AF595557

Agricultural Diversification

Editors

Claudia Di Bene
Research Centre for Agriculture and Environment (CREA-AA)
Italy

Rosa Francaviglia
Research Centre for Agriculture and Environment (CREA-AA)
Italy

Roberta Farina
Research Centre for Agriculture and Environment (CREA-AA)
Italy

Jorge Álvaro-Fuentes
Spanish National Research Council (CSIC)
Spain

Raúl Zornoza
Universidad Politécnica de Cartagena
Spain

Editorial Office
MDPI
St. Alban-Anlage 66
4052 Basel, Switzerland

This is a reprint of articles from the Special Issue published online in the open access journal *Agriculture* (ISSN 2077-0472) (available at: https://www.mdpi.com/journal/agriculture/special_issues/Agricultural_Diversification).

For citation purposes, cite each article independently as indicated on the article page online and as indicated below:

LastName, A.A.; LastName, B.B.; LastName, C.C. Article Title. *Journal Name* **Year**, *Volume Number*, Page Range.

ISBN 978-3-0365-3483-1 (Hbk)
ISBN 978-3-0365-3484-8 (PDF)

Cover image courtesy of Dr. Claudia Di Bene

Contents

About the Editors

Claudia Di Bene (Ph.D), with expertise in agronomy and soil science, has been a researcher at the Council for Agricultural Research and Economics, Research Centre for Agriculture and Environment (CREA-AA), in Rome, Italy, since 2017. In 2008, she received her PhD in Agriculture, Food, and Environment at Sant'Anna School of Advanced Studies—Pisa, Italy. Her main research topics include crop diversification strategies, low-input management, conservative agriculture, agroecology, sustainable agriculture assessment, carbon sequestration and GHG emission, soil–plant simulation models, and climate change adaptation and mitigation measures in agriculture.

Rosa Francaviglia, with expertise in agronomy, was a senior researcher at the Council for Agricultural Research and Economics, Research Centre for Agriculture and Environment (CREA-AA), in Rome, Italy, since 1981 (now retired). Her main research topics include the effects of climate change on agriculture, carbon sinks and agricultural soils, soil organic carbon simulation models, soil fertility, conservation agriculture, crop diversification, agro-environmental evaluations, soil-quality indicators, and good agro-environmental conditions (GAEC) under the EU Common Agricultural Policy.

Roberta Farina (Ph.D) has been a researcher at the Council for Agricultural Research and Economics, Research Centre for Agriculture and Environment (CREA-AA), in Rome, Italy, since 2008. She has a degree in Agricultural Science and a Ph.D in Agronomy. Her main research topics are: (i) use of ecosystem models (RothC and WinEPIC) to assess soil–crop management and climate change effects on crops yields and soil C; (ii) no-tillage, crop rotation, and cover crops effects on soil C dynamics in Mediterranean semiarid areas.

Jorge Álvaro-Fuentes (Ph.D) is a Research Scientist at the Spanish National Research Council (CSIC) in the Soil and Water Department at the Aula Dei Experimental Station (EEAD-CSIC) in Zaragoza, Spain. In 2006, he received his PhD in Soil Science and Agronomy at the University of Lleida, Spain. His main research topics include the effects of agricultural management on soil carbon and nitrogen dynamics and crop performance in Mediterranean agroecosystems.

Raúl Zornoza Belmonte (Ph.D) is full Professor at UPCT and Coordinator of Master in Advanced Techniques for Agricultural and Food Development Investigations. In 2007, he received his PhD in Soil Science from the Polytechnic University of Valencia and University Miguel Hernández (Spain). He has taught lessons in Bachelor's, Master's and Doctorate degrees about soil science, geology, innovations in agriculture, soil remediation, and soil degradation and quality. His research expertise lies in soil quality, organic matter dynamics, soil biochemistry/microbiology, and sustainable soil management for food security.

 MDPI

Editorial

Agricultural Diversification

Claudia Di Bene [1,*], Rosa Francaviglia [1], Roberta Farina [1], Jorge Álvaro-Fuentes [2] and Raúl Zornoza [3]

1 Research Centre for Agriculture and Environment, Council for Agricultural Research and Economics (CREA), 00184 Rome, Italy; r.francaviglia@gmail.com (R.F.); roberta.farina@crea.gov.it (R.F.)
2 Estación Experimental de Aula Dei, Spanish National Research Council (CSIC), 50059 Zaragoza, Spain; jorgeaf@eead.csic.es
3 Sustainable Use, Management, and Reclamation of Soil and Water Research Group (GARSA), Department of Agricultural Engineering, Universidad Politécnica de Cartagena, 30203 Cartagena, Spain; raul.zornoza@upct.es
* Correspondence: claudia.dibene@crea.gov.it

Citation: Di Bene, C.; Francaviglia, R.; Farina, R.; Álvaro-Fuentes, J.; Zornoza, R. Agricultural Diversification. *Agriculture* **2022**, *12*, 369. https://doi.org/10.3390/agriculture12030369

Received: 27 February 2022
Accepted: 4 March 2022
Published: 5 March 2022

Agricultural intensification is a highly specialized agri-food system that has contributed to raising food production worldwide due to progress in agricultural machinery and technologies, the use of improved cultivars, and external inputs such as fertilizers, irrigation, and pesticides. Nevertheless, agricultural intensification and unsustainable soil management negatively influence the environment through a decrease in air and water quality, soil erosion, depletion of soil organic matter, resistance of weeds, pests, and pathogens to pesticides, and a decrease in soil quality and agrobiodiversity. It is well-known that some changes in agricultural systems are needed for their sustainability through balancing socio-economic food production aspects with environmental goals. In this context, appropriate diversification strategies and management practices are crucial for promoting the re-design of intensive agricultural systems.

Several authors agree that crop diversity can improve crop productivity and resource use efficiency by delivering multiple ecosystem services. Coupling agricultural diversification including crop rotation, cover crops, multiple cropping and/or intercropping with low-input management strategies such as, agroecology, conservation agriculture, and organic farming contributes to increasing crop productivity and cropping system resilience in the long-term.

Despite the large scientific consensus on the potential agro-ecological and socio-economic benefits of crop diversification, some financial instruments might be necessary to favour the adoption of combined agricultural diversification strategies since the economic costs in the short-term can offset the environmental and ecological benefits in the long-term. Particularly, the re-design and diversification of agricultural intensification imply specific transition costs that must be considered by farmers and advisors in the short- and medium-term. Such costs are related to acquire new technical skills and knowledge to manage the risks due to "unknown" crops and their new market, especially in the initial implementation phases.

Therefore, research and policy must play a key role in supporting more sustainable practices for agri-food production while ensuring environmental improvements.

This Special Issue covers several topics of research relative to agricultural diversification in different parts of the world and cropping systems, where novel approaches were suggested to evaluate cropping system diversification strategies in comparison with conventional practices.

This special issue has a total of 13 research articles submitted by authors from seventeen countries: Canada, Chile, China, Ecuador, France, Germany, Italy, Lithuania, Mali, New Zealand, Niger, Poland, Singapore, Spain, Sweden, United States, and Vietnam.

The first article by Tan et al. [1] addresses the problem of poor crop productivity resulting from salinity intrusion and occasional disease outbreaks occurring in mono-cropping of rice for farmers in the Mekong Delta (Vietnam). Commonly practiced alternative farming models including the rotation of rice with fishes, shrimps, and subsidiary crops, the intensive monoculture of snakehead murrel fish and blue prawns, intercropping of blue prawn in rice paddy fields, and intercropping of blue prawn in coconut irrigation channels have been evaluated in terms of soil and water quality indicators. Among such models, the rotation of rice with different shrimp species has been demonstrated to be economically successful and ecologically sustainable, showing good adaptation to saline conditions and enabling farmers to overcome white spot disease.

The second paper by Udawatta et al. [2] examined the effects of crop management practices such as cutting height, cutting time, and the influence of plant mixture diversity on feedstock yields for bioenergy production in Missouri (USA). A monoculture of switchgrass was established along with plots that contained equal combinations of switchgrass and big bluestem. Each of these combinations also contained mixtures of native forbs and legumes seeded at various grass-to-forb ratios. The effect of species mixture was not significant on yield, while the cutting height was significant, with greater yield for the 15 cm compared to the 30 cm cutting height. However, results showed that mixtures of native warm-season grasses, forbs, and legumes are suitable for biomass production and forage crops in Missouri and can provide a source of forage during extreme summer drought conditions. When managing a forage stand using native grasses with mixtures of forbs and legumes, the frequency of cutting and timing of harvests may help to adjust costs and income potential as well as optimize equipment efficiency.

The third article by Sogoba et al. [3] addressed the cereal-cowpea intercropping practiced by smallholder farmers in Mali, a common cropping system in the Sudano-Sahelian zone of West Africa. Whether intercropping with millet or sorghum, and whatever the seasonal rainfall, the best grain yield was obtained with the wilibali (short maturing duration) variety and the best biomass yield was obtained with the sangaranka variety, which is a long-maturing duration variety. The study revealed strong trade-offs between household food opportunities and animal feeding and economic gain regarding cereal-cowpea intercropping in southern Mali. The knowledge generated revealed opportunities for alleviating some of the trade offs and achieving more promising farming decisions based on specific farm needs. Farmers selected cereal in intercropping with short maturing duration such as the wilibali variety to mainly address household food needs at specific periods corresponding to food shortages. While for those farmers prioritizing animal feeding, especially agro-pastoralists, the sangaranka variety was the best option. On the other hand, from an economic point of view, millet intercropping with cowpea was more profitable than sorghum intercropping with cowpea. Yield variability and low yields of both cereals and cowpea for all varieties combined indicated opportunities for improvement in both research and farming.

The paper by Vera-Aviles et al. [4] studied the abundance of the macro edaphic fauna and identified the beneficial families to determine the equilibrium level of the Musaceae agroecosystem in Ecuador. The mixed type of production system provides plant diversity, which favors arthropod abundance and permits lower agrochemical application without yield penalties in comparison to monocultivar systems. Within Hexapoda, the orders that presented larger populations were Collembola and Hymenoptera (based on the abundance and distribution they presented). The order Hymenoptera dominated in all of the treatments, both by its abundance and by its distribution in the studied localities, even in ecosystems with ecological imbalance. The management practices in agroecosystems can alter the community structure of pests' natural enemies, which can consequently influence their biocontrol. Since the functional composition of natural enemy communities, rather than taxonomic diversity, drive pest suppression efficiency, it is necessary to employ the functional approach to investigate the impact of management on natural enemies. Our findings showed that intraspecific diversity could be a good option to include in an IPM strategy for small and medium farmers and may help in the design of Musaceae agroecosystems to enhance the ecological regulation of pest management without putting on the farmer the constraint of management different crops.

Carton et al. [5] compared the weed suppression and yield performance between winter white lupin-triticale intercropping and lupin sole cropping throughout a set of eleven experiments during a two-year period in western France. Comparing the intercrop and the sole crop in the context of the transition to low-input crop management strategies is increasingly needed as solutions for chemical weeding are becoming scarce. In this context, results indicated that the lupin-triticale intercrop is a relevant option. Considering that a moderate lupin yield reduction can lead to a high protein yield loss, intercropping lupin with triticale does not seem to potentially perform better than sole cropping lupin regarding protein productivity on an area basis. At a broader scale, intercropping could allow an increase in lupin cropping area via increased lupin adoption by farmers due to increased weed suppression and secured total productivity. In this case, lupin intercropped with cereals could significantly contribute to the production of protein-rich grains in Europe.

Intercropping was also addressed by Munz and Reiser [6]. The agronomic optimization of intercropping systems is a challenging task given the numerous management options and the complexity of interactions between the crops and efficient methods for analyzing the influence of different management options are needed. The canopy cover of each crop in the intercropping system is a good determinant for light competition, thus influencing crop growth and weed suppression. Therefore, the study evaluated the feasibility to estimate canopy cover within an intercropping system of pea and oat based on semantic segmentation using a convolutional neural network. The network was trained with images from three datasets during early growth stages comprising canopy covers between 4% and 52%. Only the images of sole crops were used for training and then applied to images of the intercropping system. The results showed that the networks trained on a single growth stage performed best for their corresponding dataset. Combining the data from all three growth stages increased the robustness of the overall detection, but decreased the accuracy of some of the single dataset result. The accuracy of the estimated canopy cover of intercropped species was similar to sole crops and sufficient to analyze light competition.

Calvache et al. [7] examined the dynamics of water-soluble carbohydrates (WSC) use and the recovery of leaf sheaths and blades of pastures of *Bromus valdivianus* Phil. and *Lolium perenne* L. subjected to two defoliation frequencies (DFs) determined by accumulated growing degree days (AGDDs) in southern Chile. The authors also evaluated how DF influenced regrowth and accumulated herbage mass during fall. The study indicated that the leaf sheath was the principal storage organ for WSC reserves, having higher concentrations than leaf blades in fall pastures. Approximately 80% of total WSC was used during the regrowth process before WSC storage recommenced. Defoliation frequency affected WSC concentration, with longer intervals between defoliation (270 AGDDs) being preferred since the plants could recover 99% of WSC reserves and could tolerate another grazing event better. Defoliation with greater frequency (135 AGDDs) diminished the synthesis and storage of WSCs and led to slower regrowth of pasture.

The study by He et al. [8] combined statistic and economic models to evaluate the comprehensive effects of cropping systems on rice production using data collected from experimental fields. The results showed that increasing agricultural diversity through rotations, particularly potato–rice rotation (PR), significantly increased the social, economic, and ecological benefits of rice production. Yields, profits, profit margins, weighted dimensionless values of soil chemical and physical and heavy metal traits, benefits and externalities generated by PR and other rotations (e.g., fallow followed by rice (FR), and watermelon and rice rotation (WR)) were generally higher than continuous rice cropping. This suggests that agricultural diversity through rotations, particularly PR rotation, is worth implementing due to its overall benefits generated in rice production. However, due to various nutrient residues from preceding crops, fertilizer application should be rationalized to improve the resource and investment efficiency. Furthermore, the externalities (hidden ecological and social benefits/costs) generated by each of the rotation systems and the proposed ways of incenting farmers to adopt crop rotation approaches for sustainable rice production were internalized.

Dittrich et al. [9] investigated the effects of intercropping grapevine with aromatic plants using a multi-disciplinary approach. In particular, they addressed the extent to which crop diversification by intercropping impacts grapevine yield and must quality, as well as soil water and mineral nutrients (NO_3-N, NH_4-N, plant-available K and P). The experimental field was a commercial steep-slope vineyard with shallow soils characterized by a high presence of coarse rock fragments in the Mosel area of Germany. The field experiment was set up as randomized block design. Rows were either cultivated with Riesling (*Vitis vinifera* L.) as a monocrop or intercropped with *Origanum vulgare* L. or *Thymus vulgaris* L. Regarding soil moisture and nutrient levels, the topsoil (0–0.1 m) was more affected by intercropping than the subsoil (0.1–0.3 m). Gravimetric moisture was consistently lower in the intercropped topsoil. While NO_3-N was almost unaffected by crop diversification, NH_4-N, K, and P were uniformly reduced in topsoil. Significant differences in grapevine yield and quality might be dominantly attributable to climate variables, rather than to the treatments. Additionally, they also observed some insignificant yield losses due to intercropping, particularly induced by water competition. With respect to this, thyme appeared to be less competitive due to an earlier harvest date and a lower respectively shorter consumption of soil water during the crop cycle. The authors concluded that yield stabilization due to intercropping with thyme and oregano seems possible with sufficient rainfall or by irrigation.

Gecaitė et al. [10] conducted field experiments to determine yield formation regularities and plant competition effects of oat–black medick, oat–white clover, and oat–Egyptian clover relay intercropping under organic farming conditions. Oats and forage legumes were grown in mono- and intercrops. Aboveground dry matter measured at flowering, development of fruit and ripened grain, productivity indicators, oat grain yield, and nutrient content were established. The results showed that oats dominated in the intercropping systems. Oat competitive performance, which is characterized by forage legumes aboveground mass reduction compared to monocrops, was 91.4–98.9. As the oats ripened, its competitiveness tended to decline. In oat–forage legume intercropping systems, the mass of weeds was significantly lower compared to the legume monocrops. Oats and forage legumes competed for P, but N and K accumulation in biomass was not significantly affected. They concluded that, in relay intercrop, under favorable conditions, the forage legumes can easily adapt to the growth rhythm and intensity of oats and without adverse effect on their grain yield.

The challenges for food planning and policy in the regionalization of food systems, in order to shorten supply chains and develop local agriculture to feed city regions, were addressed in the article by Vicente-Vicente et al. [11]. The existing foodshed approaches enable them to assess the theoretical capacity of the food self-sufficiency of a specific region, but they struggled to consider the diversity of existing crops in a way that could be usable to inform decisions and support urban food strategies. Most studies are based on the definition of the area required to meet local consumption, obtaining a map represented as an isotropic circle around the city, without considering the site-specific pedoclimatic, geographical, and socioeconomic conditions which are essential for the development of local food supply chains. They proposed a first stage to fill this gap by combining the 'Metropolitan Foodshed and Self-Sufficiency Scenario' model, which already considers regional yields and specific land use covers with spatially explicit data on cropping patterns, soil, and topography. They used the available Europe-wide data and apply the methodology in the city region of Avignon (France), initially considering a foodshed with a radius of 30 km. Results showed that even though a theoretically-high potential self-sufficiency could be achieved for all of the food commodities consumed (>80%), when the specific pedological conditions of the area are considered, this could be suitable only for domestic plant-based products, whereas an expansion of the initial foodshed to a radius of 100 km was required for animal products to provide >70% self-sufficiency. They concluded that it is necessary to shift the analysis from the size assessment to the commodity-group–specific spatial configuration of the foodshed based on biophysical and socioeconomic features. Moreover, they discussed avenues for further research to enable the development of a foodshed assessment as a complex of complementary pieces (i.e., the 'foodshed archipelago').

The paper by Kurdyś–Kujawska et al. [12] aimed to identify the determinants of crop diversification and the impact of crop diversification on the economic efficiency of small farms in Poland. The article first provides a critical review of the literature on crop diversification, its role in stabilizing agricultural income, and its impact on economic efficiency in small farms. Secondly, the level of crop diversification was determined, and empirical research was conducted considering the economic, social, and agronomic characteristics of farms. Thirdly, the economic efficiency of farms diversifying crops was compared with farms focused on one type of production. The research material consisted of small farms participating in the Polish system of collecting and using farm accountancy data (FADN) in 2018. The level of diversification was determined using the Herfindahl-Hirschman Index. The factors influencing crop diversification were identified using the logit regression model. The Mann–Whitney U rank sum test was used to assess the significance of the differences in distributions. The research results indicated an average level of crop diversification in small farms in Poland and its regional differentiation. In addition, a statistically significant positive impact on the probability of crop diversification in small farms in Poland was found of variables such as the level of exposure of agricultural production to atmospheric and agricultural drought, the location of the farm in the frost hardiness zone, and a statistically significant negative impact of the value of its fixed assets. The existence of significant differences in the level of economic efficiency of farms diversifying crops and farms focused on one profile of agricultural production was demonstrated.

The study by Klimek-Kopyra et al. [13] assessed the effect of biochar produced from sunflower husks on soil respiration (SR), soil water flux (SWF), and soil temperature (ST), depending on its dose and different soil cover (with and without vegetation) in Poland. Moreover, the seed yield was assessed depending on the biochar fertilization. The SR, ST, and SWT were evaluated seven times in three-week intervals during two seasons over 2018 and 2019. The results indicated that the time of biochar application had a significant effect on the evaluated parameters. In the second year, significantly ($p < 0.005$) higher soil respiration (4.38 μmol s^{-1} m^{-2}), soil temperature (21.2 °C), and the level of water net transfer in the soil (0.38 m mol s^{-1} m^{-2}) compared to the first year were observed. The most effective biochar dose regarding SR and soybean yield was 60 t ha^{-1}, but a more comprehensive cost-benefit analysis is needed to recommend large-scale biochar use at this dose.

Conflicts of Interest: The authors declare no conflict of interest.

References

1. Tan, L.; Tran, T.; Loc, H. Soil and Water Quality Indicators of Diversified Farming Systems in a Saline Region of the Mekong Delta, Vietnam. *Agriculture* **2020**, *10*, 38. [CrossRef]
2. Udawatta, R.; Gantzer, C.; Reinbott, T.; Wright, R.; Robert, P.; Wehtje, W. Influence of Species Composition and Management on Biomass Production in Missouri. *Agriculture* **2020**, *10*, 75. [CrossRef]
3. Sogoba, B.; Traoré, B.; Safia, A.; Samaké, O.; Dembélé, G.; Diallo, S.; Kaboré, R.; Benié, G.; Zougmoré, B.R.; Goïta, K. On-Farm Evaluation on Yield and Economic Performance of Cereal-Cowpea Intercropping to Support the Smallholder Farming System in the Soudano-Sahelian Zone of Mali. *Agriculture* **2020**, *10*, 214. [CrossRef]
4. Vera-Aviles, D.; Suarez-Capello, C.; Llugany, M.; Poschenrieder, C.; De Santis, P.; Cabezas-Guerrero, M. Arthropod Diversity Influenced by Two Musa-Based Agroecosystems in Ecuador. *Agriculture* **2020**, *10*, 235. [CrossRef]
5. Carton, N.; Naudin, C.; Piva, G.; Corre-Hellou, G. Intercropping Winter Lupin and Triticale Increases Weed Suppression and Total Yield. *Agriculture* **2020**, *10*, 316. [CrossRef]
6. Munz, S.; Reiser, D. Approach for Image-Based Semantic Segmentation of Canopy Cover in Pea–Oat Intercropping. *Agriculture* **2020**, *10*, 354. [CrossRef]
7. Calvache, I.; Balocchi, O.; Alonso, M.; Keim, J.; López, I. Water-Soluble Carbohydrate Recovery in Pastures of Perennial Ryegrass (*Lolium perenne* L.) and Pasture Brome (*Bromus valdivianus* Phil.) Under Two Defoliation Frequencies Determined by Thermal Time. *Agriculture* **2020**, *10*, 563. [CrossRef]

8. He, D.; Ma, Y.; Li, Z.; Zhong, C.; Cheng, Z.; Zhan, J. Crop Rotation Enhances Agricultural Sustainability: From an Empirical Evaluation of Eco-Economic Benefits in Rice Production. *Agriculture* **2021**, *11*, 91. [CrossRef]
9. Dittrich, F.; Iserloh, T.; Treseler, C.; Hüppi, R.; Ogan, S.; Seeger, M.; Thiele-Bruhn, S. Crop Diversification in Viticulture with Aromatic Plants: Effects of Intercropping on Grapevine Productivity in a Steep-Slope Vineyard in the Mosel Area, Germany. *Agriculture* **2021**, *11*, 95. [CrossRef]
10. Gecaitė, V.; Arlauskienė, A.; Cesevičienė, J. Competition Effects and Productivity in Oat–Forage Legume Relay Intercropping Systems under Organic Farming Conditions. *Agriculture* **2021**, *11*, 99. [CrossRef]
11. Vicente-Vicente, J.; Sanz-Sanz, E.; Napoléone, C.; Moulery, M.; Piorr, A. Foodshed, Agricultural Diversification and Self-Sufficiency Assessment: Beyond the Isotropic Circle Foodshed—A Case Study from Avignon (France). *Agriculture* **2021**, *11*, 143. [CrossRef]
12. Kurdyś-Kujawska, A.; Strzelecka, A.; Zawadzka, D. The Impact of Crop Diversification on the Economic Efficiency of Small Farms in Poland. *Agriculture* **2021**, *11*, 250. [CrossRef]
13. Klimek-Kopyra, A.; Sadowska, U.; Kuboń, M.; Gliniak, M.; Sikora, J. Sunflower Husk Biochar as a Key Agrotechnical Factor Enhancing Sustainable Soybean Production. *Agriculture* **2021**, *11*, 305. [CrossRef]

Article

Soil and Water Quality Indicators of Diversified Farming Systems in a Saline Region of the Mekong Delta, Vietnam

Lam Van Tan [1,2,*], Thanh Tran [2] and Ho Huu Loc [3,4]

1. Department of Soil Science, College of Agriculture, Can Tho University, 3/2 Street, Ninh Kieu District, Can Tho City 900000, Vietnam
2. NTT Hi-Tech Institute, Nguyen Tat Thanh University, 300A Nguyen Tat Thanh, Ward 13, District 4, Ho Chi Minh City 70000, Vietnam; thanhtran2710@gmail.com
3. National Institute of Education, Nanyang Technological University, Singapore 637616, Singapore; huuloc20686@gmail.com
4. Faculty of Food and Environment Engineering, Nguyen Tat Thanh University, Ho Chi Minh City 70000, Vietnam

* Correspondence: lamvantan101076@gmail.com

Received: 12 December 2019; Accepted: 19 January 2020; Published: 7 February 2020

Abstract: Saltwater intrusion, a consequence of climate change and decreased water levels, has been increasingly severe in the Mekong Delta region. Thanh Phu District, Ben Tre Province, Vietnam, is a coastal region where agricultural production and local livelihood have been impaired by saltwater intrusion, resulting in the adoption of multiple coping strategies, including rotations and intercropping. This study aims to measure and evaluate soil and water quality indicators of multiple farming systems in Thanh Phu district and contributes to developing suitable cropping patterns. Soil indicators were pH, electrical conductivity, and exchangeable Na^+. Water quality characteristics include pH, salinity, dissolved N and P, alkalinity, H_2S, and chemical oxygen demand (COD). The results indicated that water pH and salinity were at suitable levels to support the growth of prawn but were below the critical level required to grow black tiger shrimp and white-legged shrimp. Water alkalinity, dissolved N, P, and COD were not constraining for the growth of shrimps. However, a significant concentration of H_2S may cause disadvantages for shrimp growth.

Keywords: blue prawn; black tiger shrimp; economic efficiency; farming systems; salinity intrusion; soil salinity; white-legged shrimp

---

1. Introduction

Salinity intrusion represents a major concern for agricultural production in coastal regions. Saline water radically alters soil conditions [1–3], severely undermines freshwater reservoirs used for consumption and irrigation and threatens the survival of freshwater living organisms and livelihoods of inhabitants residing in the affected areas. In addition, numerous aspects of farming activities, including crop harvesting, farming systems, and more importantly, crop-livestock structure, are significantly hampered by sodicity (i.e., the amount of sodium held in a soil) -induced changes in chemical and physical properties and composition of soils [4–8].

Among coastal areas of the Mekong Delta, Ben Tre province of Vietnam is a diversely-farmed region that is under significant impacts of salinity intrusion. More specifically, a quarter of the total area for agricultural production of the province amounting to 181,252 hectares is constantly affected by saltwater intrusion [9]. Saline intrusion is greatly aggravated in dry seasons where salinity with the minimum concentration of 1‰ covers an overwhelming majority of the land of Ben Tre province. The extents to which saltwater intrudes inland at the 4‰ and 1‰ thresholds are 50 and 70 km, respectively [10], causing once-fertile soil to degrade rapidly and reducing the economic efficiency of rice farming in those areas.

In response to poor crop productivity that results from salinity intrusion and occasional disease outbreaks occurring in mono-cropping of rice, farmers in the Mekong Delta have resorted to other products and adopted various diversification strategies. Commonly-practiced alternative farming models included rotation of rice with fishes, shrimps, and subsidiary crops, the intensive monoculture of snakehead murrel fish and blue prawn, intercropping of blue prawn in rice paddy fields and intercropping of blue prawn in coconut irrigation channels. Among such models, rotation of rice with different shrimp species has been demonstrated to be economically successful [11–13] and ecologically sustainable, showing good adaptation to saline conditions and enabling farmers to overcome the white spot disease [14]. Phong et al. justified the rice-shrimp intercropping model in saline soil environments by elaborating that close proximity of rice-shrimp fields to water sources, which is required for periodic water exchanges, may facilitate the flushing of salts and salt leaching through rainfalls [15]. On the other hand, *Penaeus* species, in general, and *P. monodon* (black tiger shrimp), in particular, because of their active osmoregulation against high salinity, have shown to survive well in saline ponds with frequent occurrence of salinity shocks [16]. However, high mortalities were observed in black tiger shrimp ponds with very low salinity [17].

Despite that, current coconut and rice-based diversification strategies that have been practiced heavily rely on farmers' experience, thus lacking a comprehensive scientific basis required for model refinement or extension. In addition, it is still unclear whether those farming systems will cope well with the increasing levels of salinity and not cause further degradation in soil and water quality in the coastal area in the future.

Driven by the aforementioned thrust, we conducted this study to evaluate the suitability of the popular farming models of Ben Tre with respect to the salinity intrusion. More specifically, we developed a novel set of soil and water indicators to investigate the commonly-practiced farming models in Thanh Phu district, Ben Tre province, Vietnam. The indicators were subsequently validated via comparing with the practical national standards. Methodological implication aside, the study sought to inform relevant decision-makers regarding the environmental qualities and crop suitability associated with the agricultural development of the study area.

2. Materials and Methods

2.1. Study Area

The study was carried out in Thanh Phu district, Ben Tre province, Vietnam. Meteorologically, the Thanh Phu district is a tropical monsoon climate area with two distinct seasons (dry and wet seasons). The temperature of the area is stable and averaged at around 26.6 °C annually. The maximum temperature peaks at 28.4 °C in April and falls as low as 24.3 °C in December. Temperature differences between months were minor, which is favorable for year-round cultivation. Rainfall characteristics of the district are typical to the littoral regions of the South China Sea, having the lowest annual rainfall in the Mekong Delta (1279 mm). Rainfall in the wet season (1218 mm) accounts for 95% of total annual rainfall and is starkly contrasted by that of the dry season (61 mm).

Being affected by mixed semidiurnal tides of the South China Sea, rivers in the Thanh Phu district have tidal amplitudes ranging from the maximum point of 4.1 m (from November to January) to a minimum of 2.6 m (from June to July). In the dry season, salinity intrusion occurs region-wide in which the intrusion around Ham Luong river is more severe than that around other rivers in the district. Areas that are far from the coast such as Phu Khanh and Thoi Thanh commune also suffer from salinity that lasts around 2–3 months a year. Monitoring data shows that water quality of Co Chien river is better than that of Ham Luong river. In addition, a mixed semidiurnal tidal pattern also facilitates irrigation and water supply for aquaculture in the riverbank regions.

The district can be divided into three ecological sub-regions based on salinity levels: I, II, and III as shown in Figure 1. EcoZone I (freshwater area): This area is about 11,565 hectares ranging from Binh Thanh commune to Thanh Phu town. The region is a freshwater area where two (or even three in specific areas) crops of rice are harvested within one year and are also surrounded by an anti-salinity dike line that belongs to the Project 418 of the Government. Paddy fields in this region are irrigated with fresh water. EcoZone II (brackish water area): This region has an area of about 10,000 hectares ranging from Binh Thanh commune seaward to An Dien commune. This area achieved one rice harvest per year with black tiger shrimp and white-legged shrimp cultivated intermittently. EcoZone III (saline water area): This region has an area of about 21,000 ha and specializes in extensive aquaculture of shrimp, crab, and clam hosted in mangrove forests.

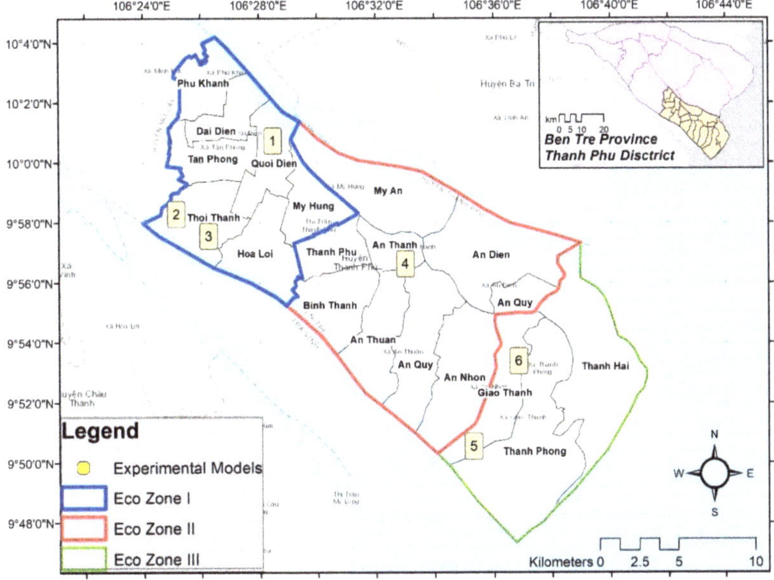

Figure 1. Administrative map of the Thanh Phu district with sampling points.

Six typical adaptive farming models were examined in this study and were described as follows Table 1:

Table 1. Sampling labels in arable areas.

Model	Description	Region	Commune	No. of Households
1	Rotation of rice (OM10252 salinity-tolerant variety) and corn (MX10 salinity-tolerant variety)	Freshwater	Quoi Dien	2
2	Intercropping of blue prawn (*Macrobrachium rosenbergii*) in coconut irrigation channels	Freshwater	Thoi Thanh	6
3	Rotation between intercropping of blue prawn in rice paddy fields (OM10252 variety) and blue prawn extensive farming	Freshwater	Quoi Dien	2
4	Rotation between intercropping of black tiger shrimp (*Penaeus monodon*) in rice paddy fields and blue prawn extensive farming	Brackish	An Thanh	2
5	Rotation between aquaculture of black tiger shrimp and white-legged shrimp (*Litopenaeus vannamei*)	Saline	Giao Thanh	2
6	Black tiger shrimp extensive farming	Saline	Giao Thanh	2

2.2. Measurement of Soil Indicators

Soil samples were collected from the farming fields for model 1 and in pond bottom sediments at a depth of 0–20 cm for other models. All soil samples were collected at the end of April (middle of the dry season). The soil in the region was illuvial soil with potential acid sulfate soil appearing at a depth of around 60 cm. Sandy soil was also found at a depth of 120 cm. The soil classification according to FAO is Endo-protho thionic GLEYSOL. In each model, soil samples from different spots were collected and then mixed. The soil analysis methods are detailed as follows Table 2:

Table 2. Methods of analyzing parameters in the soil environment.

Analytical Indicator	Reference Method
pH, EC	Extracted with distilled water, extract ratio 1:2.5 (soil: water) and measured by pH, EC meter
Nitrogen	Gianello and Bremner (1986) [18]
Phosphorus	Olsen (1954) [19]
Na^+, K^+, Mg^{2+}, Ca^{2+}	Atomic absorption spectrum (AAS)
TOC (total organic carbon)	Walkley-Black (1934) [20].
CEC (cation exchange capacity)	The measure is determined by a buffer of $BaCl_2$ 0.1 M [21]
ESP (Exchange Sodium Percentage)	The method is based on the ratio of Na^+ adsorbed and cation exchange capacity of the soil (CEC, cmol/kg): $ESP(\%) = \frac{[Na^+]}{CEC} \times 100$

2.3. Assessment of Water Environment Characteristics

Model 1 was excluded from water analysis. Time points for water collection include at the beginning (April—middle of the dry season), the middle (May-end of the dry season), and the end (June—end of the dry season) of the shrimp aquaculture period. Water samples were collected in the pond at a depth of around 20–30 cm from the surface, nearby the places in which corresponding soil samples were collected. One-liter flasks were used to collect the water samples. Prior to sample collection, the flask was washed and rinsed with pond water. Water samples were preserved according to the requirements of the analysis technique. For the H_2S indicator, water was collected close to the bottom of the pond and was added with 2 drops of zinc acetate 2N, followed by pH adjustment to 9 using NaOH [21]. Table 3 summarizes the analytical indicators of the water samples alongside the associated references.

Table 3. The methods of analysis of indicators in water.

Analytical Indicator	Reference Method
pH, EC	Hanna HI 98129
Dissolved nitrogen	Phenate method [22]
Dissolved phosphorus	Malachite Green method [23]
Total alkalinity ($CaCO_3$)	APHA (1998) method [24]
Hydrogen sulfide (H_2S)	Clesceri et al. 1998 [22]
COD	Oxidation method using permanganate kali [22]

2.4. Statistical Analysis

Microsoft Excel software is used to calculate the mean values and standard deviations between treatments (Stdev). Duncan test was used for evaluating the difference between soil and water indicators. Data were analyzed by SPSS 20.0 software.

3. Results and Discussion

3.1. Soil Characteristics in Farming Models

3.1.1. pH of Soil

The pH of soil in different farming systems is shown in Table 4 and the numbers in brackets signify the standard deviation of the measurement of each model. The differences in the standard deviations relate to various aspects in the farming practices, i.e., freshwater vs. saline water vs. brackish water; and the cropping system, i.e., shrimp, prawn, rice, or combined thereof.

Table 4. Value of soil pH in the farming models.

Models	pH
Model 1	5.90 ± 0.35
Model 2	6.24 ± 0.08
Model 3	6.68 ± 0.08
Model 4	3.82 ± 0.01
Model 5	5.65 ± 0.55
Model 6	6.39 ± 0.18

Except for model 4, all pH values ranged from 5.6 to 6.6. This pH range falls into the category of medium acidity according to Brady (1990) and is suitable for growing different types of the crop [25]. However, the highly acidic soil of model 4 is unfavorable for intercropping of aquaculture and crops [26] and could be explained by the presence of sulfidic sediments that lowered the pH of soil samples collected in dry seasons [27]. Other aquaculture systems exhibited soil pH values close to the neutrality threshold. To be specific, aquaculture models situated in freshwater and saline areas have a pH range of 6.2 to 6.6 and the model located in the brackish area had a pH value of 5.6. These elevated values are possibly due to accelerated H^+ ion exchange processes occurring in the pond environment and are in agreement with the suggested pH range (4–11) for the growth and development of cultured shrimp and aquatic organisms of Boyd (1998) [28].

In general, the pH values reported of the models are low and suitable for shrimp growth and the development of rice, corn, and coconut, albeit being slightly lower than the suggested pH range. Only model 4, which is a rotation of intercropping and extensive aquaculture, experienced a very low pH. This suggests that basal fertilization with lime should be performed before rice cultivation to improve soil pH.

3.1.2. Electrical Conductivity in Soil and Sodicity

The soil electrical conductivity (EC), a measure of salt quantity in soils, affects the growth of plants and animals. The difference in the salt concentration of the soil solution and the cell of the plant root leads to a restriction on water and nutrient uptake. Excessive soluble salts in saline soils are often associated with sodic properties, signified by high levels of sodium salt (mainly Na_2CO_3) on the absorption complex of soils. Continuous shrimp farming causes sodic soil phenomenon, hindering plant growth, disturbing the balance of water and nutrient uptake, and poses physical disadvantage [27,29]. On the other hand, EC was found to correlate strongly with ECe. To be specific, in a previously reported study analyzing 603 saline soil samples in the Mekong Delta, the ECe values were found to be strongly correlated with EC 1:2.5 (R^2 = 0.89). In addition, in comparison with ECe, EC1:2.5 was 0.41 times lower and EC1:5 was 0.28 times lower.

Table 5, which summarizes EC in examined systems, indicates that systems in the freshwater area seemed to have low conductivity with EC values ranging from 1.08 to 1.27 mS·cm^{-1}. This level is classified as low EC soil and is generally not detrimental to crop yields according to A&L Western Agricultural Laboratories guidelines. Mild salinity, as indicated by EC value of around 3.7 mS·cm^{-1}, was observed in the rotation system of black tiger shrimp-rice intercropping and extensive blue prawn farming in the brackish area [30,31]. This is also similar to those of the models in the saline area that had mildly saline soils with conductivity of about 3.36–4.99 mS·cm^{-1} [31] and consistent with the seasonality of salinity in nearby rivers and canals, which tended to show an increasingly high salinity from 8.5 to 25‰ in around April because of climate change [32,33]. In general, figures measured at farming models in brackish and saline sub-areas indicated that crop yields are probably affected at this level of conductivity. Therefore, it is necessary to select plant and animal varieties that could adapt to the soil salinity level.

Table 5. Value of soil conductivity of the models.

Models	EC (mS·cm^{-1})
Model 1	1.34 ± 0.36
Model 2	1.27 ± 0.03
Model 3	1.08 ± 0.24
Model 4	3.79 ± 0.15
Model 5	3.36 ± 0.18
Model 6	4.99 ± 0.31

3.1.3. Exchangeable Na^+ in Soil and Sodicity

The results in Table 6 showed the exchangeable Na^+ content of soil samples in different models. Models situated in the freshwater area had Na^+ values ranging from 1.08–1.37 cmol/kg. This result is higher than that of a previous study in Hau Giang province, Mekong delta, where different integrated models including shrimp-melon, shrimp-rice, and a shrimp monoculture model showed exchangeable Na^+ in the soil of 0.67; 1.80; 1.93 cmol/kg, respectively [32]. However, the percentage of exchangeable sodium (ESP) on the uptake complex in soils of freshwater models is lower than saline-sodic soil threshold (ESP < 15%), ranging from 8.50 to 10.79%, in which the rice-corn rotation model exhibited an ESP value lower than those of two other models. This is possibly due to the low accumulation of salt during dry seasons and the long leaching period (about 6 months) in wet seasons.

Table 6. Value of Na^+ exchanged and the percentage of exchangeable sodium (ESP) (%) of the models.

Models	Na^+ (cmol/kg)	ESP (%)
Model 1	1.08 ± 0.23	8.50 ± 0.23
Model 2	1.37 ± 0.05	10.79 ± 0.05
Model 3	1.14 ± 0.03	8.98 ± 0.03
Model 4	4.22 ± 0.02	33.23 ± 0.02
Model 5	4.78 ± 0.05	37.64 ± 0.05
Model 6	5.13 ± 0.42	40.39 ± 0.42

In brackish areas, model 4 (rotation of tiger shrimp-rice intercropping and monoculture of blue prawn) has an average Na^+ value of 4.22 cmol/kg, which is lower than that in the rotation aquaculture of tiger shrimp-white legged shrimp in the saline area, at 4.78 cmol/kg. The extensive aquaculture model in the saline area (model 6) exhibited a very high Na^+ value, averaged at 5.13 cmol/kg. Regarding sodicization, ESP of models in the brackish and saltwater area exceeded sodicity threshold (ESP > 15%), ranging from 33.23 to 40.19%. The maximum ESP in the saltwater is of the extensive aquaculture model (model 6 at 40.19%). In brackish farming models, the water salinity in the ponds and rice fields are reduced in the rainy season, facilitating the rotation of rice with tiger shrimp [34]. However, because of the short period for salt leaching, usually about 3 months, the soil salinity is still higher than the sodicity threshold.

In summary, indicators of models in the brackish and saline region showed clear signs of soil sodicization (ESP > 15%, pH < 8.5, EC > 4 mS·cm^{-1}) while the soil of systems in freshwater area is under sodicity threshold (ESP < 15%). Therefore, measures to leach the salt and reduce the Na^+ content in the soil are advisable to achieve productive rice farming within shrimp-rice intercropping models.

3.2. Water Characteristics in Farming Models

Water samples were collected in five farming models (model 1 was excluded from water analysis), at the beginning (April), the middle (May), and the end (June) of the shrimp farming cycle. Water samples were collected to analyze the environmental parameters such as pH, EC, dissolved nitrogen, dissolved phosphorus, alkalinity, H_2S, COD to assess the adaptability of cultured shrimp in farming models.

3.2.1. Water pH in Farming Models

pH values of water of different farming systems are presented in Figure 2. Generally, pH value is relatively high, ranging from 7.51 to 8.48 from the beginning to the end of the shrimp harvest. This pH range is in line with other pH ranges appropriate for shrimp farming, as suggested by several prior studies [35–39]. It is revealed that pH in farming systems tended to decline from the beginning to the middle of the harvest, and then rise at the end. This could be explained by the practice of applying lime at the beginning of shrimp aquaculture to rehabilitate the pond. It was reported that extreme and highly fluctuated pH might affect growth, survival, disease resistance, reproduction and nutrition of shrimps [28,40,41]. In particular, the pH of higher than 8 is reported to be highly toxic to shrimps [42].

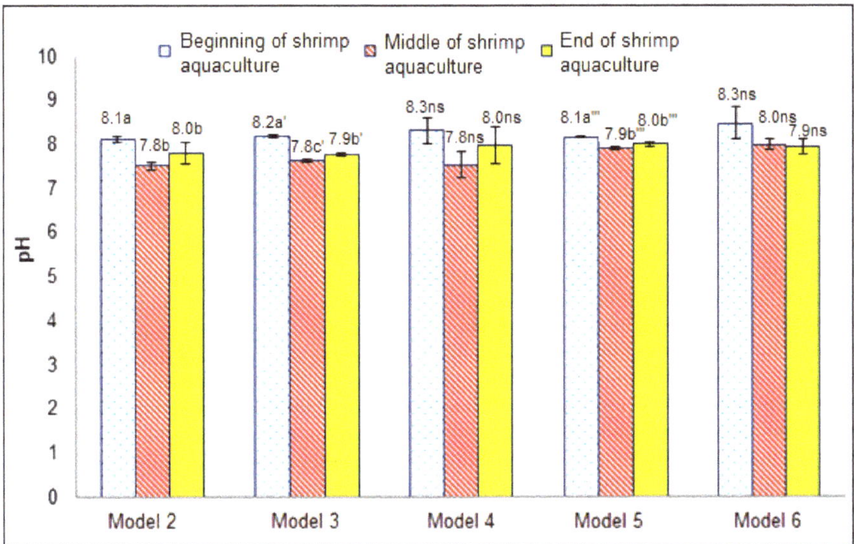

Figure 2. Variations of pH of water in examined farming systems. Notes: a and b are compared according to sampling time, the same letters within the same column or row have no significant difference. The symbols ', ''' are used to distinguish different experimental or model areas. ns: no difference (compared at a significance level of 5%).

3.2.2. Salinity of Water

The results presented in Figure 3 show that the salinity of water in the farming models varies from 2.76 to 17 mS·cm^{-1}. Two models in the freshwater area had salinity ranging from 2.76 to 5.12 mS·cm^{-1}. The rotation model of black tiger shrimp-rice intercropping and blue prawn (model 4—brackish area) had salinity ranging from 7.98 to 12.77 mS·cm^{-1}. Model of tiger shrimp-white shrimp rotation and the model of extensive shrimp (saline area) had salinity ranging from 10.3–17 mS·cm^{-1}. In general, the salinity thresholds of fresh, brackish, and saline area models are suitable for the development of blue prawn and black tiger shrimp. The salinity that is appropriate for the development of the black tiger shrimp ranges from 0 to 10 mS·cm^{-1} [34]. On the contrary, other studies suggested that the pH range for optimal growth of black tiger shrimp is 10–30 mS·cm^{-1} [40,42,43] or 15–35 mS·cm^{-1} [44].

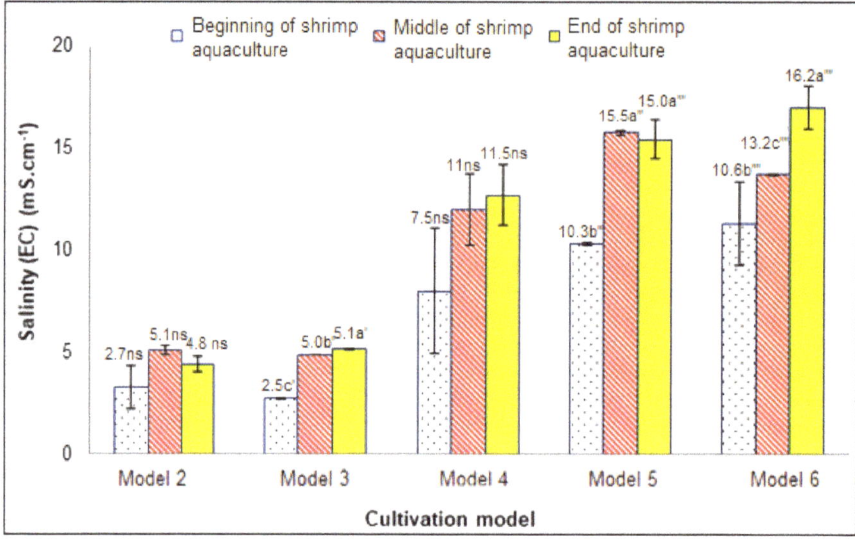

Figure 3. Variations of water salinity in farming models. Notes: a and b are compared according to sampling time, the same letters within the same column or row have no significant difference. The symbols ′, ″, ‴ are used to distinguish different experimental or model areas. ns: no difference (compared at a significance level of 5%).

Salinity seemed to be gradually increasing over the course of shrimp aquaculture, except for model 2 and 3 where the salinity increase was observable but statistically insignificant. The trend could be due to the time of sampling being at the end of the dry season, where general salinity is high. For model 3, since the model is situated in the area surrounded by anti-salinity dike, its salinity is clearly lower than those of models exposed to salinity in other subregions. It was found that salinity exceeding the tolerance limit of shrimp adversely affects the regulation, osmotic pressure, and molting of cultured shrimp, causing stunting, shell disease and resulting in low survival rate [39]. Shrimps, in general, thrive in low salinity environments yet susceptible to disease [37], which could be explained by salinity limiting the growth of some pathogenic microorganisms.

3.2.3. Soluble Nitrogen Content in Water in the Cultivation Model

The soluble protein content is another critical indicator representing nutrition in a farming pond. The excessive-high concentration of soluble protein may cause phytoplankton blooming, resulting in possible eutrophication. As suggested by Chanratchakool (2003), water with NH_4^+ nitrogen of higher than 4 mg/L is considered undesirable for shrimping farming [40]. The results presented in Figure 4 show that the content of soluble protein varied greatly from 0.30 to 0.79 mg/L. The models in the freshwater sub-region exhibited soluble protein content ranging from 0.24 to 0.65 mg/L. The content of models in brackish and saline areas ranged from 0.3 to 0.74 mg/L. This result is consistent with the threshold value of soluble protein for blue prawn, tiger shrimp, and white-legged shrimp. To be specific, Duong Nhat Long (2012) suggested the appropriate threshold for optimal development of blue prawn is NH_4^+ <1 mg/L [45]. Similarly, other studies suggested that the content of soluble nitrogen suitable for shrimp ponds could range from 0.2 to 2.0 mg/L [28,39].

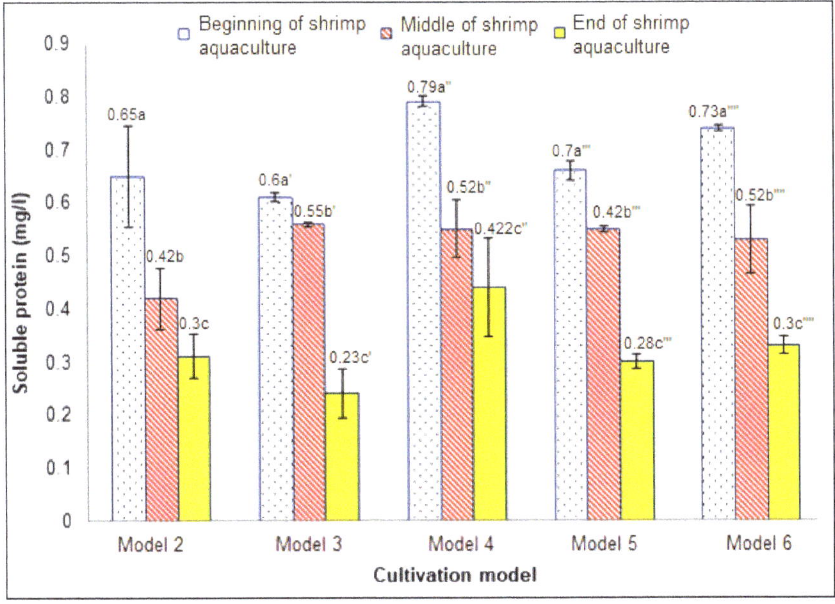

Figure 4. Variations of soluble nitrogen in farming models. Notes: a and b are compared according to sampling time, the same letters within the same column or row have no significant difference. The symbols ′, ″, ‴, ″″ are used to distinguish different experimental or model areas. ns: no difference (compared at a significance level of 5%).

For all farming systems, the content of soluble nitrogen showed a significant decline from the beginning to the end of the shrimp aquaculture. Reduced soluble protein content in the later stages could be due to the consumption of nitrogen of plants and microorganisms in the pond. In addition, submergence, waterlogging, and increased salinity could also reduce its mineralization capability and microbial activity of the soil, lowering dissolved nitrogen and soluble nitrogen. In general, the nitrogen content of the farming models is dissolved within the appropriate threshold of aquaculture ponds, not yet causing eutrophication. Since a majority (over 90%) of nitrogen content comes from shrimp feeding and metabolic products of shrimps [28], it is necessary to control protein content by observing the color and adjusting feeding to maintain the stability in water quality.

3.2.4. Soluble Phosphorus Content in Water in the Cultivation Model

The results presented in Figure 5 show that the dissolved phosphorus content ranged from 0.01 to 0.08 mg/L. The peak soluble phosphorus content was attained in the rotation aquaculture model of black tiger shrimp and white legged shrimp. Cultivation models in the fresh, brackish, and saline sub-areas all have low phosphorus content than the recommended values. To be specific, according to Nguyen Duc Hoi, the content of phosphorus dissolved in water suitable for shrimp and fish is recommended to be about 1.0 mg/L [46]. In another study, Boyd argues that the appropriate amount of dissolved phosphates in aquaculture ponds should be around 0.5 mg/L [28].

Overall, soluble phosphorus content is declining over time, and the differences are significant. This result is consistent with the study of Thai Truong Giang that the phosphorus content in pond water is low and is provided mainly through chemical fertilizers to stimulate phytoplankton to develop as a natural food source for shrimp [47]. In addition, dissolved phosphorus is also lost by uptake by phytoplankton and aquatic organisms and the phosphate uptake of the sediment sludge. As a result, it is necessary to supplement an additional amount of phosphorus to ensure the growth of shrimps in ponds [48].

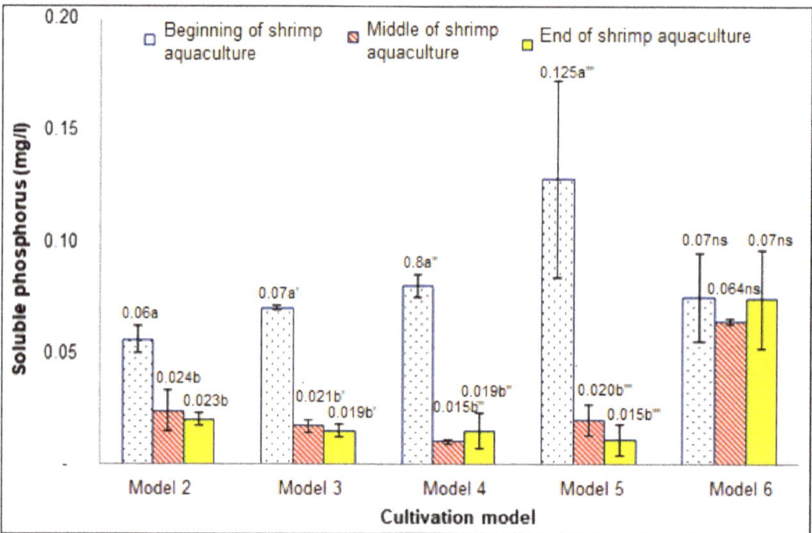

Figure 5. Variations of soluble phosphorus content in farming models. Notes: a and b are compared according to sampling time, the same letters within the same column or row have no significant difference. The symbols ', '', ''' are used to distinguish different experimental or model areas. ns: no difference (compared at a significance level of 5%).

3.2.5. Alkalinity of Water

The results presented in Figure 6 show that the alkalinity ranges from 36 to 89 mg/L. The alkalinity range of freshwater, brackish water, and saline water models was 36–88, 46–58, and 59–89 mg/L respectively. The alkalinity of water in the models is in a suitable range for aquaculture, especially blue prawn, tiger shrimp, white shrimp at the end of the crop. According to Chanratchakool et al. the alkalinity suitable for shrimp growth is in the range of 80–120 mg/L [40], which well agrees with the results of Mistein et al. obtained when monitoring the alkalinity of water in shrimp ponds in Bangladesh [49]. Studies performed in Vietnam also indicate optimal alkalinity for shrimp farming is >60 mg/L and 80–150 mg/L [50,51].

Similar to the results of soluble phosphorus content, alkalinity tended to increase over time. Alkalinity of all farming systems in the fresh, brackish, and saline areas was low at the beginning, the middle of the crop and was elevated at the end of the season. This may be due to the increase in alkalinity during the application of lime. In addition, the alkalinity of aquaculture ponds depends on the properties of pond substrate and water. For ponds in sandy soils, total alkalinity could be around 20 mg/L and ponds in limestone soils often have alkalinity above 100 mg/L [28]. Other possibilities could be due to low salinity, leading to low carbonate and bicarbonate, and underdeveloped phytoplankton, which hinders the primary production of the pond and lowers alkalinity. On the contrary, high alkalinity reduces pH fluctuations by HCO_3^- and CO_3^{2-} buffer and contributes to the assessment of acid neutralization capacity of water, expressed by HCO_3^-, CO_3^{2-}, OH^- ions [52]. In general, in comparison with the optimal range for shrimp development, the alkalinity in the models is low at the beginning [53]. Therefore, to increase the alkalinity in the water, it is necessary to apply to add lime to the early stages of stocking and after rainfalls to maintain pond quality and pH stability [51].

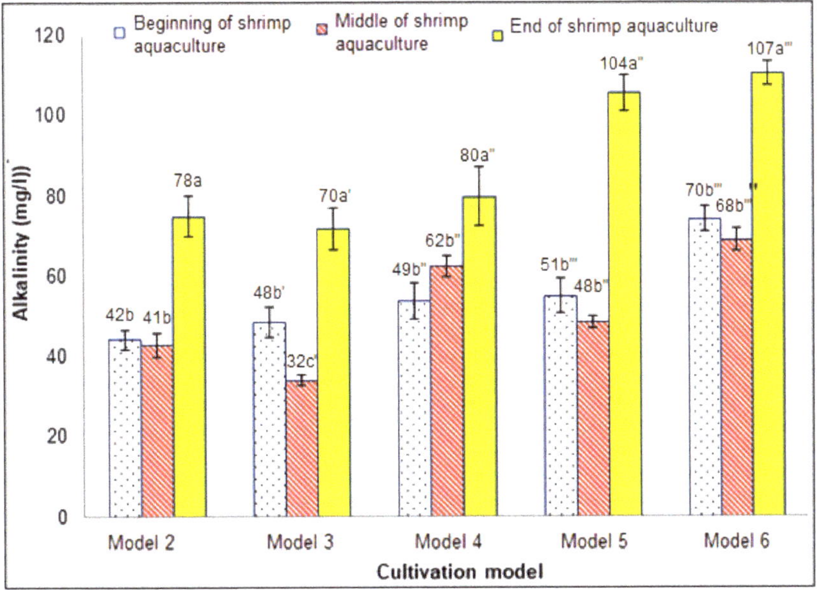

Figure 6. Variations of alkalinity in farming models. Notes: a and b are compared according to sampling time, the same letters within the same column or row have no significant difference. The symbols ′, ″, ‴ are used to distinguish different experimental or model areas. ns: no difference (compared at a significance level of 5%).

3.2.6. Hydrogen Sulfide (H_2S) Content in Pond Water

The results from Figure 7 show that the concentration of H_2S in the farming models varies from 0.1 to 0.25 mg/L, and farming systems in the brackish subregion seemed to have higher H_2S content (0.15 to 0.25 mg/L) compared to other subregions. Generally, H_2S in farming models decreased in the middle of the season and rebounded at the end of the season. This may be due to the accumulation of organic matter under anaerobic conditions, resulting in an increase in H_2S. The high concentrations of H_2S are not necessarily fatal for the shrimps but will affect the lifecycle and susceptibility to environmental conditions and diseases. The conversion of organic matter in the pond increases the reduction process in the pond bottom soil. SO_4^{2-}, derived from mineralized organic matter and from seawater, is reduced with the involvement of microorganisms to form S^{2-}, which in turn becomes H_2S [32].

The presence of H2S, even in a small amount, is harmful to the development of aquatic organisms [54–56]. Previous studies suggested that H_2S concentration in shrimp pond water should not exceed 0.03 mg/L as H_2S concentrations in the higher than 0.01 mg/L is undesirable [30] and H_2S content in the range of 0.037–0.093 mg/L has been proved to be fatal on shrimps [57]. In general, through the analysis results, concentrations of H_2S were above the appropriate threshold for the development of shrimps in all farming models, posing a high risk for most ponds in the study area. Therefore, it is necessary to prepare and rehabilitate the pond before seeding shrimps, and at the same time increase the ventilation in the pond through the implementation of the fan systems in the black tiger shrimp monoculture system.

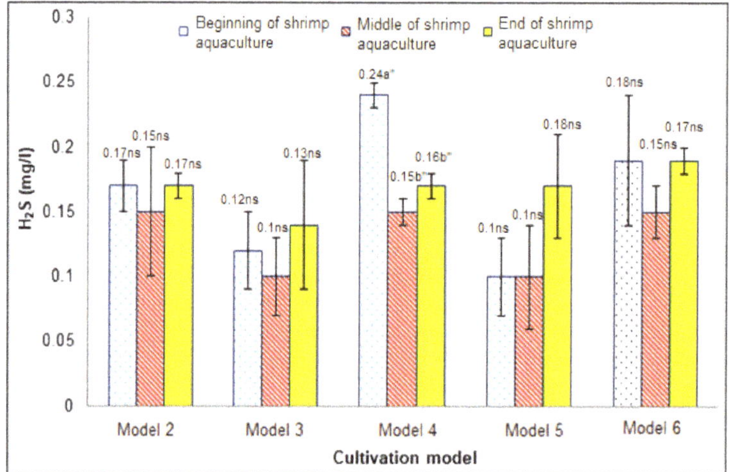

Figure 7. Fluctuations in H_2S concentration in farming models.Notes: a and b are compared according to sampling time, the same letters within the same column or row have no significant difference. The symbol ′ is used to distinguish different experimental or model areas. ns: no difference (compared at a significance level of 5%).

3.2.7. Chemical Oxygen Demand (COD) in Water

Figure 8 showed that the content of COD in water in the farming models. COD ranged from 7.10 to 880 mg/L in which the highest values were achieved at the end of the shrimp harvest. In general, COD content in the models is lower than the recommended values. According to Boyd (1998) and Smith et al. (2002), the recommend COD in the pond was about 20 mg/L [28,58]. Other COD ranges suggested in previous studies were 5–10 mg/L, 20–30 mg/L, and 15–30 mg/L [59–61].

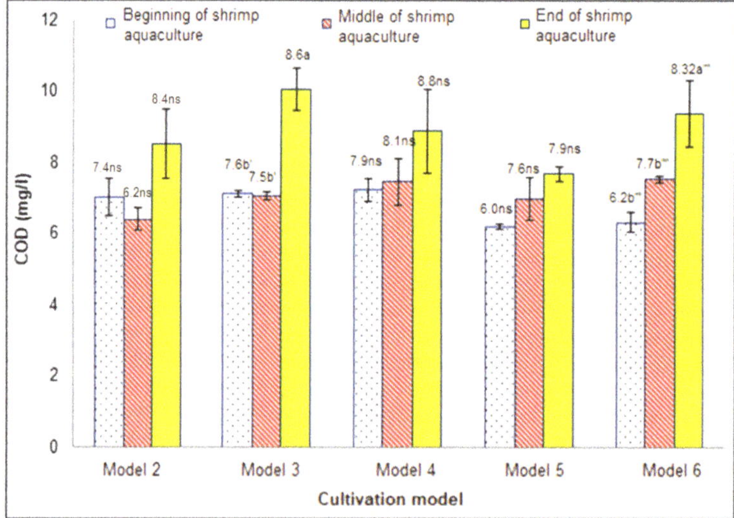

Figure 8. Fluctuations in chemical oxygen demand (COD) content in farming models. Notes: a and b are compared according to sampling time, the same letters within the same column or row have no significant difference. The symbols ′, ′′ are used to distinguish different experimental or model areas. ns: no difference (compared at a significance level of 5%).

The trend of COD content was increasing during the stages of shrimp farming, albeit the differences were statistically insignificant in some models. In most farming systems, COD levels tended to rise at the end of the season. This result is also consistent with the research results of Tat Anh Thu (2009) confirming that the COD of water increases with the increase of organic matter since COD is an indicator of organic richness in pond water [62]. COD content also showed a positive relationship with regard to the salinity of the area. This result is also consistent with the research result of Sansanayuth, explaining that high salinity hinders the mineralization rate of organic minerals, usually leading to lower COD content [63].

Overall, COD values in the farming pond are at a low level in comparison with the recommended values, which is indicative of favorable development of blue prawn, black tiger shrimp, and white shrimp.

4. Conclusions

Many water and soil quality indicators were measured in different farming systems in a saline district in the Mekong Delta. Results showed that the water environment in aquaculture ponds is generally favorable for the development of the three shrimp species, reflected by suitable pH, nitrogen, phosphorous content, and COD. Salinity is optimal for the development of blue prawn but falls within lower thresholds for white legged and black tiger shrimps. In addition, water in the surveyed models has low alkalinity and excessively high H_2S content. Both of which adversely affect the shrimp growth. Regarding soil quality, soil pH is suitable for shrimp aquaculture. However, exchangeable Na^+ was high and soil in models in the brackish and saline region was sodicized. Key remediation strategies derived from these results should involve soil leaching, increasing the water alkalinity and lowering H_2S content in aquaculture ponds.

Author Contributions: Investigation, L.V.T., T.T., and H.H.L.; writing—original draft, L.V.T. and T.T. All authors have read and agreed to the published version of the manuscript.

Funding: This research received no external funding.

Conflicts of Interest: The authors declare no conflict of interest.

References

1. Cline, W.R. *Global Warming and Agriculture: Impact Estimates by Country*; Peterson Institute: Washington, DC, USA, 2007.
2. Ministry of Natural Resource and Environment of Vietnam—MONRE. *Scenarios for Climage Change and Sea Water Rise in Vietnam*; Vietnam resource; Environment and Maps Publisher: Ha Noi, Vietnam, 2012.
3. Solomon, S.; Manning, M.; Marquis, M.; Qin, D. *Climate Change 2007-The Physical Science Basis: Working Group I Contribution to the Fourth Assessment Report of the IPCC*; Cambridge University Press: Cambridge, UK; New York, NY, USA, 2007.
4. Akter, M.; Oue, H. Effect of Saline Irrigation on Accumulation of Na+, K+, Ca2+, and Mg2+ Ions in Rice Plants. *Agriculture* **2018**, *8*, 164. [CrossRef]
5. Mavi, M.S.; Marschner, P.; Chittleborough, D.J.; Cox, J.W.; Sanderman, J. Salinity and sodicity affect soil respiration and dissolved organic matter dynamics differentially in soils varying in texture. *Soil Biol. Biochem.* **2012**, *45*, 8–13. [CrossRef]
6. Tripathi, S.; Kumari, S.; Chakraborty, A.; Gupta, A.; Chakrabarti, K.; Bandyapadhyay, B.K. Microbial biomass and its activities in salt-affected coastal soils. *Biol. Fertil. Soils* **2006**, *42*, 273–277. [CrossRef]
7. Pan, C.; Liu, C.; Zhao, H.; Wang, Y. Changes of soil physico-chemical properties and enzyme activities in relation to grassland salinization. *Eur. J. Soil Biol.* **2013**, *55*, 13–19. [CrossRef]
8. Hamdi, L.; Suleiman, A.; Hoogenboom, G.; Shelia, V. Response of the Durum Wheat Cultivar Um Qais (*Triticum turgidum* subsp. *durum*) to Salinity. *Agriculture* **2019**, *9*, 135. [CrossRef]
9. Ben Tre People's Committee. *Report on Socio-Economic Situation, Security and Defense in 2012 and Directions for 2013*; Ben Tre Statistical Office: Ben Tre, Vietnam, 2012.

10. Ben Tre People's Committee. *Report on Actual State of Environment in Ben Tre province in 2011*; Ben Tre Statistical Office: Ben Tre, Vietnam, 2011.
11. Phuong, N.T.; Minh, T.H.; Tuan, N.A. Overall of Black Tiger Shrimp Farming Models in Mekong Delta. Conference on Development of Coastal Aquatic Resources. Master's Thesis, Department of Environmental Management - Faculty of Environment and Natural Resources Nong Lam University, Ho Chi Minh City, Vietnam, 4 August 2008.
12. Nguyen, H.Q.; Korbee, D.; Ho, H.L.; Weger, J.; Thi Thanh Hoa, P.; Thi Thanh Duyen, N.; Dang Manh Hong Luan, P.; Luu, T.T.; Ho Phuong Thao, D.; Thi Thu Trang, N.; et al. Farmer adoptability for livelihood transformations in the Mekong Delta: A case in Ben Tre province. *J. Environ. Plan. Manag.* **2019**, *62*, 1603–1618. [CrossRef]
13. Ho, H.L.; Nguyen, T.H.D.; Nguyen, T.C.; Kim, N.I.; Shimizu, Y. Integrated evaluation of ecosystem services in prawn-rice rotational crops, Vietnam. *Ecosyst. Serv.* **2017**, *26*, 377–387.
14. Srinivasan, M.; Khan, S.A.; Rajagopal, S. Culture of prawn in rotation with shrimp. *NAGA* **1997**, *20*, 21–23.
15. Phong, N.D.; My, T.V.; Nang, N.D.; Tuong, T.P.; Phuoc, T.N.; Trung, N.H. *Salinity Dynamics and its Implications for Cropping Patterns and Rice Performance in Rice-Shrimp Farming Systems in My Xuyen and Gia Rai*; Sustainability of Rice-Shrimp Systems in the Mekong Delta, Vietnam, ACIAR Technical Report; University of Sydney: Sydney, Australia, 2002.
16. Dall, W.H.B.J.; Hill, B.J.; Rothlisberg, P.C.; Sharples, D.J. *The Biology of the Penaeidae*; Academic Press Publisher: Cambridge, MA, USA, 1990.
17. Minh, T.H.; Jackson, C.J.; Hoa, T.T.T.; Ngoc, L.B.; Preston, N. *Growth and Survival of Penaeus Monodon in Relation to the Physical Conditions in Rice-Shrimp Ponds in the Mekong Delta*; Sustainability of rice-shrimp systems in the Mekong Delta, Vietnam, ACIAR Technical Report; University of Sydney: Sydney, Australia, 2002.
18. Gianello, C.; Bremner, J.M. Comparison of chemical methods of assessing potentially available organic nitrogen in soil. *Commun. Soil Sci. Plant Anal.* **1986**, *17*, 215–236. [CrossRef]
19. Olsen, S.R. *Estimation of Available Phosphorus in Soils by Extraction with Sodium Bicarbonate*; U.S. Dept. of Agriculture: Washington, DC, USA, 1954.
20. Nelson, D.W.; Sommers, L.E. *Total Carbon, Organic Carbon, and Organic Matter 1. Methods of Soil Analysis. Part 2. Chemical and Microbiological Properties*; Purdue University: West Lafayette, IN, USA, 1982; pp. 539–579.
21. Houba, V.J.C.; Van Der Lee, J.J.; Novzamsky, I.; Walinga, I. *Soil and Plant Analysis. Aseries of Syllabi, Part 5: Soil Analysis Procedures*; Department of Soil Science and Plant Nutrition, Wageningen Agriicultural University: Gelderland, The Netherland, 1988.
22. Clesceri, L.S.; Greenberg, A.E.; Eaton, A.D. *Standard Methods for the Examination of Water and Wastewater*; American Public Health Association: Washington, DC, USA, 1998; pp. 4–415.
23. Hens, M. Aqueous Phase Speciation of Phosphorus in Sandy Soils. Ph.D. Thesis, Katholieke Universiteit Leuven, Leuven, Belgium, 1999.
24. American Public Health Association. *Standard Methodsfor the Examination of Water and Wastewter*, 20th ed.; American Public Health Association, American Waterworks Association and Water Environment Federation: Washington, DC, USA, 1998.
25. Brady, N.C. *The Nature and Properties and Soils. AK Ghosh*; Printing-Hall of India Pvt. Ltd.: New Delhi, India, 1990; p. 383.
26. Avnimelech, Y.; Lacher, M. On the Role of Soil in the Maintenance of Fish Pond's Fertility. In *Hyperophic Ecosystems*; Barica, J., Mur, L.R., Eds.; W. Junk BV Publisher: The Hague, The Netherland, 1980; pp. 251–255.
27. Ngo-Hoang, D.L.; Thuy, N.T.T. Local Knowledge Ben Tre's Coastal on Resilience to Climate Change Background Study (August 27, 2019). Available online: https://ssrn.com/abstract=3367754 (accessed on 13 November 2019). [CrossRef]
28. James, E.M. *Water Quality for Pond Aquaculture*; Research and Development Series No.43; International Center for Aquaculture and Aquatic Environment, Alabama Agricultural Experiment Station, Auburn University: Auburn, AL, USA, 1998; pp. 789–811.
29. Guong, V.T. *Assessing Soil and Water Quality and Proposing Appropriate Use Measures for Shrimp Farming Models in My Xuyen District, Soc Trang Province (Vietnamese Version)*; Department of Science and Technology of Soc Trang Province: Soc Trang, Vietnam, 2003.

30. Hung, N.N. Measurement of EC and conversion for salinity evaluation scale of saline soil of rice-shrimp farming in Mekong Delta (Vietnamese). *J. Agric. Rural Dev.* **2010**, *5*, 41–45.
31. Lamond, R.E.; Whitney, D.A. *Management of Saline and Sodic Soils*; Retrieved 10.1; Kansas State University Agricultural Experiment Station and Cooperative Extension Service: Manhattan, KS, USA, 2020.
32. Centre for Meteorological Forecast of Ben Tre. *Report on Salinity Instrusion in Ben Tre Province to 2012*; Centre for Meteorological Forecast of Ben Tre: Ben Tre, Vietnam, 2012.
33. Ben Tre Department of Resources and Environment. *Report on Environmental Context in 2011*; Ben Tre Department of Resources and Environment: Ben Tre, Vietnam, 2011.
34. Hoa, N.M. Additional sources of potassium into the soil in rice-intensive farming systems in the Mekong Delta. *J. Soil Sci.* **2006**, *24*, 62–65.
35. Turnbull, P.C.J.; Funge-Smith, S.; Macrea, I.; Slimswan, C. *Health Management in Shrimp Pond*; Agricultural Publisher: Ho Chi Minh city, Vietnam, 2002.
36. Long, D.N. *Giant Freshwater Shrimp (Macrobrachium Rosenbergi) Farming Techniques in Coconut Garden Ditches and in Rice Fields in Ben Tre Province*; Can Tho University, Department of Freshwater Aquaculture—Department of Fisheries: Can Tho, Vietnam, 2012.
37. Nho, N.T.; Cuong, T.K.; Diep, L.M.; Ne, V.T.; Lien, N.T.; Hau, N.T.M. *Questions About Black Tiger Shrimp*; Publisher of Agriculture: Ho Chi Minh City, Vietnam, 2002.
38. Hung, N.P. *Cultivation Techniques of Shrimp—Rice Models*; Bac Lieu Department of Agriculture and Rural Development, Center for Agriculture and Fishery Extension: Bac Lieu, Vietnam, 2012.
39. Ayers, R.S.; Westcot, D.W. *Water Quality for Agriculture*; FAO Irrigation and Drainage Paper. No. 29. Rev. 1; FAO: La Jolla, CA, USA, 1985.
40. Chanratchakool, P. *Health Management in Shrimp Ponds*; Aquatic Animal Health Research Institute: Bangkok, Thailand, 1998; ISBN 9789747604511.
41. Thu, T.A. Surveying the Quality of Soil, Water Environment in Aquaculture and Nutrient Accumulation in Aquaculture Ponds in Two Districts of Vinh Chau and My Xuyen in Soc Trang Province. Master's Thesis, Department of Soil Science, Faculty of Agriculture, Can Tho University, Can Tho, Vietnam, 2010.
42. Hung, N.N. Evaluation on methods of phosphorus analysis on corn field in Mekong Delta. *J. Soil Sci.* **2009**, *32*, 62–66. (In Vietnamese)
43. Kungvankij, P.; Chua, T.E.; Pudadera, B.J., Jr.; Corre, K.G.; Borlongan, E.; Tiro, L.B., Jr.; Talean, G.A. Shrimp Culture: Pond Design, Operation and Management. 1986. Available online: http://www.fao.org/3/AC210E/AC210E00.htm (accessed on 20 November 2019).
44. Whetstone, J.M.; Treece, G.D.; Browdy, C.L.; Stokes, A.D. *Opportunities and Constraints in Marine Shrimp Farming*; Southern Regional Aquaculture Center: Washington, DC, USA, 2000.
45. Long, D.N. Technique for Farming of Blue Prawn (Macrobrachium Rosenbergi) in Coconut Irrigation Canals and in Rice Paddy Fields in Ben Tre. Master's Thesis, Aquaculture Department, Faculty of Agricuture, Can Tho University, Can Tho, Vietnam, 2012.
46. Hoi, N.D. *Management of Water Quality in Aquaculture Pond*; Institute for Aquatic Farming: Bac Ninh, Vietnam, 2000.
47. Giang, T.T. Survey on Some Chemical and Physical Properties of Soil and Water of Specialized Systems of Shrimp and Rice Shrimp on Acid Sulfate and Non-Alum Soil in Thoi Binh, Cai Nuoc and Dam Doi districts, Ca Mau. Master's Thesis, Department of Soil Science, Faculty of Agriculture, Can Tho University, Can Tho, Vietnam, 2003.
48. Te, B.Q. *Intensive Shrimp Farming Ensures Food Hygiene and Safety Under the GAP Model*; Ministry of Agriculture and Rural Development, Research Institute for Aquaculture 1: Ha Noi, Vietnam, 2009.
49. Milstein, A.; Islam, M.S.; Wahab, M.A.; Kamal, A.H.M.; Dewan, S. Characterization of water quality in shrimp ponds of different sizes and with different management regimes using multivariate statistical analysis. *Aquac. Int.* **2005**, *13*, 501–518. [CrossRef]
50. Hoa, T.V.; Dom, T.V.; Khiem, D.V. *Technique for Intensive Farming of Black Tiger Shrimp in 101 Frequently Asked Questions in Aquacultural Production*; Youth Publisher: Ho Chi Minh, Vietnam, 2002.
51. Hung, N.P. *Technique for Farming of Shrimp-Rice Integrated Model*; Centre for Agricultural and Aquacultural Extension, Bac Lieu Department of Agricultural and Rural Development: Bac Lieu, Vietnam, 2012.
52. Lazur, A. *Growout Pond and Water Quality Management. JIFSAN (Joint Institute for Safety and Applied Nutrition) Good Aquacultural Practices Program*; University of Maryland: College Park, MD, USA, 2007.

53. Wurts, W.A.; Durborow, R.M. Interactions of pH, Carbon Dioxide, Alkalinity and Hardness in Fish Ponds. 1992. Available online: https://appliedecology.cals.ncsu.edu/wp-content/uploads/SRAC-0464.pdf (accessed on 25 November 2019).
54. Schwedler, T.E.; Tucker, C.S.; Beleau, M.H. Non-infectious diseases. In *Channel Catfish Culture*; Tucker, C.S., Ed.; Elsevier: New York, NY, USA, 1985; Volume 15, pp. 497–541.
55. Vismann, B. Sulfide species and total sulfide toxicity in the shrimp Crangon crangon. *J. Exp. Mar. Biol. Ecol.* **1996**, *204*, 141–154. [CrossRef]
56. Gopakumar, G.; Kuttyamma, V.J. Effect of hydrogen sulphide on two species of penaeid prawns Penauesindicus (H. Milne Edwards) and Metapenauesdobsoni (Miers). *Bull. Environ. Contam. Toxicol.* **1996**, *57*, 824–828. [CrossRef] [PubMed]
57. Kutty, M.N. *Site Selection for Aquaculture: Chemical Features of Water*; Lectures presented at ARAC [African Regional Aquaculture Centre] for the Senior Aquaculturists Course. Port Harcourt (Nigeria); Microfiche no: 297785. RAF/82/009 ARAC/87/WP/12(9); Establishment of African Regional Aquaculture Centre: Aluu, Nigeria, 1987; 40p.
58. Smith, D.M.; Burford, M.A.; Tabrett, S.J.; Irvin, S.J.; Ward, L. The effect of feeding frequency on water quality and growth of the black tiger shrimp (Penaeus monodon). *Aquaculture* **2002**, *207*, 125–136. [CrossRef]
59. Te, B.Q. *Intensive Shrimp Farming to Ensure Food Safety Following GAP Model*; Institute for Aquacultural Research, Vietnam Ministry for Agriculture and Rural Development: Ha Noi, Vietnam, 2009.
60. Cat, L.V. *Treatment of Nitrogen- and Phosphorus-Rich Wastewater*; Institute of Science and Technology of Vietnam, Natural Science and Technology Publisher: Ha Noi, Vietnam, 2006.
61. Phu, Q.T. *Textbook on Analysis of Water Quality and Pond Environment*; Department of Aquaculture, Can Tho University: Can Tho, Vietnam, 2004.
62. Thu, T.A. Survey on Soil and Water Quality for Aquaculture and Nutrition Accumulation in Aquaculture Ponds in Vinh Chau and My Xuyen Dictrict, Soc Trang Province. Ph.D. Thesis, Can Tho University, Can Tho, Vietnam, 2009.
63. Sansanayuth, P.; Phadungchep, A.; Ngammontha, S.; Ngdngam, S.; Sukasem, P.; Hoshino, H.; Ttabucanon, M.S. Shrimp pond effluent: Pollution problems and treatment by constructed wetlands. *Water Sci. Technol.* **1996**, *34*, 93–98. [CrossRef]

Article

Influence of Species Composition and Management on Biomass Production in Missouri

Ranjith P. Udawatta [1,2,*], Clark J. Gantzer [1], Timothy M. Reinbott [3], Ray L. Wright [3], Robert A. Pierce II [1] and Walter Wehtje [1]

1 School of Natural Resources, University of Missouri, Columbia, MO 65211, USA; gantzerc@missouri.edu (C.J.G.); PierceR@missouri.edu (R.A.P.II); WehtjeW@missouri.edu (W.W.)
2 The Center for Agroforestry, School of Natural Resources, University of Missouri, Columbia, MO 65211, USA
3 University of Missouri Bradford Research Center, 4968 Rangeline Road, Columbia, MO 65201-8973, USA; reinbottT@missouri.edu (T.M.R.); wrightr@missouri.edu (R.L.W.)
* Correspondence: UdawattaR@missouri.edu; Tel.: +573-882-4347; Fax: +573-882-1977

Received: 23 January 2020; Accepted: 11 March 2020; Published: 13 March 2020

Abstract: Perennial biofuel crops help to reduce both dependence on fossil fuels and greenhouse gas emissions while utilizing nutrients more efficiently compared to annual crops. In addition, perennial crops grown for biofuels have the potential to produce high biomass yields, are capable of increased carbon sequestration, and are beneficial for reducing soil erosion. Various monocultures and mixtures of perennial grasses and forbs can be established to achieve these benefits. The objective of this study was to quantify the effects of feedstock mixture and cutting height on yields. The base feedstock treatments included a monoculture of switchgrass (SG) and a switchgrass:big bluestem 1:1 mixture (SGBBS). Other treatments included mixtures of the base feedstock with ratios of base to native forbs plus legumes of 100:0, 80:20, 60:40, and 20:80. The study was established in 2008. Biomass crops typically require 2 to 3 years to produce a uniform stand. Therefore, harvest data were collected from July 2010 to July 2013. Three harvest times were selected to represent (1) biomass for biofuel (March), (2) forage (July), and (3) forage and biomass (October). Annual mean yields varied between 4.97 Mg ha^{-1} in 2010 to 5.56 Mg ha^{-1} in 2011. However, the lowest yield of 2.82 Mg ha^{-1} in March and the highest yield of 7.18 Mg ha^{-1} in July were harvested in 2013. The mean yield was 5.21 Mg ha^{-1} during the 4 year study. The effect of species mixture was not significant on yield. The cutting height was significant ($p < 0.001$), with greater yield for the 15 cm compared to the 30 cm cutting height. Yield differences were larger between harvest times during the early phase of the study. Yield difference within a harvest time was not significant for 3 of the 10 harvests. Future studies should examine changes in biomass production for mixture composition with time for selection of optimal regional specific species mixtures.

Keywords: big blue stem; Cave in rock; claypan; forbs; legumes

---

1. Introduction

Bioenergy acts (Biomass Research and Development Act 2000, Energy Policy Act 2005, and Energy Independence and Security Act 2007) and the Farm Bills of 2002, 2008, and 2014 have promoted renewable energy production, mandating 136 billion L production of biofuel by 2022. Biofuels play a role in helping reduce dependence on fossil fuels and greenhouse gas emissions [1]. Midwestern United States monocultures of perennial grasses have been promoted as a potential crop for biomass production. Perennial grasses use nutrients more efficiently compared to annual crops and produce high dry matter yields while reducing soil erosion and increasing carbon sequestration [2,3]. Planting perennial grasses provides numerous ecosystem services, including the reduction of non-native species [4]. Tilman suggests that marginal or retired Conservation Reserve Program (CRP) lands have potential for bioenergy production to avoid land competition with food production [5,6]. This idea supports the general consensus from Missouri agricultural producers that biofuel production will primarily be implemented on less productive soils, which are typically used for livestock production or are currently enrolled in the CRP [7]. Despite the benefits of growing bioenergy crops, challenges exist for producing sufficient feedstock to meet the 2022 production target.

US corn grain yield has increased by 0.12 Mg ha^{-1} annually since 1955, thanks to improvements in genetics and management [8]. Currently, more than 37% of the US corn crop is being used for ethanol production [9]. The highest corn production to date occurred in 2016 with 0.36 10^9 Mg of production on 38 10^6 ha of land [9]. Using an average corn-to-ethanol conversion rate, 2016 corn production could have produced 15.4 billion L of ethanol [10]. Although corn feedstock is an important contributor, other cellulosic feedstocks will be required to meet the 136 billion L target in 2022.

Typically, monocultures of switchgrass (*Panicum virgantum* L.), a native warm-season grass, or *Miscanthus x giganteus*, an introduced species, have been used for biomass production for cellulosic ethanol production. These two species have the potential to produce enough ethanol to offset one-fifth of the US fuel use if planted on ~9% of US cropland [11]. These species have been identified as potential high-biomass-producing species on water- and nutrient-limited eroded soils [12]. In a monoculture, switchgrass yields could vary between 5 and 20 Mg ha^{-1} in the US and are determined by weather, soil, ecotype (upland/lowland), and management [13–16]. Data from 39 sites across 17 states showed biomass yields of ~9 and ~13 Mg ha^{-1} for upland and lowland ecotypes, respectively [16].

Recent studies showed that a diversified cropping system may produce equal or greater biomass for fuel production compared to monocrop systems [17,18]. A yield from a diverse mixture of native grasses and forbs was ~240% greater than for monocrop yield after a decade [17]. In a 5 year study conducted across nine locations, Jungers [18] observed greatest yields with a mixture of four and eight species both with and without added nitrogen. Over time, biomass yield remained constant for these diverse mixtures.

Combination of native legumes and forbs with native warm-season grasses can decrease the need for N fertilizer and thus provide more options for livestock producers (e.g., grazing or biomass production for fuel). Tilman [5,6,17] highlighted the benefits that result from diverse combinations of native grasses and forbs, which create a portfolio effect whereby one species' potential lack of performance is compensated by other species. Such mixtures can also provide additional financial incentives through enrolment in conservation programs (Ex, Cover Crop Standard Practice 340; https://efotg.sc.egov.usda.gov/references/public/IN/340_Cover_Crop.pdf). Establishing diverse mixtures of native forbs, legumes, and warm-season grasses on agricultural land can also provide environmental benefits such as habitat for bobwhite quail, grassland birds, and other wildlife requiring springtime successional vegetation for food and cover [19,20]. Several studies have described the importance of grazer interactions and ecological benefits as well as how grazing alters plant community species composition and impacts nutrient cycles [21,22]. A combination of native grasses, forbs, and legumes can improve soil and water conservation and provide habitats for pollinator species [23,24]. Establishing and managing diverse mixtures of native plants for biomass and forage crops reintroduces a species matrix adapted to a region's climate and soil conditions [25].

In contrast to previously mentioned studies, significant variations in biomass yield across landscapes because of factors including soil type and weather conditions have been reported [26,27]. Studying feedstock yields for five years across topography in North Dakota, South Dakota, and Nebraska, Schmer [28] reported inconsistent relationships between switchgrass yields and topographic attributes. In Iowa and Minnesota, the effect of a field's position within a particular landscape on biomass yield was inconsistent [29,30].

Biomass yield used for feedstock is also influenced by cropping practice and management [2,3]. The timing of harvest impacts yield quality parameters including moisture and ash contents and other traits [13,31–33]. Studying switchgras harvest times in Iowa, Vogel [34] recorded optimum feedstock yields for R3 to R5 (panicle fully emerged from boot to postanthesis) maturity stages. Others have shown yield responses to N fertilizers. No nitrogen treatment showed increasing yields up to the fourth year, while yields declined in the last sampling for the 160 kg ha^{-1} treatment in a 16 year study [35]. Angelini [36] concluded that fertilization mostly affects the initial four years of crop growth and then declines afterwards. The effects of management and soil fertility or nutrient supply are becoming increasingly important for the development of an efficient feedstock production strategy that can also provide other ecosystem services such as improved soil conservation as well as enhanced pollinator and wildlife habitat. Cutting heights influence the amount of residual vegetation that is available for use by a variety of grassland wildlife species for protective cover during the winter and for nesting habitats during spring [20]. Research suggests that the development of new switchgrass cultivars based on local ecotypes will also provide increased opportunities for the production of biomass and improved ecological services [37,38].

Considerable inconsistencies exist between the effects management and species composition on feedstock yields. Few studies have been conducted in claypan soils to examine possible differences in biomass yields for feedstock as influenced by factors such as soil fertility and crop management practices. Claypan soils (Major Land Resource Area 113) are characterized by a dense, impermeable clay horizon with very low hydraulic conductivity and greater runoff potential, thus potentially removing large amounts of sediment and nutrients from agricultural watersheds [39]. The objectives of this study were to examine the effects crop management practices such as cutting height, cutting time, and the influence of plant mixture diversity on feedstock yields for four years. A monoculture of switchgrass was established along with plots that contained equal combinations of switchgrass and big bluestem (*Andropogon gerardii* Vitman). Each of these combinations also contained mixtures of native forbs and legumes seeded at various grass-to-forb ratios to address these study objectives.

2. Materials and Methods

The study was established at the University of Missouri's Bradford Research Center (MU BREC) in 2008. The study area consisted of a corn–soybean rotation prior to the establishment of the biomass plots, and soybean was the crop harvested during the previous year. This site represents a claypan soil (Mexico silt loam 0%–2% slope; Fine, smectitic, mesic Aeric Vertic Epiaqualfs) of central Missouri. The primary climate–soil–plant community classification is Claypan Summit Prairie (ecological site ID: R113XY001MO) see http://esis.sc.egov.usda.gov).

Feedstock treatments included a monoculture of switchgrass (SG), a switchgrass and big bluestem 1:1 mixture (SGBBS), and these grasses planted with varying ratios of native forbs and legumes. The switchgrass variety used was Cave in Rock. The grass to forb and legume species ratios used in the plantings were 100:0, 80:20, 60:40, and 20:80. The selection of native species, legumes, and forbs in the mixtures were based on ability to fix nitrogen and provide forage while enhancing plant diversity of the stand and improving wildlife habitat [40,41]. The legume species were Partridge pea (*Chamaecrista fasciculate* Michx.), Illinois bundle flower (*Desmanthus illinoensis* (Michx.) MacMill. ex B.L. Rob & Fernald), Showy tick trefoil (*Desmodium candense* (L.) DC.), Roundhead lespedeza (*Lespedeza capitata* Michx.), Slender lespedeza (*Lespedeza virginica* (L.) Britton), and Sensitive briar (*Mimosa quadrivalvis var. nuttallii*). Forbs were: Ashy sunflower (*Helianthus mollis* Lam.), Purple coneflower (*Echinacea purpurea* (L.) Moench), Plains coreopsis (*Coreopsis palmate* Nutt.), Maximimillian Sunflower (*Helianthus maximiliani* Schrad.), wild bergamot *(Monarda fistulosa* L.), and Oxeye sunflower (Heliopsis helianthoides (L.) Sweet).

Plots were established using a Hege plot drill during the fall of 2008. Broadleaf weeds were controlled with 2–4-D within the monoculture plots at full label rate. Monoculture grasses were fertilized with 36 kg (80 lbs) of nitrogen ha^{-1} rate according to soil test recommendations each year. Seeding rates were 50 seeds per 93 cm^2. The study design consisted of four replicated blocks; thus, the design had four replications. Each plot within a block was 9.1m wide and 15m long. Each plot was divided into two subplots for the two cutting height treatments of 15 and 30 cm. Each of these subplots were completely harvested at the specified treatment height in March, July, and October. Total yield was determined by harvesting with a forage harvester constructed for the study. Dry weight of biomass for each harvested plot was determined by oven drying at 50 °C for a minimum of 3 days. Dry matter biomass yields for each year and treatment were analyzed using SAS 9.2 [42] PROC GLM MIXED to determine individual mixture, year, harvest time, and harvest height treatments on yields as described by Steel [43]. Regressions were analyzed to evaluate the effect of cutting time (month and month plus year) on yields and reported cutting time and yield relationships.

3. Results and Discussion

3.1. Rainfall and Temperature

Annual precipitation varied from the normal 30 y mean value of 1083 mm by 6%, −28%, −38%, and −13% for 2010, 2011, 2012, and 2013 (Figure 1, Table 1). Precipitation deviated by 15%, −33%, −36%, and −9% from the normal for 1 March to 31 October for 2010, 2011, 2012, and 2013. Rainfall amounts differed by 17%, −41%, −68%, and −60% from the long-term mean in those years during the most productive growth period of 1 June to 1 September. A normal year receives 335 mm of rain during the most productive growth period, compared to 393, 197, 106, and 133 mm during the study. In 2010, 2011, 2012, and 2013, rainfall amounts were below the long-term mean for 9, 10, 10, and 9 months. Among four years, 2012 was the driest and significant crop failures occurred in the county.

Monthly weather generally followed the long-term pattern until May in 2010 (Figure 1). July and September rainfall amounts were greater than the long-term monthly values. Sufficient rainfall during the first nine months of 2010 might have helped good growth of grass, better survival, and productive growth. However, three months (October, November, and December) of the year had amounts less than 50% of the long-term monthly values (Figure 1). The low rainfall in the last three months of 2010 and early 2011 may have influenced the soil moisture recharge and the plant growth in subsequent years.

Lower rainfall amounts in 2011, 2012, and 2013 caused large cumulative deficits within a year compared to the cumulative 30 year mean. The cumulative deficit was 289, 417, and 142 on December 31 of 2011, 2012, and 2013. The 2011 cumulative rainfall was below the normal for the entire year, while it was below normal from May to December in 2012. Rainfall in April and May of 2013 caused greater cumulative rainfall amounts than the long-term values. Lower rainfall amounts after June 2013 created a rainfall deficit during the last six months of 2013.

Maximum monthly temperature values were similar to 30 year monthly maximum values in 2010 (Figure 1). In 2011, monthly maximum values were above normal for March, May, June, July, October, November, and December. For 2012, only September and October had lower monthly maximum values than the 30 year monthly maximum values. Maximum March and July temperatures were 7 °C greater than the 30 year monthly maximum values. The last year of the study (2013) had favorable temperature conditions for plant growth (Figure 1).

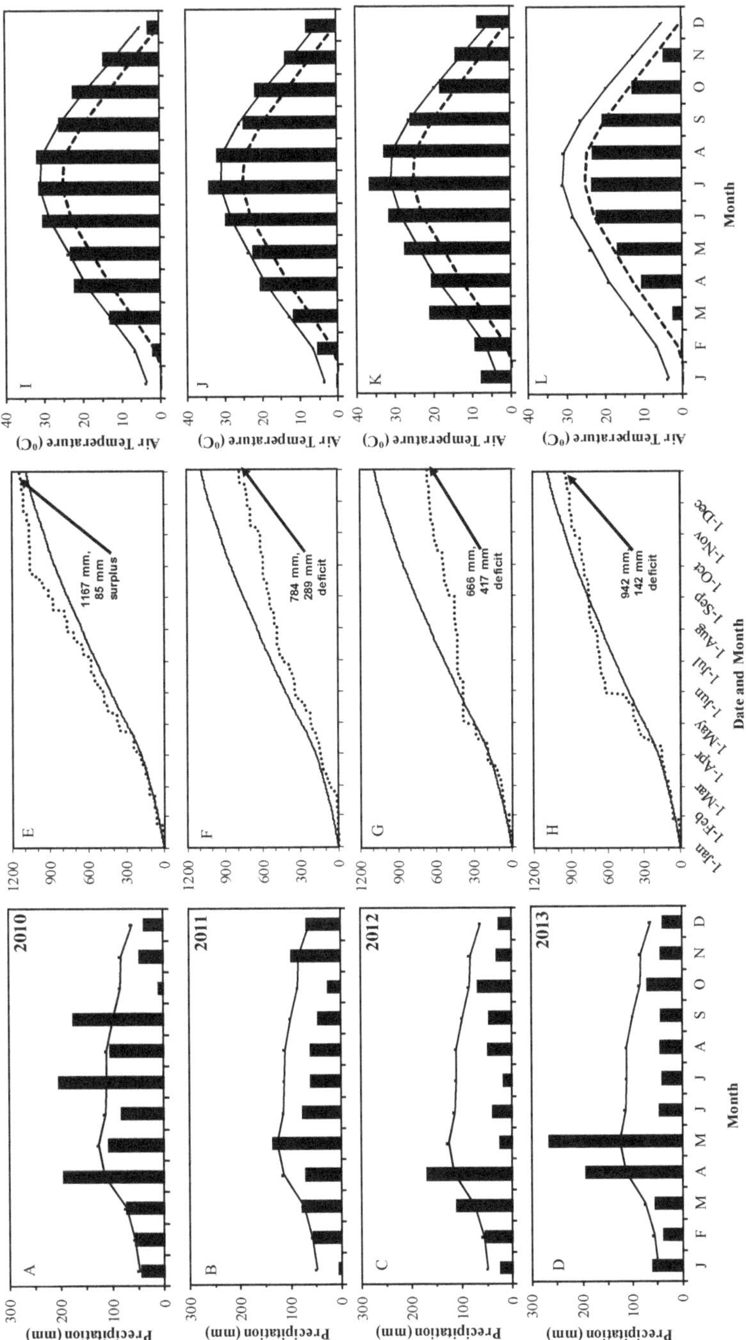

Figure 1. Monthly rainfall distribution (bars) and monthly 30 year mean (line) for 2010, 2011, 2012, and 2013 (**A–D**); cumulative long-term and annual cumulative rainfall for 2010, 2011, 2012, and 2013 (**E–H**); and monthly maximum temperature (bars), monthly 30 year maximum (black line) and monthly 30 year mean (broken line) for 2010, 2011, 2012, and 2013 (**I–L**) at Bradford Research Center, University of Missouri.

Table 1. Annual, 1 March to 31 October, and 1 June to 1 September rainfall amounts and deviations from the long-term means during the study period at the Bradford Research Center, University of Missouri, USA.

Rainfall Category	Long-Term	2010	2011	2012	2013
Annual (mm)	1083	1149	784	666	942
Percent deviation from the normal precipitation		6	−28	−38	−13
1 March to 31 October (mm)	832	959	554	529	760
Percent deviation from the normal precipitation		15	−33	−36	-9
1 June to 1 September (mm)	335	393	197	106	133
Percent deviation from the normal precipitation		17	−41	−68	-60

3.2. Biomass Yield and Weather

Annual mean yields during the study varied between 4.97 Mg ha^{-1} in 2010 to 5.56 Mg ha^{-1} in 2011. However, the lowest yield of 2.82 Mg ha^{-1} in March and the highest yield of 7.18 Mg ha^{-1} in July were observed in the same year, 2013 (Figure 2). Forage yields might have reflected the effects of growth responses during the early phase and the effects of weather. Our biomass yields were lower when compared with 10 Mg ha^{-1} harvested after three years [44]. These lower yields can be attributed to early phase of the experiment, dry weather conditions, and soil water deficit (Figure 1). In a metadata analysis Wullschleger [16] reported switchgrass yield ranging from 1 to 40 Mg ha^{-1}, with the majority of data points within the 10–14 Mg ha^{-1} range. Our yields were within the ranges observed in other areas in the country.

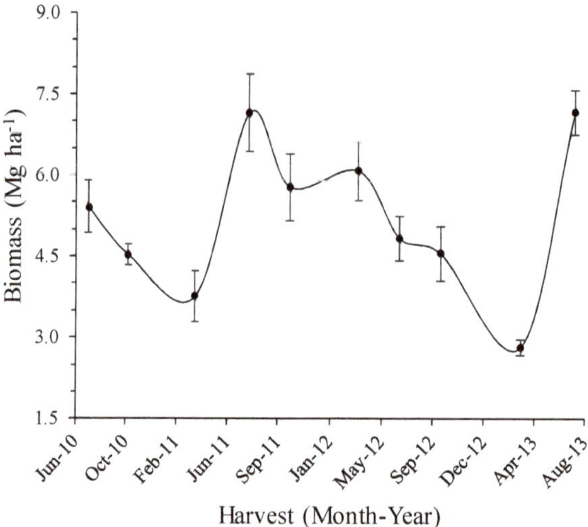

Figure 2. Distribution of mean biomass yields for the eight mixture treatments for 2010, 2011, 2012, and 2013 by harvest time at the Bradford Research Center of University of Missouri, Columbia, Missouri, USA.

Biomass yields fluctuated during the study period (Figure 2). Biomass yield decreased by 16% from the first harvest to the second harvest, and another 17% decrease occurred between the second and third harvests. The fourth harvest had the second largest (7.15 Mg ha^{-1}) yield during the study, a 90% increase from the third harvest. Favorable growth conditions including above normal rainfall in 2010 and temperature conditions might have helped support plant growth before the fourth harvest (Figure 1). Similarly to our results, high precipitation and favorable temperatures produced 39 Mg ha^{-1} yields on switchgrass cultivar Alamo [16,45]. Dragoni [26] observed no yield differences in *Arundo donax* L. in single- and double-harvest systems in Italy over two years. However, biomass yields were lower in the second year for both harvest systems. Compared to our study, they had received sufficient rain during their study. In their study, the double-cut harvest had lower yield in the second harvest of the second year, while single-cut harvest had no substantial yield reduction. Biomass yields also varied by cutting time. The mean March yield was 69% of the July yield during the study. July yield was the highest with 6.14 Mg ha^{-1}. October yield was 4.95 Mg ha^{-1}, about 81% of the July yield. A regression using the cutting month as the independent variable and biomass yield as the dependent variable showed a nonsignificant (p = 0.191) linear relationship with a 0.37 coefficient of determination (r^2). A regression model with year and month improved the r^2 (0.45), although not significantly (p = 0.66).

Lower rainfall amounts and higher temperature conditions likely caused poor plant growth and lower yields after the fourth harvest of July 2011. The yield decline was 19% from the fourth to the fifth harvest. Subsequent forage yields were smaller than the previous yields for the next five harvests until March 2013 (Figure 2). March 2013 was the lowest yield during the 4 year study (2.82 Mg ha^{-1}). Soil moisture deficit and above normal maximum temperatures likely affected the plant growth from July 2011 to March 2013 (Figure 1). Despite a 417 mm cumulative deficit by December 31 of 2012, the April–May precipitation and favorable temperature conditions in 2013 might have helped better plant growth and biomass yield thus increased to 7.18 Mg ha^{-1} in July 2013, a 155% increase from the March 2013 yield. Muir [46] observed that March to August rainfall in Stephenville, TX highly correlated with biomass yields. Similar to our results, in a 4-year study, Lee [47] correlated their biomass yields to April and May precipitation in South Dakota. The two yield increments, 3rd to 4th harvest and 9th to 10th harvest, suggested that rainfall was the main factor that controlled the biomass yields during the study. The surplus of 85 mm by December 2010 and the second surplus between March and May 2013 may have contributed to the recorded largest forage yields during the study.

Below normal rainfall and severe drought conditions significantly decreased biomass yields of all treatments. The yield increase in the fourth harvest (July 2011) can be attributed to early rain events in 2011 and favorable growth conditions. The reduction in yields in October 2011 could have been due to the lower rainfall and thus soil moisture limitations. Below normal rainfall during this growth period reduced yield. Severe drought, low soil moisture status, and extreme temperatures of 2012 reduced the subsequent yields in 2012. Similar to our results, Wullschleger [16] reported that biomass yield varied by temperature and rainfall. In their metadata analysis, biomass yield increased with increasing temperatures up to 14 °C and decreased. Sufficient rainfall during the growing season and favorable temperatures are critical factors for biomass yields [16,48–50]. Additionally, flexibility of harvesting can help to address weather patterns and bioenergy market for optimum benefits [51]. Other studies have shown that crop maturity negatively affected methane yields, while juvenile traits were detrimental for thermochemical processes but beneficial for anaerobic digestion [52–54].

3.3. Mixture Composition and Yields

Our two main mixtures showed slightly different patterns of yield during the study (Figure 3). The SG mixtures had three prominent yield peaks, while the SGBBS mixtures had only two peaks. The first two mean peak yields were also larger for SG mixtures compared to SGBBS mixtures. This might indicate that SG responded better to favorable conditions compared to BBS. Similarly to our results, Jefferson [55] reported greater yield potential for SG across a latitudinal gradient compared to other species. However, the differences in yields among mixtures were not significant. Generally, all eight combinations followed the same pattern. The initial three harvests showed continuously declining yields. The fourth, sixth, and eighth yields were larger than the third, fourth, and seventh yields for most mixtures. The difference between mixtures were the smallest for 9th and 10th harvests. The 9th and 10th harvest occurred in 2013 after three years of growth. Species that were not suitable for the site and non-competitive species might have disappeared by this time and the yields would have come from the surviving few species. Each treatment may have been well established with surviving species during the fourth year. Figure 3 also shows variable yield differences among mixtures for the first three years of data collection, which supported this hypothesis, as those yields consisted of poorly performing species mixtures occupying the soil and space.

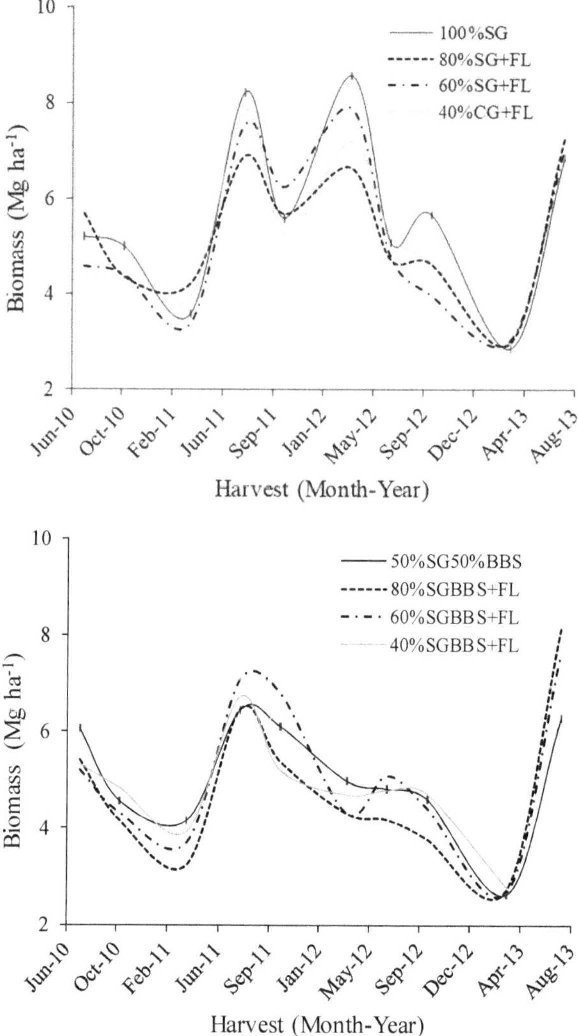

Figure 3. Distribution of mean biomass yields for the eight mixture treatments (100% switchgrass (SG), 80% SG with forbs and legumes, 60% SG with forbs and legumes, 20% SG with forbs and legumes, 50% SG with 50% big blue stem (BBS), 80% SGBBS with forbs and legumes, 60% SGBBS with forbs and legumes, 20% SGBBS with forbs and legumes) for 2010, 2011, 2012, and 2013 at the Bradford Research Center of University of Missouri, Columbia, Missouri, USA.

Many studies have reported increased biomass yields, and some with greater than 50% increases, with polycultures as compared to monocultures [5,17,56,57]. However, increased biomass yields with polyculture in long-term studies are inconclusive. Tilman [17] showed increased biomass yields with stand maturity for polycultures versus monocultures. In contrast, others have not observed similar increases in polyculture yields with stand maturity across various environments [58–61]. Our study in the Midwest of the USA was conducted during a time with below normal precipitation, a severe drought in 2012, and above normal temperature conditions. We cannot determine whether the differences in polyculture and monoculture were influenced by stand maturity or weather conditions. Studies that evaluate biomass yields for polycultures and monocultures on environmental gradients may be needed to determine site-suitable mixtures to meet the energy independence from biofuel, as data are lacking in the literature.

3.4. Cutting Height and Biomass Yield

The height of cutting had a significant effect on biomass yield ($p < 0.0001$; Figure 4). More biomass was harvested from the 15 cm cutting compared to the 30 cm cutting height. The 4 year means for the 15 cm and 30 cm height treatments were 5.98 and 4.43 Mg ha^{-1} for the study, respectively (Table 2). Similar differences were observed for yields by cutting heights within each year.

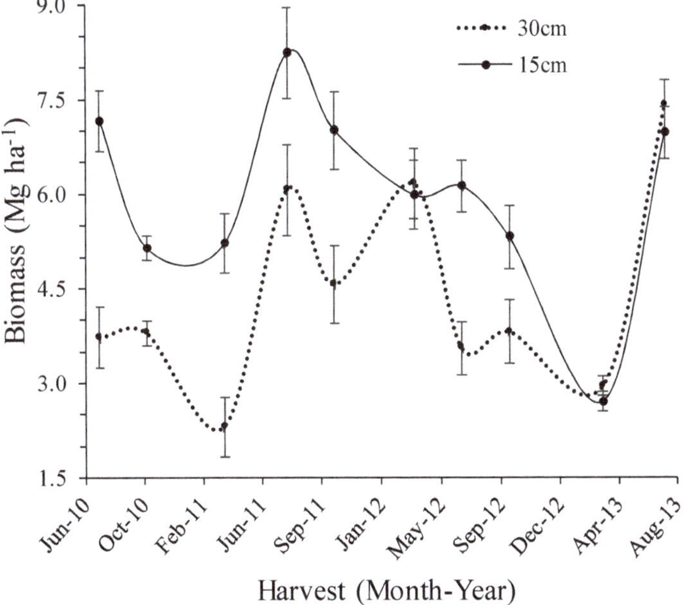

Figure 4. Distribution of mean biomass yields for 15 cm and 30 cm cutting height treatments for 2010, 2011, 2012, and 2013 at the Bradford Research Center of University of Missouri, Columbia, Missouri, USA.

Table 2. Biomass yields by harvest time, change (as a percentage of previous harvest), annual mean for 15 cm and 30 cm cutting treatments at the Bradford Research Center, University of Missouri, Columbia, Missouri, USA.

	Harvest		**15 cm**			**30 cm**	
Harvest Time	**Number**	**Yield**	**Change**	**Annual**	**Yield**	**Change**	**Annual**
		Mg ha^{-1}	**%**	**Mg ha^{-1}**	**Mg ha^{-1}**	**%**	**Mg ha^{-1}**
July 2010	1	7.16			3.72		
October 2010	2	5.14	−28	6.15	3.79	2	3.76
March 2011	3	5.21	1		2.30	−39	
July 2011	4	8.23	58		6.06	163	
October 2011	5	7.00	−15	6.81	4.56	−25	4.31
March 2012	6	5.99	−14		6.16	35	
July 2012	7	6.12	2		3.54	−42	
October 2012	8	5.31	−13	5.80	3.81	7	4.5
March 2013	9	2.70	−49		2.94	−23	
July 2013	10	6.96	158	4.83	7.40	152	5.17
Study period				5.98			4.43

The difference between biomass yields for 15 cm and 30 cm cutting heights (15 cm yield minus 30 cm yield) varied between −0.44 Mg ha^{-1} for July 2013 and 2.91 Mg ha^{-1} in March 2011. Yield differences were much larger during the early phase of the study (Figure 4, Table 2). The yield difference within a harvest time was not significant for 3 (6th, 9th, and 10th) of the 10 harvests. The first harvest of 2012 and both March and July harvests of 2013 had slightly larger yields for the 30 cm cutting height than the 15 cm cutting height, although these differences were not significant.

The 15 cm cutting height treatment consistently produced greater yields in 2010 and 2011. Only two cutting heights in 2012 had greater yields for the 15 cm cutting height as compared to the 30 cm cutting height. The greatest yields for 15 cm (6.81 Mg ha^{-1}) and 30 cm (5.17 Mg ha^{-1}) cutting heights were observed in 2011 and 2013, respectively. For both cutting height treatments, similar yield increases were found for July 2013 harvest, with an average increase of 154% (158% and 152%), compared to the March 2013 yields. The 30 cm treatments showed the greatest increase, with a 163% increase between the March and October harvests in 2011.

The two cutting height treatments did not respond in the same fashion. In some years, it reduced the subsequent biomass yield. For example, biomass yields declined from October 2010 to March 2011, July 2011 to October 2011, March 2012 to July 2012, and October 2012 to March 2013 for the 30 cm height treatment (Table 2, Figure 3). These declines ranged from −23% to −42%. The 15 cm treatments showed yield declines for July 2010 to October 2010 and July 2011 to March 2013, with declines ranging from −13% to −49%. The 15 cm treatment recorded reduced yield for five times as compared to four times for the 30 cm treatment. Although overall yields were greater for the 15 cm treatment, the number of times with reduced yields was lower for the 30 cm treatment.

Food reserves and greater amount of biomass left in the field may have contributed to these differences. In a pruning height study, Tipu [62] found greater *Leucaena leucocephala* yields for higher cutting heights, with significantly greater number of branches, lengths of branches, and leaves per branch. A grazing study in Mongolia suggested taller cutting heights grazing land management, although initial yields were greater for shorter cutting heights [63]. These yield changes might indicate the effect of stored nutrients that can have an effect on subsequent growth of cutting height treatments. We cannot explain the mixture effects on resilience, as we did not estimate the mixture compositions in each year and at the end of the study.

3.5. Management Implications for Biomass Production

The effect of mixtures of the two main grass species combinations was unexpected, as there was no significant yield difference were observed between mixtures. However, the mixtures dominated by switchgrass had slightly more biomass production. Since the study did not evaluate the changes in mixture composition with time for a longer period (>10 years), we are unable to comment on reseeding frequency for the maintenance of a mixture.

The study highlighted the importance of cutting height. The 15 cm cutting height generated more biomass. However, during the last two harvests, the 15 and 30 cm cutting heights produced similar yields. This may suggest increased resilience and adaptability of the 15 cm cutting versus the 30 cm cutting. This study emphasized the importance of long-term evaluation of management, as the yields were almost identical in the fourth year. During the 4 year study period, rainfall was below normal in three years and temperatures were extremely high in two years. Weather factors influenced this 4 year study's results.

Landowners may consider the establishment of biomass crops to avoid yield decreases near riparian buffers and to protect soils and water resources from erosion. Integration of economically valuable perennial species into biomass strips can help to generate additional income while improving soil, water, and wildlife habitats [64]. Biomass crop rows near the streams may also qualify for other conservation practices where landowners may minimize expenses and generate income.

4. Conclusions

In this 4 year plot study, we evaluated the biomass yields that resulted from the use of monocultures of native warm season grasses and varying mixtures of native forbs and legumes, with three cuttings conducted each year and at two cutting heights. Yields declined from the early cutting to subsequent cuttings in 2010 and 2012 when averaged across all mixtures, cutting times, and heights. In 2011 and 2013, yields increased from March to July, and declined in September for 2011. Whether or not mixtures were used was not significant, which indicates that the integration of native forbs and legumes with native warm-season grasses did not negatively influence biomass or forage production.

During this study, mid-Missouri experienced levels of annual precipitation that were well below the long-term mean, which influenced yield. However, results showed that mixtures of native warm-season grasses, forbs, and legumes are suitable for biomass production and forage crops in Missouri and can provide a source of forage during extreme summer drought conditions. This diversity of vegetation can also be managed to benefit a variety of wildlife in Missouri. Plots with varying ratios of mixtures generated acceptable yields compared with plots that utilized monocultures of native grasses and generally required fewer inputs, such as applications of nitrogen fertilizer, after initial seeding and establishment.

These results emphasize the importance of selecting site-suitable species for production, environmental, and economic benefits. Although cutting height was a major determinant of crop yields during the first three years after establishment, those differences disappeared during the last year of the study. Landowners who expect long-term benefits from these stands may have to sacrifice the initial forage yields that result from short cutting heights until the third or fourth year after establishment. However, landowners can optimize the value of using mixtures of native forbs and legumes with warm-season grasses by altering the timing of a harvest to take advantage of various markets, whether through cutting for biomass production in the late fall or spring or by haying or grazing for a livestock forage during the summer season. These are important considerations in managing a forage stand using native grasses with mixtures of forbs and legumes. The frequency of cutting and timing of harvests may help to adjust costs and income potential as well as optimize equipment availability.

Author Contributions: Investigation: R.P.U., C.J.G., T.M.R., R.A.P.II; Project Administration: T.M.R., R.A.P.II, W.W., R.L.W.; Methodology: T.M.R., R.A.P.II, W.W., R.L.W.; Formal analysis: R.P.U., C.J.G.; Original draft preparation: R.P.U., C.J.G.; Writing: R.P.U., C.J.G; Writing, reviewing, and editing: R.P.U., C.J.G., T.M.R., R.A.P.II, R.L.W. All authors have read and agreed to the published version of the manuscript.

Funding: This research was funded by NRCS Grant Number 0023235 NRCS 69-3A75-9-141Timothy Reinbott Titled "Demonstrating alternative methods of biofuel production to enhance farm profitability while improving wildlife habitat and soil and water conservation."

Conflicts of Interest: The authors declare no conflict of interest.

References

1. U.S. DOE (Department of Energy) 2009. Available online: http://www1.eere.energy.gov/biomass/biomass (accessed on 31 March 2017).
2. Karp, A.; Shield, I. Bioenergy from plants and the sustainable yield challenge. *New Phytol.* **2008**, *179*, 15–32. [CrossRef]
3. Zegada-Lizarazu, W.; Elbersen, H.W.; Cosentino, S.L.; Zatta, A.; Alexopoulou, E.; Monti, A. Agronomic aspects of future energy crops in Europe. *Biofuels Bioprod. Bioref.* **2010**, *4*, 674–691. [CrossRef]
4. Soper, J.M.; Raynor, E.J.; Wienhold, C.; Schacht, W.H. Evaluating composition and conservation value of roadside plant communities in a grassland biome. *Environ. Manag.* **2019**, *63*, 789–803. [CrossRef] [PubMed]
5. Tilman, D.; Hill, J.; Lehman, C. Carbon-negative biofuels from low-input high-diversity grassland biomass. *Science* **2006**, *314*, 1598–1600. [CrossRef] [PubMed]
6. Tilman, D.; Socolow, R.; Foley, J.A.; Hill, J.; Larson, E.; Lynd, L.; Pacala, S.; Reilly, J.; Searchinger, T.; Somerville, C.; et al. Beneficial biofuels–The food, energy, environment trilemma. *Science* **2009**, *325*, 270–271. [CrossRef] [PubMed]
7. Gelfand, I.; Sahajpal, R.; Zhang, X.; Izaurralde, R.C.; Gross, K.L.; Robertson, G.P. Sustainable bioenergy production from marginal lands in the, US Midwest. *Nature* **2013**, *493*, 514–517. [CrossRef]
8. Nielson, R.L. Historical Corn Grain Yields for Indianan and the US 2012. Available online: https://www.agry.purdue.edu/ext/corn/news/timeless/yieldtrends.html (accessed on 1 June 2017).
9. NASS.USDA.GOV. Available online: usda.nass.usda.gov/Charts_and_Maps/graphics/2017 (accessed on 10 August 2017).
10. Bothast, R.J.; Schlicher, M.A. Biotechnological processes for conversion of corn into ethanol. *Appl. Microbiol. Biotechnol.* **2005**, *67*, 19–25. [CrossRef]
11. Heaton, E.A.; Dohleman, F.G.; Long, S.P. Meeting US biofuel goals with less land: The potential of Miscanthus. *Glob. Chang. Biol.* **2008**, *14*, 2000–2014. [CrossRef]
12. Wright, L.; Turhollow, A. Switchgrass selection as a "model "bioenergy crop: A history of the process. *Biomass Bioenergy* **2010**, *34*, 851–868. [CrossRef]
13. Adler, P.R.; Sanderson, M.A.; Boateng, A.A.; Weimer, P.J.; Jung, H.J.G. Biomass yield and biofuel quality of switchgrass harvested in fall or spring. *Agron. J.* **2006**, *98*, 1518–1525. [CrossRef]
14. Gunderson, C.A.; Davis, E.B.; Jager, H.I.; West, T.O.; Perlack, R.D.; Brandt, C.C.; Wullschleger, S.D.; Baskaran, L.M.; Wilkerson, E.G.; Downing, M.E. Exploring Potential, US Switchgrass Production for Lignocellulosic Ethanol. 2008 Technical Manuscript ORNL/TM-2007/183. Available online: http://bioenergy.ornl.gov (accessed on 1 April 2016).
15. Vogel, K.P.; Mitchell, R.B. Heterosis in switchgrass: Biomass yield in swards. *Crop Sci.* **2008**, *48*, 2159–2164. [CrossRef]
16. Wullschleger, W.S.; Davis, E.B.; Borsuk, M.E.; Gunderson, C.A.; Lynd, L.R. Biomass production in switchgrass across the united states: Database description and determinants of yield. *Agron. J.* **2010**, *102*, 1158–1168. [CrossRef]
17. Tilman, D.; Reich, P.B.; Knops, J.M.H. Biodiversity and ecosystem stability in a decade-long grassland experiment. *Nature* **2006**, *441*, 629–632. [CrossRef] [PubMed]
18. Jungers, J.M.; Clark, A.T.; Betts, K.; Mangan, M.E.; Sheaffer, C.C.; Wyse, D.L. Long-term biomass yield and species composition in native perennial bioenergy cropping systems. *Agron. J.* **2015**, *107*, 1627–1640. [CrossRef]
19. Meehan, T.D.; Hurlbert, A.H.; Gratton, C. Bird communities in future bioenergy landscapes of the Upper Midwest. *Proc. Natl. Acad. Sci. USA* **2010**, *107*, 18533–18538. [CrossRef]
20. Pierce, R.A., II; Reinbott, T.; Wright, R.; White, B.; Potter, L. *Establishing and Managing Early Successional Habitats for Wildlife on Agricultural Lands*; MU Extension Publication MP: Missouri, MO, USA, 2010; Volume 907, p. 20.
21. McNaughton, S.J. Ecology of grazing ecosystem: The Serengeti. *Ecol. Monogr.* **1985**, *55*, 259–294. [CrossRef]
22. Burkepile, D.E. Comparing aquatic and terrestrial grazing ecosystems: Is the grass really green? *Oikos* **2013**, *122*, 306–312. [CrossRef]

23. Gardiner, M.A.; Tuell, J.K.; Isaacs, R.; Gibbs, J.; Ascher, J.S.; Landis, D.A. Implications of three biofuel crops for beneficial arthropods in agricultural landscapes. *Bioenergy Res.* **2010**. [CrossRef]
24. Haaland, C.; Naisbit, R.E.; Bersier, L.F. Sown wildflower strips for insect conservation: A review. *Insect Conserv. Divers.* **2011**, *4*, 60–80. [CrossRef]
25. Packard, S.; Mutel, C.F. (Eds.) *The Tallgrass Restoration Handbook: For Prairies, Savannas, and Woodlands*; Society for Ecological Restoration; Island Press: Washington, DC, USA, 1997; p. 463.
26. Dragoni, F.; Nasso, N.N.; O Di Tozzini, C.; Bonari, E.; Ragaglini, G. Aboveground yield and biomass quality of giant reed (arundo donax l.) as affected by harvest time and frequency. *Bioenergy Res.* **2015**, *8*, 1321–1331. [CrossRef]
27. Hao, X.; Thelen, K.; Gao, J. Spatial variability in biomass yield of switchgrass, native prairie, and corn at field scale. *Agron. J.* **2016**, *108*, 548–558. [CrossRef]
28. Schmer, M.R.; Mitchell, R.B.; Bogel, K.P.; Schacht, W.H.; Marx, D.B. Spatial and temporal effects on switchgrass stands and yield in the Great Plains. *Bioenergy Res.* **2010**, *3*, 159–171. [CrossRef]
29. Thelemann, R.; Johnson, G.; Sheaffer, C.; Banerjee, S.; Cai, H.; Wyse, D. The effect of landscape position on biomass crop yield. *Agron. J.* **2010**, *102*, 513–522. [CrossRef]
30. Wilson, D.M.; Heaton, E.A.; Schulte, L.A.; Gunther, T.P.; Shea, M.E.; Hall, R.B.; Headlee, W.L.; Moore, W.L.; Boersma, N.N. Establishment and short-term productivity of annual and perennial bioenergy crops across a landscape gradient. *Bioenergy Res.* **2014**, *7*, 885–898. [CrossRef]
31. Lewandowski, I.; Heinz, A. Delayed harvest of miscanthus—Influences on biomass quantity and quality and environmental impacts of energy production. *Eur. J. Agron.* **2003**, *19*, 45–63. [CrossRef]
32. Tahir, M.H.N.; Casler, M.D.; Moore, K.J.; Brummer, E.C. Biomass yield and quality of reed canarygrass under five harvest management systems for bioenergy production. *Bioenergy Res.* **2011**, *4*, 111–119. [CrossRef]
33. Kering, M.K.; Butler, T.J.; Biermacher, J.T.; Guretzky, J.A. Biomass yield and nutrient removal rates of perennial grasses under nitrogen fertilization. *Bioenergy Res.* **2012**, *5*, 61–70. [CrossRef]
34. Vogel, K.P.; Brejda, J.J.; Walters, D.T.; Buxton, D.R. Switchgrass biomass production in the Midwest, USA: Harvest and nitrogen management. *Agon. J.* **2002**, *94*, 413–420. [CrossRef]
35. Monti, A.; Zegada-Lizarazu1, W. Sixteen-year biomass yield and soil carbon storage of giant reed (*Arundo donax* L.) grown under variable nitrogen fertilization rates. *Bioenergy Res.* **2016**, *9*, 248–256. [CrossRef]
36. Angelini, L.G.; Ceccarini, L.; Bonari, E. Biomass yield and energy balance of giant reed (Arundo donax L.) cropped in central Italy as related to different management practices. *Eur. J. Agron.* **2005**, *22*, 375–389. [CrossRef]
37. Yang, Z.; Shen, Z.; Tetreault, H.; Johnson, L.; Friebe, B.; Frazier, T.; Huang, L.; Burklew, C.; Zhang, X.; Zhao, B. Production of autopolyploid lowland switchgrass lines through in vitro chromosome doubling. *Bioenergy Res.* **2014**, *7*, 232–242. [CrossRef]
38. Johnson, L.C.; Olsen, J.T.; Tetreault, H.; DeLaCruz, A.; Bryant, J.; Morgan, T.J.; Knapp, M.; Bello, N.M.; Baer, S.G.; Maricle, B.R. Intraspecific variation of a dominant grass and local adaptation in reciprocal garden communities along a, US Great Plains' precipitation gradient: Implications for grassland restoration with climate change. *Evol. Appl.* **2015**, *8*, 705–723. [CrossRef]
39. Anderson, S.H. Claypan and its Environmental Effects. In *Encyclopedia of Agrophysics*; Encyclopedia of Earth Sciences Series; Gliński, J., Horabik, J., Lipiec, J., Eds.; Springer: Dordrecht, The Netherlands, 2011.
40. USDA NRCS 2016. Available online: https://www.nrcs.usda.gov/wps/portal/nrcs/detail/mo/plantsanimals/?cid=stelprdb1083060 (accessed on 10 June 2018).
41. Sudkamp, S.; Chapman, R.; Pierce, R.A., II. *Quail Friendly Plants of the Midwest*; University of Missouri Extension MP: Missouri, MO, USA, 2008; Volume 903, p. 121.
42. SAS. *SAS User's Guide, Statistics*; Version 5; SAS Institute: Cary, NC, USA, 2010.
43. Steel, R.G.D.; Torrie, J.H.; Dickey, D.A. *Principles and Procedures of Statistics: A Biometrical Approach*; Mcgraw Hill Series in Probability and Statistics; Mcgraw Hill: New York, NY, USA, 1997; p. 672.
44. Heaton, E.; Voigt, T.; Long, S.P. A quantitative review of comparing the yields of two candidate C-4 perennial biomass crops in relation to nitrogen, temperature and water. *Biomass Bioenergy* **2004**, *27*, 21–30. [CrossRef]
45. Kiniry, J.R.; Tischler, C.R.; van Esbroeck, G.A. Radiation use efficiency and leaf, CO2 exchange for diverse C4 grasses. *Biomass Bioenergy* **1999**, *17*, 95–112. [CrossRef]
46. Muir, J.P.; Sanderson, M.A.; Ocumpaugh, W.R.; Jones, R.M.; Reed, R.L. Biomass production of Alamo switchgrass in response to nitrogen, phosphorus, and row spacing. *Agron. J.* **2001**, *93*, 896–901. [CrossRef]

47. Lee, D.K.; Boe, A. Biomass production of switchgrass in central South Dakota. *Crop Sci.* **2005**, *45*, 2583–2590. [CrossRef]
48. Sanderson, M.A.; Reed, R.L.; Ocumpaugh, W.R.; Hussey, M.A.; Van Esbroeck, G.; Read, J.C.; Tischler, C.R.; Hons, F.M. Switchgrass cultivars and germplasm for biomass feedstock production in Texas. *Bioresour. Technol.* **1999**, *67*, 209–219. [CrossRef]
49. Berdahl, J.D.; Frank, A.B.; Krupinsky, J.M.; Carr, P.M.; Hanson, J.D.; Johnson, H.A. Biomass yield, phenology and survival of diverse cultivars and experimental strains in western North Dakota. *Agron. J.* **2005**, *97*, 549–555. [CrossRef]
50. Parrish, D.J.; Fike, J.H. The biology and agronomy of switchgrass for biofuels. *Crit. Rev. Plant. Sci.* **2005**, *24*, 423–459. [CrossRef]
51. Monti, A.; Bezzi, G.; Pritoni, G.; Venturi, G. Long-term productivity of lowland and upland switchgrass cytotypes as affected by cutting frequency. *Bioresour. Technol.* **2008**, *99*, 7425–7432. [CrossRef]
52. Nizami, A.S.; Korres, N.E.; Murphy, J.D. Review of the integrated process for the production of grass biomethane. *Environ. Sci. Technol.* **2009**, *43*, 8496–8508. [CrossRef] [PubMed]
53. Prochnow, A.; Heiermann, M.; Plöchl, M.; Linke, B.; Idler, C.; Amon, T.; Hobbs, P.J. Bioenergy from permanent grassland—A review: 1. Biogas. *Bioresour. Technol.* **2009**, *100*, 4931–4944. [CrossRef] [PubMed]
54. Prochnow, A.; Heiermann, M.; Plöchl, M.; Amon, T.; Hobbs, P.J. Bioenergy from permanent grassland—A review: 2. Combustion. *Bioresour. Technol.* **2009**, *100*, 4945–4954. [CrossRef] [PubMed]
55. Jefferson, P.G.; McCaughey, W.P. Switchgrass (*Panicum virgatum* L.) cultivar adaptation, biomass production, and cellulose concentration as affected by latitude of origin. *ISRN Agron.* **2012**, *2012*, 763046. [CrossRef]
56. Khalsa, J.; Fricke, T.; Weisser, W.W.; Weigelt, A.; Wachendorf, M. Effects of functional groups and species richness on biomass constituents relevant for combustion: Results from a grassland diversity experiment. *Grass Forage Sci.* **2012**, *67*, 569–588. [CrossRef]
57. Jungers, J.M.; Sheaffer, C.C.; Lamb, J.A. The effect of nitrogen, phosphorus, and potassium fertilizers on prairie biomass yield, ethanol yield, and nutrient harvest. *Bioenergy Res.* **2015**, *8*, 279–291. [CrossRef]
58. Cardinale, B.J.; Nelson, K.; Palmer, M.A. Linking species diversity to the functioning of ecosystems: On the importance of environmental context. *Oikos* **2000**, *91*, 175–183. [CrossRef]
59. Adler, P.R.; Sanderson, M.A.; Weimer, P.J.; Vogel, K.P. Plant species composition and biofuel yields of conservation grasslands. *Ecol. Appl.* **2009**, *19*, 2202–2209. [CrossRef]
60. Johnson, M.V.; Kiniry, J.R.; Sanchez, H.; Polley, H.W.; Fay, P.A. Comparing biomass yields of low-input high-diversity communities with managed monocultures across the central United States. *Bioenergy Res.* **2010**, *3*, 353–361. [CrossRef]
61. Griffith, A.P.; Epplin, F.M.; Fuhlendorf, S.D.; Gillen, R. A comparison of perennial polycultures and monocultures for producing biomass for biorefinery feedstock. *Agron. J.* **2011**, *103*, 617–627. [CrossRef]
62. Tipu, S.U.; Hossain, K.L.; Islam, M.O.; Hossain, M.A. Effect of pruning height on shoot biomass yield of *Leucaena leucocephala*. *Asian J. Plant Sci.* **2006**, *5*, 1043–1046.
63. Baatar, B.; Svavarsdóttir, K.; Aradóttir, Á.L.; Ægisdóttir, H.H. *Effects of Cutting Height and Frequency on Yield in a Mongolian Rangeland*; Land Restoration Training Programme: Reykjavík, Iceland, 2008.
64. Schulte, L.A.; Niemi, J.; Helmers, M.J.; Liebman, M.; Arbuckle, J.D.; James, D.E.; Kolka, R.K.; O'Neal, M.E.; Tomer, M.D.; Tyndall, J.C.; et al. Prairie strips improve corn–soybean croplands. *Proc. Natl. Acad. Sci. USA* **2017**, *114*, 11247–11252. [CrossRef] [PubMed]

Article

On-Farm Evaluation on Yield and Economic Performance of Cereal-Cowpea Intercropping to Support the Smallholder Farming System in the Soudano-Sahelian Zone of Mali

Bougouna Sogoba [1,2,*], Bouba Traoré [3,4], Abdelmounaime Safia [1], Oumar Baba Samaké [2], Gilbert Dembélé [2], Sory Diallo [4], Roger Kaboré [2], Goze Bertin Benié [1], Robert B. Zougmoré [5] and Kalifa Goïta [1]

1 Centre D'Applications et de Recherches en TéléDétection (CARTEL), Université de Sherbrooke, Sherbrooke, QC J1K 2R1, Canada; abdelmounaime.a.safia@usherbrooke.ca (A.S.); goze.bertin.benie@usherbrooke.ca (G.B.B.); Kalifa.Goita@USherbrooke.ca (K.G.)
2 ONG AMEDD (Association Malienne d'Éveil au Développement Durable) Darsalam II-Route de Ségou, Rue 316 porte 79, BP 212 Koutiala, Mali; oumar.samake@ameddmali.org (O.B.S.); gilbert.dembele@ameddmali.org (G.D.); agrisahel@yahoo.fr (R.K.)
3 International Crops Research Institute for the Semi-Arid Tropics (ICRISAT-Niger), BP 12404 Niamey, Niger; B.Traore@cgiar.org
4 Institut D'Economie Rurale (IER)-Mali, BP 438 Bamako, Mali; diallo.sory@yahoo.fr
5 CGIAR Research Program on Climate Change, Agriculture and Food Security (CCAFS), ICRISAT West and Central Africa Office, BP 320 Bamako, Mali; R.Zougmore@cgiar.org
* Correspondence: bougouna.sogoba@usherbrooke.ca

Received: 20 April 2020; Accepted: 6 June 2020; Published: 9 June 2020

Abstract: Cereal-cowpea intercropping has become an integral part of the farming system in Mali. Still, information is lacking regarding integrated benefits of the whole system, including valuing of the biomass for facing the constraints of animal feedings. We used farmers' learning networks to evaluate performance of intercropping systems of millet-cowpea and sorghum-cowpea in southern Mali. Our results showed that under intercropping, the grain yield obtained with the wilibali (short maturing duration) variety was significantly higher than the yield obtained with the sangaranka (long maturing duration) variety whether with millet (36%) or sorghum (48%), corresponding, respectively, to an economic gain of XOF (West African CFA franc) 125 282/ha and XOF 142 640/ha. While for biomass, the yield obtained with the sangaranka variety was significantly higher by 50% and 60% to that of wilibali with an economic gain of XOF 286 526/ha (with millet) and XOF 278 516/ha (with sorghum). Total gain obtained with the millet-cowpea system was significantly greater than that obtained with the sorghum-cowpea system by 14%, and this stands irrespective of the type of cowpea variety. Farmers prefer the grain for satisfying immediate food needs instead of economic gains. These results represent an indication for farmer's decision-making regarding cowpea varieties selection especially for addressing household food security issues or feeding animals.

Keywords: intercropping; cropping systems; Sub-Saharan Africa; millet and sorghum; diversification

1. Introduction

In Mali, millet (*Pennissetum glaucum* (L.) R. Br.) and sorghum (*Sorghum bicolor* (L.) Moench) represent 1/3 of all crops and contribute mainly to the food security of the population, especially in rural areas [1,2]. Cowpea (*Vigna unguiculata* (L.) Walp.) is largely produced by farm households as a staple food crop, and with 22 to 30% protein content, it has become a major source of low-cost nutrition for the urban and rural poor who cannot afford meat and milk products [3]. Cowpea varieties are divided into early- (wilibali variety) or medium-maturing types (korobalen variety); highly grain-productive and late-maturing types; and high fodder production types (sangaranka variety). The planting date and pattern of cowpea plants vary from farmer to farmer, and the plants occupy 30 to 50% of the land area in each field [4].

Major constraints for farming systems in the region are related to high inter- and intra-annual rainfall variability resulting in recurrent droughts [5] and secondly to years of crop nutrient-mining and limited organic or inorganic resupply [6]. A high diversity of farming systems between agro-ecological and socioeconomic environments [7] and poor resource endowments of households limit options and opportunities to address specific production constraints [8]. Furthermore, many projections on West Africa's future climate prognosticate adverse impacts that are likely to lead to productivity crises unless sustainable solutions are in place. It is estimated that crop growing periods in West Africa may shorten by an average of 20% by 2050, causing a 40% decline in cereal yields and a reduction in cereal biomass for livestock [9].

Crop diversification including intercropping in this region reduces the risk of crop failure for smallholder farmers [10] by improving productivity per unit of land when compared with those of sole cropping systems [11]. This is especially true within low input, subsistence-oriented, agro-pastoral land use systems in the Sudano-Sahelian zone of West Africa [12].

Cereal-cowpea intercropping has long been practiced by smallholder farmers and has become part of the common cropping system. The traditional system of intercropping consists of mixing and planting cereals and cowpea seeds on the same hill, resulting in important inter-specific competition and low yields of the component crops. Although sole cropping of cowpea is profitable, farmers grow cowpea within a mixed cropping system because it fits well into the low input labor-intensive tradition of growing crops in the region [13] and favors greater yield on a given piece of land [14].

The land equivalent ratio index (LER) of cereal-cowpea intercropping in the region usually has a value greater than 1, indicating no detrimental competition between both crops [15].

However among several studies comparing sole cereals cropping [16], sole cowpea cropping [17], or cereal-cowpea intercropping [18], there are limited results that take into account the integrated benefits of the system, including valuing of the biomass, which has become important and widespread in the cities. Information on the monetary value of biomass in the system is scanty and less informative for supporting traders. From that perspective, whether with cereals or with cowpea, biomass has become as important as grain for human consumption.

In this study, we used the farmers' learning networks in partnership with researchers to evaluate the performance of intercropping systems of millet-cowpea and sorghum-cowpea in southern Mali. Our specific objectives were to: (i) evaluate cereal-cowpea system performance, (ii) analyze the rotation effect of cereal-cowpea intercropping, and (iii) identify economic benefits of cereal-cowpea intercropping.

2. Materials and Methods

2.1. Study Area

The study was conducted in the Soudano-Sahelian zone of southern Mali, covering the commune of Tominian (13.2857° N, 4.5908° W) and Yorosso (12.3548° N, 4.7782° W). The rainy season lasts from June to October with rainfall peaks in August. In 2016, mean seasonal rainfall in the region was 1005 mm while in 2017 seasonal rainfall was 861 mm. The number of rainy days was 51 in 2016 compared to 45 in 2017. The dry season includes a relatively cold period from November to February and a hot period lasting from March to May. The mean maximum temperature is 34 °C during the rainy season and 40 °C during the hot dry period.

Vegetation in the region is savannah with trees and shrubs, mainly from a natural regeneration system, and cultivated lands are mainly characterized by parks of Vitellaria paradoxa (shea nut tree),

Parkia biglobosa (néré), and Adansonia digitata (baobab). The mean population density is 16.4 inhabitants per km^2 with a mean of 8 persons per household [19].

Cropping land is spatially dispersed and the largest share is allocated to cereal production. Sorghum (*Sorghum bicolor* (L.) Moench) and millet (*Pennisetum glaucum* (L.) R.Br.) are the main crops, representing, respectively, 38% and 32% of the cultivated area, but maize (*Zea mays* L.) is also important, covering 12%. Cereals are grown in a two- or three-year rotation with cotton (*Gossypium hirsutum* L.). Fertilizer and pesticides are mainly applied to cotton and maize. Millet and sorghum usually do not receive fertilizer but benefit in the crop rotation from previous fertilizer applications to cotton or maize.

Cattle is a key component of the mixed crop-livestock farming systems in the study area. Eighty per cent of farmers own at least one pair of oxen, a cultivator, and a seeder, and use animal traction for soil preparation, weeding, and sowing [3].

The soils are mainly Ferric Lixisols with low clay content (<10%) in the topsoil. Soils are in general moderately acidic with a pH of around 5–6 [20] and with low nutrient holding capacity and low organic matter content [21]. The fertilizer application rates recommended by agricultural research and extension services have generally proven too costly for smallholder farmers. In addition, they involve a high financial risk, which is a major factor driving decision making for smallholder farmers [22].

2.2. Field Experimentation

Experimentation was the last phase in a series of four activities focusing mainly on biophysical characterization of farm fields, farmer's dialogue on the cereal-cowpea intercropping system, technical organization, and cropping system selection by the respective farmers.

A total of 159 trials including 76 with millet and 83 with sorghum, both intercropped with cowpea, were conducted in 2016 and 2017 in 108 villages. The experimental design for each trial was arranged in a randomized block design with 4 treatments based on 3 improved cowpea varieties and the farmers' local cowpea variety. The selection of cowpea varieties was oriented towards the farmers' objectives, which were mainly based on earliness of production and availability of biomass for animal feeding. The same varieties were simultaneously tested at all sites.

The intercropping system was designed by the community based on a previous study [15] and on the farmers' experience. The implemented intercropping system consisted of 2 rows of cereals (millet or sorghum) followed by 2 rows of cowpea varieties (Table 1). Each farmer selected either millet or sorghum in combination with cowpea varieties.

Table 1. System characterization.

Crop	Variety	Time to Maturity (days)	Duration
Cowpea	Wilibali	60–65	Short
Cowpea	Korobalen	70–75	Medium
Cowpea	Sangaranka	90–100	Long
Cowpea	Local	60–70 and 90–100	Short and long
Sorghum	Jakumbè (CSM63E)	90–100	Medium
Millet	Toroniou	90–100	Medium

Field plot size for each treatment was 100 m^2. For cereals (millet and sorghum) and cowpea varieties (wilibali and korobalen), the inter-row distance was 0.75 m, with a within-row plant distance of 0.4 m because of erected stem character and small space occupation rate. For sangaranka and local varieties, within-row plant distance was 0.8 m and the inter-row distance was 0.75 m. These varieties are creeping crops with large space occupation rates. The distance between adjacent cereal and cowpea rows was 0.75 m. All crops were thinned (2 plants/hole) at 15 days after planting to achieve the recommended planting densities.

Planting dates mostly occurred in June. Weeding was carried out before 20 days after planting and again between 30 and 40 days after planting (Table 2), i.e., weeding was completed twice for each field.

Based on the national recommendation and the farmers' common practice, an average of 100 kg/ha of diammonium phosphate (18-46-0) was applied between 15 and 20 days after planting. To protect crops from enemies, particularly cowpea, water-based Neem [23] was spread between 35 and 45 days after planting (DAP) for the first application and between 50 and 55 DAP for second application.

Table 2. Cropping management (days after planting, DAP) under cowpea intercropping with millet and sorghum in southern Mali.

Crop	Year	Planting Date	1st Weeding (DAP)	2nd Weeding (DAP)	Fertilizer Application (DAP)	Biopesticide Treatment 1 (DAP)	Biopesticide Treatment 2 (DAP)
Millet	2016	12/07 ± 8.2	19.0 ± 6.6	34.3 ± 7.9	21.9 ± 5.7	36.0 ± 6	53.8 ± 15
Sorghum		15/07 ± 6.7	19 ± 10.6	35.83 ± 13.5	21.56 ± 9.5	46.30 ± 18.7	58.00 ± 22
Millet	2017	13/07 ± 7.3	18.0 ± 6.4	35.5 ± 8.4	18.3 ± 6.8	35.4 ± 10.6	49.7 ± 9.6
Sorghum		12/07 ± 6	17.8 ± 12.6	34.54 ± 12.3	16.34 ± 5.8	40.60 ± 19.1	55.88 ± 19.2

Rotation effect was determined based on the crop cultivation calendar for the previous three years. In total, the effects of three types of rotation, i.e., cereal-cereal, cereal-legume, and cereal-cotton, on yield were analyzed using an unbalanced design regression model. The cereal consisted of millet, sorghum, or maize while the legume consisted of groundnut or cowpea.

2.3. *Measurement*

The timing of different operations including planting, weeding, harvesting, and fertilizing was recorded by field technicians. Crop physiology status such as flowering and maturity dates was also collected. At crop maturity, farmers harvested the total area of the plot with the assistance of the researchers. Mature millet and sorghum plants were harvested following the local practice of cutting the panicles and bagging. Legume pods were harvested when mature. Biomass of all crops was weighed at the plot, and a sub-sample was taken for weighing. Millet ears, sorghum panicles, and legume pods were dried on a clean floor at the homestead and were threshed and hand-winnowed; legume pods were shelled by hand. Grains were weighed and grain sub-samples were taken and weighed as well. All sub-samples (grain and biomass) were dried and re-weighed to determine dry weights in kg/ha.

2.4. *Statistical Analysis*

Because of the varied number of experiments per village, across villages, and per year, we used an unbalanced design using the GenStat regression model for the variables mean separation. Firstly, ANOVA was performed to separately evaluate the simple effect of cowpea grain yield with cowpea biomass yield under intercropping with millet and sorghum. Secondly, for the purpose of economic analysis, we compared yield of grain to biomass under intercropping with each of the two cereals and their respective interactions with varieties. Treatment structure consisted of either grain or biomass variables for varieties per crop and their respective interactions with the year, representing the annual rainfall effect. Villages were considered as replicate. Significant means were separated using average standard error of difference (SED). We also used Box plots for capturing the distribution of variables.

2.4.1. System Gain

To determine system economic gain per hectare, we used a gross margin (GM) analysis model that is equal to the difference between total revenue (TR) and total variable cost (TVC) and is expressed as follow:

$$GM\ (\pi) = \sum TR - \sum TVC \tag{1}$$

Total revenue means the total market price of production per hectare multiplied by the crops' yields (grain or biomass) while TVC includes mainly input costs such as insecticide, fertilizer, and ploughing. The system economic gain was expressed in West African CFA franc (XOF).

2.4.2. Farmers' Ranking of Cowpea Varieties

A total of 30 farmers (18% of the total) participated to the prioritization of the cereal-cowpea intercropping systems using a paired comparison scaling method. Each farmer was requested to provide a weighted score for cowpea grain, biomass, and total income. The respective scores were multiplied by the number of scores for each cowpea variety to obtain a total weighted score that was then divided by the total number of respondents to obtain the weighted mean score (WMS). Rank order was given according to the WMS values.

3. Results

3.1. Yield of Grain and Biomass of Cowpea under Intercropping with Millet and Sorghum

Grain yield distribution of cowpea varieties under intercropping with millet showed that in the 25% trial, yields of korobalen and sangaranka were less than 100 kg/ha while in the 75% trial, yields were below 400 kg/ha (Figure 1). In contrast, in the 25% trial for wilibali, yields of korobalen and sangaranka were less than 200 kg/ha while in the 75% trial, yields were below 500 kg/ha. Grain yield distribution for the local variety varied from 100 kg/ha to 500 kg/ha. Statistical analysis of cowpea grain yield showed that the best yield was obtained with the wilibali variety, which was significantly higher than that of the the sangaranka variety with a difference of + 150 kg/ha (Table 3). This result did not change over years or with varieties.

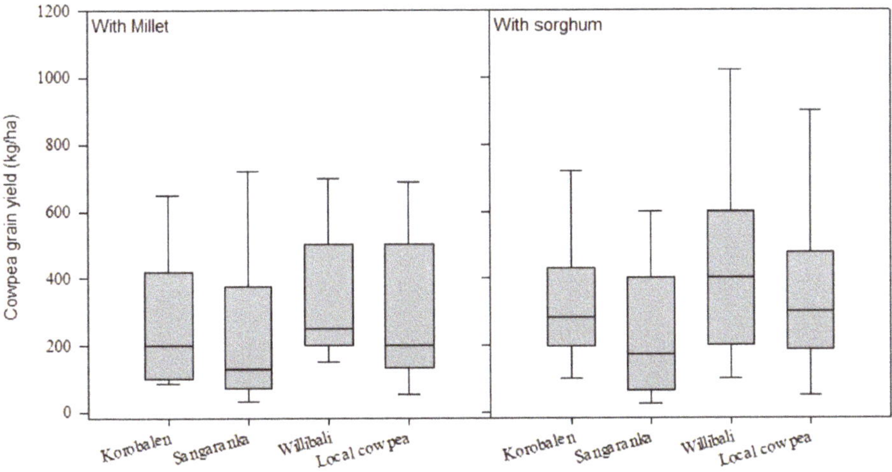

Figure 1. Cowpea grain yield under intercropping with millet and sorghum in southern Mali.

For cowpea biomass, there was great variability depending on varieties (Figure 2). Distribution showed that 75% of the biomass yield of korobalen and the local variety was below 1400 kg/ha, while for the sangaranka yield, distribution was higher and varied from 900 kg/ha to 2200 kg/ha. Cowpea biomass yield obtained with wilibali varied less (500 kg to 1000 kg/ha). Statistical analysis indicated that the best biomass yield was obtained with sangaranka, which was significantly higher than that obtained with korobalen, wilibali, or the local variety, with a difference of 482 kg/ha, 820 kg/ha, and 721 kg/ha, respectively (Table 3). Comparing grain yield to biomass showed that biomass yield of cowpea was statistically higher than that of grain yield ($p < 0.05$). Interaction between grain and biomass yield with cowpea varieties was significant, indicating that performance of grain or biomass yield depends on cowpea varieties. Thus, under intercropping with millet, the cowpea grain yield obtained with the wilibali variety was statistically higher to that with sangaranka while its biomass yield was significantly higher compared to that of wilibali.

Table 3. Cowpea yield (kg/ha) under intercropping with millet and sorghum in southern Mali.

	Cowpea Yield with Millet Intercropping				**Cowpea Yield with Sorghum Intercropping**			
	DF	**Grain**	**DF**	**Biomass**	**DF**	**Grain**	**DF**	**Biomass**
Village	36	-	39	-	38	-	38	-
Cowpea variety	3	-	3	-	3	-	3	-
Korobalen		310.3		1154		382.6		1125
Sangarakan		261.1		1637		253.2		1592
Wilibali		417.6		815		475.5		659
Local cowpea		317.2		913		386.1		1012
v.r		2.69		13.58		4.38		17.78
p-value		0.04		0.001		0.003		0.001
SED		54.53		140.6		66.79		126.9
Year 2016	1	337.9		1171	1	416.8		1212
Year 2017		319.5		1068		331.9		964
P-value Year		0.79		0.57		0.22		0.10
Interaction of cowpea and Year	3	0.78		0.95	3	0.23		0.51
p-value cowpea grain vs. biomass			0.001			0.001		

DF: Degrees of Freedom.

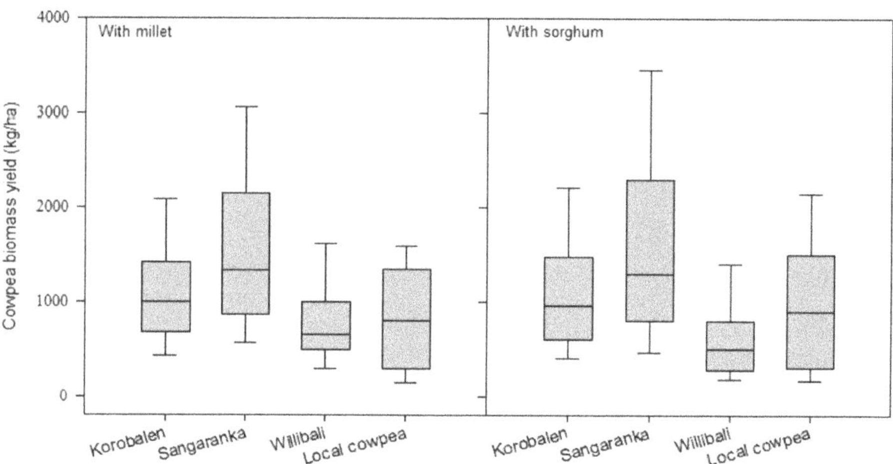

Figure 2. Cowpea biomass yield (kg/ha) under intercropping with millet and sorghum in southern Mali.

With sorghum intercropping (Figure 1), cowpea yield distribution was similar to that with millet except in the 75% trial, where the yield of wilibali was lower, 600 kg/ha, compared to 500 kg/ha with millet. Grain yield obtained with the wilibali variety was 25% and 15% significantly higher than the yields obtained with the sangaranka and local varieties (Table 3). Interaction effect of cowpea yield under intercropping with sorghum and year was not significant. For cowpea biomass (Figure 2) when intercropped with sorghum, the yield obtained with sangaranka was significantly higher by 30% and 25%, respectively, for wilibali and korobalen and by 45% for the local variety. Interaction effect of year with cowpea varieties was not significant, indicating that difference in cowpea biomass is not related to a particular year.

3.2. *Grain and Biomass Yield of Millet and Sorghum under Intercropping with Cowpea Varieties*

Grain yield of millet and sorghum varied similarly irrespective of cowpea varieties. With cowpea varieties, in the 25% trial, millet and sorghum grain yield was less than 300 kg/ha while 50% of yield was between 300 kg/ha and 900 kg/ha (Figure 3). Millet grain yields were statistically similar with a mean of 577 kg/ha regardless of intercropping with cowpea varieties (Table 4). For millet biomass (Figure 4), 50% of the biomass yield varied from 1500 kg/ha to 6300 kg/ha with a mean of 4033 kg/ha. The biomass yield difference within cowpea varieties was not significantly different while the year effect was significant ($P < 0.05$) whether with grain or with biomass yield. In 2016, millet grain yield was 636 kg/ha and was higher by 20% to that of 2017, while biomass yield was 5415 kg/ha in 2016 and was higher by 48% to that of 2017. For sorghum, grain yield was 621 kg/ha and yield difference under intercropping with cowpea varieties were not statistically significant ($P > 0.05$) (Table 4). For sorghum biomass, mean yield was 2923 kg/ha and differences under cowpea varieties was statistically significant. However, year effect was significant for biomass yield. In 2016, mean sorghum biomass yield was 4298 kg/ha and higher by 63% to that obtained in 2017.

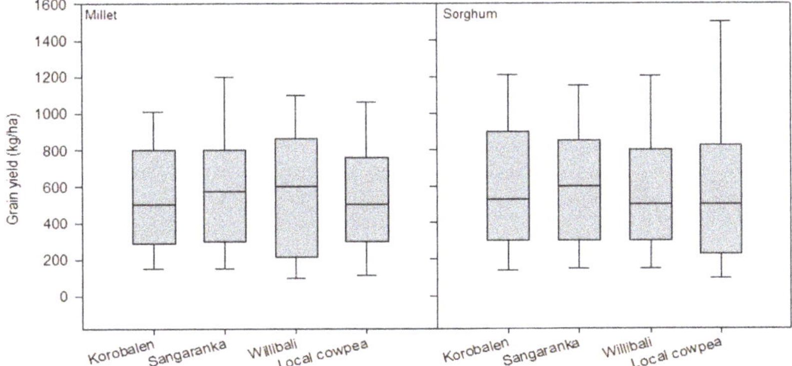

Figure 3. Grain yield of millet and sorghum under intercropping with different varieties of cowpea in southern Mali.

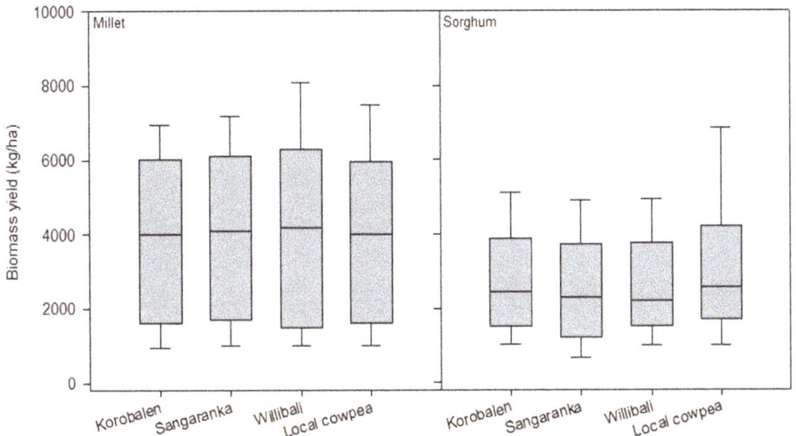

Figure 4. Biomass yield (kg/ha) of millet and sorghum under intercropping with different cowpea varieties in southern Mali.

Table 4. Yield (kg/ha) of millet and sorghum under intercropping with cowpea varieties in southern Mali.

	Millet Yield with Cowpea Intercropping				Sorghum Yield with Cowpea Intercropping			
	DF	**Grain**	**DF**	**Biomass**	**DF**	**Grain**	**DF**	**Biomass**
Village	37	-	31	-	37	-	37	-
Cowpea variety	3	-	3	-	3	-	3	-
Korobalen		561.5		3916		616.7		2793
Sangarakan		612.9		4061		638.8		2821
Wilibali		596.5		4117		617.3		2773
Local cowpea		538.9		4040		612		3308
v.r		1.33		0.28		0.09		3.02
p-value		0.27		0.84		0.96		0.03
SED		40.98		228		57.01		206
Year 2016	1	635.7		5415	1	632.9		4298
Year 2017		511.5		2795		610.3		1563
p-value Year		0.01		0.001		0.75		0.001
Interaction of crop and Year	3	0.37		0.44	3	0.64		0.678
p-value grain vs. biomass			0.001				0.001	

DF: Degrees of Freedom.

3.3. *Effect of Crop Rotation on Intercropping*

Separation of the effects of crop rotation and/or inter-cropping from continuous cereal showed that there are additional benefits to crop yield if crops are rotated with cash crop or at least intercropped with legume crops. For sorghum, the yield obtained after cereal-legume and cotton-cereal rotation was significantly higher than the yield obtained with cereal-cereal rotation (Table 5). For the systems with millet and sorghum biomass, the yield obtained after cotton-cereal and cereal-legume rotation was significantly higher than the yield obtained after the cereal-cereal rotation.

Table 5. Performance of millet and sorghum under intercropping with cowpea and according to type of rotation in 37 villages.

Rotation	**Millet Yield from Intercropping with Cowpea**		**Sorghum Yield from Intercropping with Cowpea**	
	Grain	**Biomass**	**Grain**	**Biomass**
Cereal-Cereal	529.3	4808	448	2445
Cereal-Legume	624.8	5359	694.6	2871
Coton-Cereal	588.1	6470	711.9	3689
P-value	0.227	0.001	0.001	0.001
SED	72.69	504.5	82.18	335.3

3.4. *Economic Gains*

3.4.1. Gains with Cowpea Grain and Biomass under Intercropping with Millet and Sorghum

Results showed that grain gain per hectare with cowpea varieties under intercropping with millet varied accordingly. Higher gain with cowpea grain was obtained with millet-wilibali system (XOF 125 282/ha) which was significantly higher than gain obtained with korobalen, sangaranka and local variety by respectively 26%, 37% and 24% (Table 6).

For cowpea biomass under intercropping with millet, gain obtained was statistically different ($p < 0.05$). Mean biomass gain obtained (XOF 286 526/ha) with sangaranka was significantly higher than those obtained with the korobalen, wilibali and local variety respectively by 30%, 50% and 44 % (Table 6).

Table 6. Gain /ha from grain and biomass of cowpea under intercropping with millet and sorghum. The values are expressed in actual currency of West African, the Franc CFA (XOF).

	Cowpea Gain with Millet Intercropping				Cowpea Gain with Sorghum Intercropping			
	DF	**Grain**	**DF**	**Biomass**	**DF**	**Grain**	**DF**	**Biomass**
Village	36	-	39	-	38	-	38	-
Cowpea variety	3	-	3	-	3	-	3	-
Korobalen		93,085		201,908		114,782		196,880
Sangarankan		78,341		286,526		75,972		278,516
Wilibali		125,282		142,559		142,640		115,284
Local cowpea		95,160		159,726		115,821		177,179
v.r.		2.69		13.58		4.38		17.78
p-value		0.04		0.001		0.005		0.001
SED		16,683		24,851		20,038		21,658
Year 1	1	101,371	1	204,889	1	125,027	1	212,039
Year 2		95,839		186,971		99,561		168,648
p-value year		0.79		0.58		0.22		0.10
Interaction of cowpea and Year	3	0.78	3	0.62	3	0.23	3	0.51
p-value grain vs. biomass			0.001				0.001	
p-value inter. grain and biomass cowpea variety			0.001				0.001	

DF: Degrees of Freedom.

By comparing gain obtained with cowpea grain to that of biomass under intercropping with millet, results show that gain with biomass (XOF 195 791/ha) was significantly greater than that of grain by 49%, corresponding to a difference of XOF 95 480/ha. However, this difference varied according to the intercropping systems which is due to the significant effect of interaction between gain from grain and biomass according to cowpea varieties. As consequence, greater biomass gain was obtained with the system millet-sangaranka while it has low gain from grains and alternatively best gain from grains was obtained with the system mil-wilibali while it has low biomass gain.

With sorghum, greater grain gain was obtained with the system sorghum-wilibali (XOF 142 640/ha) which was significantly higher ($p < 0.05$) than the gains obtained with the sorghum-sangaranka, sorghum-korobalen and local variety by 47%, 20%, and 19%, respectively (Table 6). On the other hand, with the biomass, greater gain was obtained with the sorghum-sangaranka which was statistically higher than the gain obtained with the sorghum-wilibali, sorghum local variety and sorghum-korobalen, by 59%, 36% and 29%, respectively.

By comparing the two variables grain and biomass under intercropping with sorghum, results showed that mean gain of XOF 187 649/ha obtained with biomass was significantly greater than that obtained with grain by 38% corresponding to a difference of XOF 71 025 /ha.

As with millet system, there was a significant effect of the interaction between grain and biomass gains based on cowpea varieties under intercropping with sorghum. Thus, greater biomass gain was obtained with the sorghum-sangaranka while it has the lowest gain from grains. Moreover, greater gain from grain was obtained with the sorghum-wilibali system while it has the lowest biomass gain.

3.4.2. Gain with Millet and Sorghum in Intercropping with Cowpea

Results showed that gain obtained with millet grain as well biomass in intercropping with cowpea was not statistically significant whatever cowpea variety (Table 7). However, gain obtained with biomass was significantly greater than that obtained with the grain by 75%, regardless cowpea variety. With regards to sorghum, yield was not significant unlike for biomass where system sorghum-wilibali had lowest gain. Gain obtained with biomass was greater than that obtained with the grain by 74% and this stands whatever cowpea varieties.

Table 7. Gain of grain and biomass/ha for millet and sorghum under intercropping with cowpea varieties. The values are expressed in actual currency of West African, the Franc CFA (XOF).

	Millet Gain/ha under Intercropping with Cowpea				Sorghum Gain/ha under Intercropping with Cowpea			
	DF	Grain	DF	Biomass	DF	Grain	DF	Biomass
Village	37	-	24	-	37	-	37	-
Cowpea variety	3	-	3	-	3	-	3	-
Korobalen		88,874		301,989		96,825		349,087
Sangarankan		96,490		299,626		100,286		352,600
Wilibali		94,299		276,184		96,923		346,667
Local cowpea		85,150		286,654		96,092		413,483
v.r		1.28		0.40		0.09		3.02
p-value		0.28		0.75		0.96		0.03
SED		6503		25,811		8951		25,806
Year 2016	1	100,419		263,817	1	99,367	1	537,295
Year 2017		80,832		297,775		95,814		195,343
p-value year		0.02		0.78		0.75		0.001
Interaction of cowpea and Year	3	0.34	3	0.08	3	0.64	3	0.67
p-value grain vs. biomass		0.001				0.001		

DF: Degrees of Freedom.

3.4.3. Total Economic Gain per System

By comparing the two systems, total gain obtained with millet-cowpea system was significantly greater than that obtained with sorghum-cowpea system by 14% corresponding to a difference of XOF 123 676/ha and this stands irrespective the type of cowpea variety (Table 8). For both systems millet and sorghum, total gain varied significantly from year to year. In 2016, for millet-cowpea system, mean gain was XOF 1124389/ha and was 39% higher than that obtained in 2017. For sorghum-cowpea system mean gain in 2016 was XOF 954 739/ha and was 49% higher than that of 2017.

Table 8. Total gain/ha of the system millet-cowpea and sorghum-cowpea. The values are expressed in actual currency of West African, the Franc CFA (XOF).

	DF	Total Gain/ha (Millet and Cowpea)	DF	Total Gain/ha (Sorghum and Cowpea)
Village	28	-	36	-
Cowpea variety	3	-	3	-
Korobalen		879,428		737,843
Sangaranka		994,643		800,579
Wilibali		863,508		691,568
Local cowpea		853,056		737,606
v.r		1.95		2.13
p-value		0.12		0.09
SED		63,516		49,408
Year 2016	1	1124,389	1	954,739
Year 2017		686,145		484,574
p-value Year		0.001		0.001
Interaction of crop and Year	3	0.19	3	0.29
p-value total gain of millet vs. sorghum	1		0.009	
p-value inter. income of millet and sorghum and cowpea variety	3		0.81	

DF: Degrees of Freedom.

4. Discussion

4.1. Yield Variation under Intercropping with Cereal

Although cowpea is of vital importance to the livelihoods of most Malian farmers, we found that whether with millet or sorghum, cowpea yields were low, and 75% of the yields were less than 500 kg/ha. This result is similar to that of [24], supporting that cowpea grain yields in farmers' fields can be below 300 kg/ha.

In the study area, soil fertility is low, including low organic carbon and especially P deficiency, which may limit cowpea yield through growth limitation and impaired pod formation and N fixation [25]. We found high variability in yield whether with millet, sorghum, or cowpea varieties. This can be due to agricultural practices variability, which may depend on farm resource endowment status. A farm with appropriate equipment can benefit more from the first rain for earlier planting, while a delay in planting, especially with a low resource farm type, may result in significant yield penalty. Furthermore, variability in soil fertility management across the region can also result in yield variations as can biotic factors such as the presence of trees, which varied from 10 to 40 trees per farm ha depending on field topographic position [26].

4.2. Cowpea Varieties under Intercropping

By comparing cowpea varieties under intercropping, the best yield was obtained with the wilibali variety, whether with millet or sorghum and whatever the year, and this was mainly due to the shortness of time to crop maturity. This variety can be harvested in as little as 60–80 days and therefore can avoid the seasonal late water-stress that mostly occurs in September.

With the potential of a short growing season, the wilibali variety enables households to have grains for consumption or sale during the "hungry period", especially when grain reserves from the previous cereal harvests were reduced and current crops are still not ready to be harvested. On the other hand, the best biomass yield was obtained with sangaranka, because the long duration of maturity time maximized the thermal temperature sum. With the high biomass yielding potential, the sangaranka variety offers opportunities for animal feeding, especially in the zones where grazing has become increasingly rare due to the expansion of cropping fields [27] and the poor quality of grazing [27].

Given local farming constraints, each of the two products grain or biomass offers opportunities for each farmer. Thus, farmers with less sufficient financial or technical means (land, equipment, etc.) for farming and whose primary objective is for food for their families can select the wilibali variety. In contrast, farmers with sufficient technical background, means for farming, and with many animals can select the sangaranka variety because of the high potential biomass production for animal feeding.

Regarding millet and sorghum, we found that grain yields varied similarly whatever the cowpea variety and there was no difference among cereal grains and biomass yields due to cowpea varieties. In similar regions intercropped with a legume, cereal grain yields may increase up to 55% compared to cereal alone [28] through improvement of the soil moisture due to soil covering, which limits evapotranspiration [29]. However, research has demonstrated that in some cases, intercropping may reduce cereals yields by 10% due to increased competition for resources [30]. This points out challenges related to setting adapted management strategies, in particular, planting date offsets between the main and secondary crops depending on the start and variability of the seasonal rainfall.

4.3. Cereal-Cowpea Rotation

For farmers, selection of cereals to consider in the rotation depends on the current fertility level as of the soil as well as on the households' capacity to produce organic manure [31]. Beyond grain for human consumption and fodder for animals feeding, the cowpea system plays an important role in soil fertility by maintaining and improving nutrient availability [32]. We found that cereal grain and biomass yield obtained after cereal-legume and cotton-cereal rotation were higher compared to that obtained with cereal-cereal rotation. Enhanced cereal yield following legume planting can be attributed to enhanced phosphorus (P) nutrition for cereals through improving soil chemical P availability and microbiologically increased P uptake [32]. Cereal-legume rotation contributes to soil P restoration and nitrogen (N) availability, especially in acidic soils, which are found in most of Sahelian, where P was found to be a major constraint to crop growth [33]. With a crop rotation system, soil bacterial communities have greater species diversity than under continuous cultivation with the same crop [33].

Cotton-cereal rotation represents 35 to 40% of the cropping system in southern Mali [34]. In the study area, cotton was introduced as an alternative source of cash for farmers, but also to allow other crops to benefit from the system. Our results show greater yields of millet and sorghum after cotton-cereal rotation, which is certainly due to the residual fertilizer effects [35]. This result indicates the importance of cotton in achieving food security for smalholder farmers. Furthermore, cotton provides access to fertilizer through credit schemes from cotton companies, to which farmers would not have access otherwise, and which are crucial for sustained crop productivity [35]. The result also reflects the need of direct application of mineral fertilizer on millet or sorghum. With application of only 3 g as a microdose, the yield of millet and sorghum increased by 70% and 52%, respectively [36]. The onus is on policymakers and extension workers to promote the use of the microdosing technique under cereal cropping, especially in the regions where cotton is driving the system.

4.4. Economic Performance of the Cereal-Cowpea System and Farmers' Perceptions

We found that whether intercropping with millet or sorghum, the greatest gain for grain was obtained with the wilibali variety while greatest gain for biomass was obtained with the sangaranka variety. This is mainly due to the highest grain and biomass yield obtained, respectively, with wilibali and sangaranka varieties. However, this gain can be subject to variation depending particularly on market opportunities regarding the price variation from ± 20 to 30% across the same year for cereal and cowpea grain in the region [37].

The results represent an indication for farmer's decision-making regarding cowpea varieties selection, especially for addressing house food security issues or feeding animals. Furthermore, although cowpea biomass gain is greater than cowpea grain, the farmers' choice is usually geared towards grains for satisfying immediate food needs. This is supported by farmers' preferential classification (Table 9), under which grains and biomass come as a priority before immediate economic gain. Selling cowpea grain is not a priority for farmers, but it occurs, especially when there is surplus production because of a good rainfall pattern or when there is a social emergency requiring cash. Profitability of the cereal-cowpea production system depends mainly on farm size, family labor, seed access and quality, as well as fertilizer and crop protection strategies [38].

Table 9. Farmer's evaluation and selection of technology.

	Karobalen	Sangaranka	Willibaly	Local Cowpea	Noted	Rank for Grain and Biomass
Biomass	12.48	12.39	6.63	11.13	10.50	II
Grain	12.03	14.39	24.63	13.65	17.02	I
Gross margin	0.83	0.88	1.40	0.89	1.04	III
Total noted	25.35	27.67	32.66	25.66		
Rank	III	II	I	III		

Our results show that by comparing the two systems, the total gain obtained with the millet-cowpea system was significantly greater than that obtained with the sorghum-cowpea system, and this stands whatever the type of cowpea variety. This is explained by the millet biomass, which we found to be 28% greater than that of sorghum. However, variation of biomass between millet and sorghum may depend on the variety and the date of planting [16]. A variety with a long maturing duration with an earlier planting date may produce more biomass with higher revenue. While a short maturing duration variety may result in low biomass revenue even with an earlier planting date.

In the cereal system of southern Mali, attribution of crop per surface does not only depend on satisfying a household's food needs or revenue but may also rely on food preferences based on the cultural education [39].

5. Conclusions

Whether intercropping with millet or sorghum and whatever the seasonal rainfall, the best grain yield was obtained with the wilibali (short maturing duration) variety and the best biomass yield was obtained with the sangaranka variety, which is a long-maturing duration variety. The study revealed strong trade-offs between household food opportunity and animal feeding and economic gain regarding cereal-cowpea intercropping in southern Mali. The knowledge generated revealed opportunities for alleviating some of the trade-offs and achieving more promising farming decisions based on specific farm needs. Farmers selected cereal in intercropping with short maturing duration such as the wilibali variety to mainly address household food needs at specific periods corresponding to food shortages. While for those farmers prioritizing animal feeding, especially agro-pastoralists, the sangaranka variety was the best option. On the other hand, from an economic point of view, millet intercropping with cowpea is more profitable than sorghum intercropping with cowpea. Yield variability and low yields of both cereals and cowpea for all varieties combined indicates opportunities for improvement in both research and farming.

Author Contributions: Conceptualization, B.S.; Formal analysis, B.T.; Investigation, G.D., S.D., R.K., and G.B.B.; Project administration, O.B.S.; Supervision, R.Z. and K.G.; Validation, A.S.; Writing—original draft, B.S. and B.T.; Writing—review and editing, R.B.Z. and K.G. All authors have read and agree to the published version of the manuscript.

Funding: This research was funded by the McKnight foundation, grant number 14-239.

Acknowledgments: We acknowledge and thank the McKnight foundation for supporting the farmer's knowledge project in Mali and giving us the opportunity to conduct this research. Support from the Natural Science and Engineering Research Council (NSEC) of Canada is fully acknowledged. The authors also acknowledge the support of the Climate Change, Agriculture and Food Security (CCAFS) Program of the CGIAR (for details visit, https://ccafs.cgiar.org/donors). Thanks also to the three reviewers for their detailed comments that helped us to improve the manuscript.

Conflicts of Interest: The authors declare no conflict of interest.

References

1. FAO. *Corporate Statistical Database (FAOSTAT), 2018*; FAO: Rome, Italy, 2019.
2. Shewry, P.R. Advances in Sorghum and Millet Research Special Issue. *J. Cereal Sci.* **2019**, *85*, A1. [CrossRef]
3. Sanogo, O.M. Participatory Evaluation of the Dairy Cow Supplementation Technologies in a Country Setting at Mali (Koutiala)/ Évaluation Participative Des Technologies de Supplementation Des Vaches Laitières En Milieu Paysan Au Mali (Koutiala). *Can. J. Dev. Stud.* **2010**, *31*, 91–106. [CrossRef]
4. FAO. *West African Catalogue of Plant Species and Varieties*; FAO: Rome, Italy, 2008.
5. Traore, B.; Corbeels, M.; van Wijk, M.T.; Rufino, M.C.; Giller, K.E. Effects of Climate Variability and Climate Change on Crop Production in Southern Mali. *Eur. J. Agron.* **2013**, *49*. [CrossRef]
6. Jones, A.; Breuning-Madsen, H.; Brossard, M.; Dampha, A.; Deckers, J.; Dewitte, O.; Gallali, T.; Hallett, S.; Jones, R.; Kilasara, M. Soil Atlas of Africa. Luxembourg. *Publ. Off. Eur. Union* **2013**. Available online: https://op.europa.eu/en/publication-detail/-/publication/f57a0bdf-94d1-11e5-983e-01aa75ed71a1 (accessed on 9 June 2020).
7. Giller, K.E.; Tittonell, P.; Rufino, M.C.; van Wijk, M.T.; Zingore, S.; Mapfumo, P.; Adjei-Nsiah, S.; Herrero, M.; Chikowo, R.; Corbeels, M.; et al. Communicating Complexity: Integrated Assessment of Trade-Offs Concerning Soil Fertility Management within African Farming Systems to Support Innovation and Development. *Agric. Syst.* **2011**, *104*, 191–203. [CrossRef]
8. Falconnier, G.N.; Descheemaeker, K.; Van Mourik, T.A. Understanding Farm Trajectories and Development Pathways: Two Decades of Change in Southern Mali Understanding Farm Trajectories and Development Pathways: Two Decades of Change in Southern Mali. *F. Crop. Res.* **2015**. [CrossRef]
9. Zougmoré, R.; Partey, S.; Ouédraogo, M.; Omitoyin, B.; Thomas, T.; Ayantunde, A.; Ericksen, P.; Said, M.; Jalloh, A. Toward Climate-Smart Agriculture in West Africa: A Review of Climate Change Impacts, Adaptation Strategies and Policy Developments for the Livestock, Fishery and Crop Production Sectors. *Agric. Food Secur.* **2016**, *5*, 26. [CrossRef]
10. Challinor, A.; Wheeler, T.; Garforth, C.; Craufurd, P.; Kassam, A. Assessing the Vulnerability of Food Crop Systems in Africa to Climate Change. *Clim. Chang.* **2007**, *83*, 381–399. [CrossRef]

11. Rusinamhodzi, L.; Corbeels, M.; Nyamangara, J.; Giller, K.E. Maize–Grain Legume Intercropping Is an Attractive Option for Ecological Intensification That Reduces Climatic Risk for Smallholder Farmers in Central Mozambique. *F. Crop. Res.* **2012**, *136*, 12–22. [CrossRef]
12. Ayantunde, A.A.; Turner, M.D.; Kalilou, A. Participatory Analysis of Vulnerability to Drought in Three Agro-Pastoral Communities in the West African Sahel. *Pastoralism* **2015**, *5*. [CrossRef]
13. Kamara, A.Y.; Omoigui, L.O.; Kamai, N.; Ewansiha, S.U.; Ajeigbe, H.A. *Improving Cultivation of Cowpea in West Africa*; Burleigh Dodds Science Publishing: Cambridgeshire, UK, 2018.
14. Lithourgidis, A.S.; Dordas, C.A.; Damalas, C.A.; Vlachostergios, D.N. Annual Intercrops: An Alternative Pathway for Sustainable Agriculture. *Aust. J. Crop Sci.* **2011**, *5*, 396–410.
15. Falconnier, G.N.; Descheemaeker, K.; Mourik, T.A.V.; Giller, K.E. Unravelling the Causes of Variability in Crop Yields and Treatment Responses for Better Tailoring of Options for Sustainable Intensification in Southern Mali. *F. Crop. Res.* **2016**, *187*, 113–126. [CrossRef]
16. Traore, B.; van Wijk, M.T.; Descheemaeker, K.; Corbeels, M.; Rufino, M.C.; Giller, K.E. Evaluation of Climate Adaptation Options for Sudano-Sahelian Cropping Systems. *F. Crop. Res.* **2014**, *156*, 63–75. [CrossRef]
17. Takim, F.O. Advantages of Maize-Cowpea Intercropping over Sole Cropping through Competition Indices. *J. Agric. Biodivers. Res.* **2012**, *1*, 53–59.
18. Maman, N.; Dicko, M.; Abdou, G.; Kouyate, Z.; Wortmann, C. Pearl Millet and Cowpea Intercrop Response to Applied Nutrients in West Africa. *Agron. J.* **2017**, *109*, 2333–2342. [CrossRef]
19. Instat. *Annuaire Statistique Du Mali 2018*; Instat: Bamako, Mali, 2018.
20. Hazelton, P.; Murphy, B. *Interpreting Soil Test Results: What Do All the Number Means?* CSIRO Publishing: Collingwood, VIC, Australia, 2007.
21. Vanlauwe, B.; Descheemaeker, K.; Giller, K.E.; Huising, J.; Merckx, R. Integrated Soil Fertility Management in Sub-Saharan Africa: Unravelling Local Adaptation. *SOIL Discuss.* **2014**, 1239–1286. [CrossRef]
22. Abdoulaye, T.; Sanders, J.H. Stages and Determinants of Fertilizer Use in Semiarid African Agriculture: The Niger Experience. In *Agricultural Economics*; Blackwell Publishing Inc.: Malden, MA, USA, 2005; Volume 32, pp. 167–179.
23. Uddin, R.O., II; Abdulazeez, R.W. Comparative Efficacy of Neem (Azadirachta Indica), False Sesame (Ceratotheca Sesamoides) Endl. and the Physic Nut (Jatropha Curcas) in the Protection of Stored Cowpea (Vigna Unguiculata) L. Walp against the Seed Beetle Callosobruchus Maculatus (F.). *Ethiop. J. Environ. Stud. Manag.* **2013**, *6*, 827–834. [CrossRef]
24. Ajeigbe, H.A.; Singh, B.B.; Adeosun, J.O.; Ezeaku, I.E. Participatory On-Farm Evaluation of Improved Legume-Cereals Cropping Systems for Crop-Livestock Farmers: Maize-Double Cowpea in Northern Guinea Savanna Zone of Nigeria. *Afr. J. Agric. Res.* **2010**, *5*, 2080–2088.
25. Adu-Gyamfi, J.J.; Aigner, M.; Gludovacz, D. Variations in Phosphorus Acquisition from Sparingly Soluble Forms by Maize and Soybean in Low-and Medium-P Soils Using P-32. *Proc. Int. Plant Nutr. Colloquium XVI* **2009**, 1008–1016.
26. Cuny, P.; Sorg, J.-P. Forêt et Coton Au Sud Du Mali: Cas de La Commune Rurale de Sorobasso. *BOIS FORETS DES Trop.* **2003**, *276*, 17–31.
27. Turner, M.D.; Ayantunde, A.A.; Patterson, K.P.; Patterson, E.D., III. Livelihood Transitions and the Changing Nature of Farmer–Herder Conflict in Sahelian West Africa. *J. Dev. Stud.* **2011**, *47*, 183–206. [CrossRef]
28. Trail, P.; Abaye, O.; Thomason, W.E.; Thompson, T.L.; Gueye, F.; Diedhiou, I.; Diatta, M.B.; Faye, A. Evaluating Intercropping (Living Cover) and Mulching (Desiccated Cover) Practices for Increasing Millet Yields in Senegal. *Agron. J.* **2016**, *108*, 1742–1752. [CrossRef]
29. Sissoko, F.; Affholder, F.; Autfray, P.; Wery, J.; Rapidel, B. Wet Years and Farmers' Practices May Offset the Benefits of Residue Retention on Runoff and Yield in Cotton Fields in the Sudan–Sahelian Zone. *Agric. Water Manag.* **2013**, *119*, 89–99. [CrossRef]
30. Rusinamhodzi, L.; Makumbi, D.; Njeru, J.M.; Kanampiu, F. Performance of Elite Maize Genotypes under Selected Sustainable Intensification Options in Kenya. *F. Crop. Res.* **2020**, *249*, 107738. [CrossRef] [PubMed]
31. Blanchard, M.; Coulibaly, K.; Bognini, S.; Dugue, P.; Eric, V. Diversity in the Quality of Organic Manure Produced on Farms in West Africa: What Impact on Recommendations for the Use of Manure? *Biotechnol. Agron. Soc. Environ.* **2014**, *18*, 512–523.

32. Bationo, A.; Ntare, B.R. Rotation and Nitrogen Fertilizer Effects on Pearl Millet, Cowpea and Groundnut Yield and Soil Chemical Properties in a Sandy Soil in the Semi-Arid Tropics, West Africa. *J. Agric. Sci.* **2000**, *134*, 277–284. [CrossRef]
33. Alvey, S.; Bagayoko, M.; Neumann, G.; Buerkert, A. Cereal/Legume Rotations Affect Chemical Properties and Biological Activities in Two West African Soils. *Plant Soil* **2001**, *231*, 45–54. [CrossRef]
34. De la Croix, D. *L'agriculture Familiale Des Zones Cotonnières d'Afrique de l'ouest: Le Cas Du Mali*; AFD: Paris, France, 2004.
35. Ripoche, A.; Crétenet, M.; Corbeels, M.; Affholder, F.; Naudin, K.; Sissoko, F.; Douzet, J.M.; Tittonell, P. Cotton as an Entry Point for Soil Fertility Maintenance and Food Crop Productivity in Savannah Agroecosystems-Evidence from a Long-Term Experiment in Southern Mali. *F. Crop. Res.* **2015**, *177*, 37–48. [CrossRef]
36. Ibrahim, A.; Abaidoo, R.C.; Fatondji, D.; Opoku, A. Fertilizer Micro-Dosing Increases Crop Yield in the Sahelian Low-Input Cropping System: A Success with a Shadow. *Soil Sci. Plant Nutr.* **2016**, *62*, 277–288. [CrossRef]
37. RESIMAO. *Bulletin Mensuel Sur Les Marches Agricoles*; Mai 2017-N°009; RESIMAO: Madina, Saudi Arabia, 2017.
38. Mohammed, U.S.; Mohammed, F.K. Profitability Analysis of Cowpea Production in Rural Areas of Zaria Local Government Area of Kaduna State, Nigeria. *Int. J. Dev. Sustain.* **2014**, *3*, 1919–1926.
39. Dunkel, F.V. *Incorporating Cultures' Role in the Food and Agricultural Sciences*; Academic Press: Cambridge, MA, USA, 2017.

Article

Arthropod Diversity Influenced by Two *Musa*-Based Agroecosystems in Ecuador

Daniel Vera-Aviles [1,*], Carmita Suarez-Capello [1], Mercè Llugany [2], Charlotte Poschenrieder [2], Paola De Santis [3,4] and Milton Cabezas-Guerrero [5]

1 Facultad de Ciencias Agrarias, Universidad Técnica Estatal de Quevedo, Quevedo 120509, Ecuador; csuarez@uteq.edu.ec
2 Laboratory of Plant Physiology, Universitat Autònoma de Barcelona, Bellaterra, 08193 Barcelona, Spain; merce.llugany@uab.cat (M.L.); Charlotte.Poschenrieder@uab.cat (C.P.)
3 Research Centre Bioversity International, Via dei Tre Denari 472/a, 00054 Maccarese (Fiumicino), Italy; p.desantis@cgiar.org
4 Department Environmental Biology, Università Sapienza, Ple Aldo Moro, 5 00185 Rome, Italy
5 Facultad de Ciencias Pecuarias. Universidad Técnica Estatal de Quevedo, Quevedo 120509, Ecuador; mcabezas@uteq.edu.ec
* Correspondence: dvera@uteq.edu.ec

Received: 7 April 2020; Accepted: 7 June 2020; Published: 18 June 2020

Abstract: Banana and plantain (*Musa* spp.) are very important crops in Ecuador. Agricultural production systems based on a single cultivar and high use of external inputs to increase yields may cause changes in the landscape structure and a loss in biodiversity. This loss may be responsible for a decrease in the complexity of arthropod food webs and, at the same time, related to a higher frequency and range of pest outbreaks. Very little is known either about the ecological mechanisms causing destabilization of these systems or the importance of the diversity of natural enemies to keep pests under control. Few studies have focused on this issue in tropical ecosystems. Here, we address this problem, comparing two *Musa*-based agroecosystems (monocultivar and mixed-species plantations) at two sites in Ecuador (La Maná and El Carmen) with different precipitation regimes. The diversity of soil macro fauna, represented by arthropods, was established, as indicators of the abovementioned disturbances. Our ultimate goal is the optimization of pest management by exploring more sustainable cropping systems with improved soil quality. Arthropod abundance was higher in the mixed system at both localities, which was clearly associated with the quality of the soils. In addition, we found Hymenoptera species with predatory or parasitic characteristics over the pests present in the agroecosystems under study. These highly beneficial species were more abundant at the locality of La Maná. The mixed type of production system provides plant diversity, which favors beneficial arthropod abundance and permits lower agrochemical application without yield penalties in comparison to the monoculture. These findings will help in the design of *Musa*-based agroecosystems to enhance pest control.

Keywords: monocultivar; mixed-species plantation; biodiversity; arthropod; soil; on-farm biodiversity indicators

---

1. Introduction

At present, biodiversity is being lost at an unprecedented rate due to human activities. Research has devoted a great deal of effort to assess the importance of biodiversity for the functioning and stability of agroecosystems and for the provision of environmental services. Pest management has become more efficient on numerous occasions, as a valuable environmental service provided by biodiversity. However, this is threatened by human activity [1,2].

It is well known that agricultural production systems are intensified by enhanced use of external inputs to increase yields, causing a change in the landscape structure. These intense cropping systems are prone to losing their biodiversity and become destabilized. Studies carried out by Michalko and Košulič [3] showed that components of biodiversity peaked at different levels of canopy openness on oak forest stands. Therefore, the restoration and suitable forest management of such conditions will retain important diversification of habitats. Diversity is an index composed of two variables: the abundance of species (or groups of species) and the equitability (or uniform distribution of individuals between groups) [4]. Michalko and Košulič [3] suggested that the permanent presence of small-scale improvements could be suitable conservation tools to prevent the general decline of woodland biodiversity in the intensified landscape, and proposed that all the suggested improvements would promote species diversity, conservation aspects, and functional diversity. A loss in plant diversity decreases the complexity of arthropod food webs that could be related to the observed higher frequency and range of pest outbreaks [5]. However, little is known about the ecological mechanisms that result in this destabilization or the importance of the diversity of the natural enemies to keep the pest under control [6]. In this context, it is essential to identify those species/orders/families that, with their presence or absence, are important indicators of the quality of the agroecosystem under study. The soil microbiota represents the largest group of terrestrial animal organisms; among them, the phylum Arthropoda is considered one of the most important for man; despite its small size, they are visible to the naked eye, have a relatively short life cycle and fulfill numerous environmental services [7].

In the last decade, the diversity of the different production systems has been a matter of concern due to changes in ecosystems by human activities in agriculture, livestock and forestry [8]. These actions cause a great alteration in the processes of configuring the habitat of the organisms occupying that environment, causing negative effects on the diversity of the macro fauna and altering the balance between the ecosystem, the soil and the plants [9]. We suppose that all the suggested improvements would promote species diversity, conservation aspects, and functional diversity. These are essential to ecosystem functions and the restoration of forest environments in landscapes under intense human land use.

The decrease in diversity and abundance of the edaphic macro fauna caused by human activity has generated concerns that have stimulated the development of research on the impact of different types of tillage on soil biota. A clear example is described by Arroyo and Iturrondobeitia [10], who evaluated the diversity of arthropods in forests and different monocultivar agricultural systems, finding high values of species richness in forest areas, in contrast with the agroecosystem receiving fertilization and a general management of the crop. These kinds of studies are very scarce in the tropics, especially those related to agricultural crops.

Banana and plantain monocultivar plantations in Ecuador have been established in areas where primary forest has been eroded. A feature of these tropical soils is their dependence on the biomass and species diversity of the forest that covers it. Once the protective cover of the forest is eliminated, the productivity and fertility per unit area decreases.

Banana monocultivar are affected by many pests due to their weak ecological balance [11]. This drastically reduces the yield after the first two cropping years. For this reason, the banana (*Musa balbisiana*) producers need large areas of land and the consequent expansion to compensate for the fall in production per hectare. Extensive commercial banana crops are governed by technological standards aimed towards the intense production of fruit for export. Plantains (*Musa acuminata*), instead, besides being an important export crop, are also an essential staple food of coastal Ecuador and many other countries along the tropics, where diversity in respect to pest and disease pressures exists. Although farmers try to apply basically the same management practices as banana enterprises, plantain is in the hands of medium and small farmers with limited access to resources and who use lower levels of technology.

There are different approaches to tackle those problems. One approach intensively used is integrated pest management (IPM) strategies, focusing on using agronomic management techniques to reduce pesticide use, but IPM concentrates on modifying the environment around predominantly modern cultivars and has tended to exclude the potential of using within-crop diversity through genetic mixtures (crop variety mixtures) for example, or the planned deployment of different varieties in the same production environment. A diverse genetic basis of resistance (e.g., crop variety mixtures) is beneficial for the farmer because it allows a more stable management of pest and disease pressure than what a monocultivar system allows.

Intraspecific biodiversity increases resistance to pests in crops, developing more biological support, thus, ensuring production. Greater intraspecific biodiversity improves the biota in the soil, creating synergy with the crop. The mixture of natural and human selection gives the particularity of the environment for different agroecosystems. It requires agricultural species with genetic characteristics that adapt to the different environments. The most palpable case is that of maize, conducted in different climatic zones under different constrains [12].

There is very little information and research work of this kind in the region, therefore, the development of a project in progress under an agreement between the Technical University of Quevedo, Ecuador and Bioversity International [13] was used to conduct this study in order to determine possible bioindicators of the use of intraspecific biodiversity in two banana production systems on the Ecuadorian coast. The present work aimed at developing a study of the *Arthropoda* population over two Musa agroecosystems (single and mixed Musa cultivar plantations, at two sites in Ecuador (La Maná and El Carmen)), to establish its value as a bioindicator under Ecuadorian conditions.

2. Materials and Methods

2.1. Site Description

The study was conducted at two sites in Ecuador with different soil characteristics and precipitation regimes (Table 1) in 2014. La Maná site is located in the province of Cotopaxi (00°53′43″ S; 79°11′05″ O) at 354 m altitude. El Carmen site is located in Manabí province (00°16′14″ S; 79°29′12″ O) at 250 m altitude.

Table 1. Soil and weather characteristics of both sites and agricultural systems studied during 2014.

		La Maná		EL Carmen	
Soil Characteristic		**Mixed**	**Monocultivar**	**Mixed**	**Monocultivar**
Texture		Sandy loam	Sandy loam	Silty loam	Silty loam
pH		5.9	5.5	6.4	6.0
Organic matter (%)		4.3	2.9	4.0	3.2
NH_4 (ppm)		26	24	27	24
Weather Variables [1]		**Rainy Season**	**Dry Season**	**Rainy Season**	**Dry Season**
Precipitation (mm)		2589	398	2084	461
Temperature (°C)	max	29.2	27.5	30.3	28.7
	min	20.4	19.4	20.4	19.2
	med	24.8	23.4	25.3	23.9
RH (%)		89.2	87.3	87.6	86.9
Light hours		231	383	328	447

[1] Source: National Institute of Meteorology and Hydrology (INAMHI), San Juan Station (La Maná) and El Carmen Station (Manabí).

2.2. Experimental Design

The experimental sites consisted of two plots of 1 ha each. One plot is for traditional monocultivar system and the other for the mixed-species. The plot with the traditional system was established more than 30 years ago and was based on the cultivar Orito (Musa acuminata AB type of genome) in La Maná site, and cultivar Barraganete (Musa balbisiana AA type genome) in El Carmen site. The other plot corresponded to a mixed Musaceae system based on 12 different Musa cultivars, planted in 2009 (Table 2). Each cultivar was represented by subplots of 24 plants, with borders of plants of the local cultivar. Each subplot was repeated three times and randomly distributed in the ha. A spacing of 3 × 3 m (1111 plants ha^{-1}) was adopted in all plots.

Table 2. *Musa* ecotypes used in the mixed-species plot according to their genome.

Banana: *Musa Acuminata* (AA)	Plantain: *Musa Balbisiana* (AB)
Orito (AA)	Barraganete (AAB)
Gros Michel (AAA)	Maqueño Verde (AAB)
Guineo Jardin (AAA)	Dominico (AAB)
Filipino (AAA)	Dominico Harton (AAB)
Williams (AAA)	Dominico Negro (AAB)
	Dominico Gigante (AAB)
	Limeño (AAB)

2.3. Agronomic Management in the Experimental Plots

In both sites, the traditional systems were established by deep tillage practices, while tillage was minimum in the mixed systems. The main annual field operations are summarized as follows:

- Biweekly leaf pruning, eliminating folded senescent and dead leaves, as well as necrotic parts of leaves that have less than 30% necrosis.
- Shoots were eliminated every two months, selecting only one vigorous basal sucker as a replacement for the next generation.
- Peeling of the banana plant or elimination of dry leaf sheath from the pseudostem, every month during the rainy season and every two months in the dry season.
- Fruit harvesting every two weeks during rainy season and every three weeks for the dry season.
- Chime of corms from harvested plants.
- Fertilization (65 kg ha^{-1} of N, 45 kg ha^{-1} of P_2O_5 and 156 kg ha^{-1} of K_2O) distributed at the beginning and end of the rainy season.
- Manual weed controls every month in the rainy season and every two months in the dry season, after evaluating the species present.
- Chemical weed control, two applications/year of glyphosate (2 L ha^{-1}) in the monocultivar system, one in the rainy season and another at the beginning of the dry season. The mixed plot received only one application during the rainy season. Throughout the dry season, manual weeding was performed every two months in both systems.
- Chemical control of *Cosmopolites sordidus* was applied once a year with 10 g $plant^{-1}$ of Alodrin RB only in the monocultivar system.

2.4. Arthropod Abundance and Identification

Arthropod samples were taken randomly from both (mixed and monocultivar) sites (La Maná and El Carmen) and two seasons of the year (rainy and dry). To collect the most representative taxes, two different trap systems were used: a pitfall trap consisting of a 1 L capacity vessel, buried at ground level during 72 h, to catch organisms falling into the container filled with water and liquid detergent. The second type of traps were "Chromatic" traps, which consists of yellow plastic plates (18 cm diameter) placed for 3 h on the floor, with a solution of water and liquid soap. A total of 80 samples were taken—40 for each site—and these, in turn, subdivided into 20 subsamples for each production system and 10 for each type of trap used. The collected individuals were taken to the university laboratory, where they were quantified and classified up to the taxonomic level of the order, dividing them by phylum, class (insect, arachnid, myriapod, mollusks and annelids) and orders. The indicators of biodiversity of individuals that constitute the soil macrobiota were determined for the total of arthropods found, which were described according to criteria of Moreno [14].

To determine the abundance of each arthropod taxon (order), two sampling periods were carried out, one between January and May during the rainy season and another between June and November during the dry season.

Arthropods were identified using taxonomic keys [15] or arriving at the level of orders and family and as far as possible; they were differentiated between beneficial and harmful specimens.

2.5. Statistical Analysis

Based on the total structure and number of the arthropod community collected, comparison of means of the abundance of orders or groups of arthropods were performed and a multifactorial analyses of variance (ANOVA) was carried out with factors: site (El Carmen y La Mana), season (rainy and dry season) and type of cultural system (mixed and monocultivar). Calculations were done using the R Commander program [15]; values of each plot were previously transformed to $\sqrt{(x + 1)}$.

In order to determine the abundance, richness, diversity (Simpson indexes), similarity (Jaccard coefficient) and equitability (J) of the soil macro fauna by site and production system, the PAST statistical program [16] was used.

The Simpson diversity index was calculated as $\lambda = \Sigma pi^2$, where pi = proportional abundance of species i, that is, the number of individuals of species i divided by the total number of individuals in the sample during the time period considered. This index shows the probability that two individuals taken at random from a sample are of the same species. It is strongly influenced by the importance of the most dominant species [16].

To determine the relationships between factors and arthropod abundances, a Principal Component Analysis (PCA) was performed, in such a way that the distribution of the sites and agricultural systems was visualized both in the rainy and dry season, taking into account the characteristic uncorrelated environmental factors, which explain much of the original total variability.

3. Results and Discussion

The influence of the three factors (mixed and monocultivar agricultural systems; season and sites) on the abundance of different arthropod orders are shown in Table 3. The insect orders, Coleoptera and Diptera, were significantly influenced by the agricultural system, while site was a determinant factor for the orders Hymenoptera, and Prostigmata ($p > 0.01$). Orders Collembola, Hemiptera and Orthoptera were represented in all factors. The season of the year had highly significant influence on the majority of the orders found. There were no significant differences between the two cropping systems and the localities for the orders Araneae, Diptera, Hymenoptera, and Spirobolida. These orders have great adaptation capacity to diverse environments. The long life-cycles of some of these species could be responsible for their presence throughout the year [17]. In addition, some species influence the transformation of biodegradable waste, especially the organic matter deposited on the soil surface, incorporating it into the edaphic system, through the tunnels and channels that the coleoptera excavate. This facilitates infiltration and aeration of the soil [18].

Table 3. Influence of agricultural system, season and site on the abundance of arthropod populations. The values result from the factorial analysis of variance when comparing these factors. Values were transformed to $\sqrt{(x + 1)}$ before the analysis.

Order	Factor	Square Mean	F	*p*
Araneae	Site	0.09	1.40	0.240
	Season	11.05	168.96	0.000
	Agricultural system	0.01	0.17	0.686
	Site × Season	0.00	0.06	0.804
	Site × Agricultural system	0.08	1.19	0.280
	Season × Agricultural system	0.22	3.39	0.070

Table 3. *Cont.*

Order	Factor	Square Mean	F	p
Coleoptera	Site	0.62	12.79	0.001
	Season	4.36	89.16	0.000
	Agricultural system	0.67	13.74	0.000
	Site × Season	0.09	1.85	0.178
	Site × Agricultural system	0.30	6.22	0.015
	Season × Agricultural system	1.07	21.89	0.000
Collembola	Site	576.04	247.12	0.000
	Season	0.74	0.32	0.575
	Agricultural system	103.40	44.36	0.000
	Site × Season	5.35	2.30	0.134
	Site × Agricultural system	12.92	5.54	0.021
	Season × Agricultural system	2.97	1.27	0.263
Diptera	Site	0.72	1.64	0.203
	Season	14.77	33.72	0.000
	Agricultural system	4.76	10.87	0.002
	Site × Season	010	0.24	0.629
	Site × Agricultural system	0.12	0.28	0.601
	Season × Agricultural system	0.18	0.41	0.526
Hemiptera	Site	7.07	49.38	0.000
	Season	0.52	3.60	0.062
	Agricultural system	3.18	22.24	0.000
	Site × Season	3.76	26.25	0.000
	Site × Agricultural system	0.62	4.33	0.041
	Season × Agricultural system	0.21	1.48	0.227
Hymenoptera	Site	16.07	11.16	0.001
	Season	183.74	127.62	0.000
	Agricultural system	5.20	3.61	0.061
	Site × Season	33.49	23.26	0.000
	Site × Agricultural system	1.85	1.28	0.261
	Season × Agricultural system	3.13	2.17	0.145
Orthoptera	Site	8.06	29.27	0.000
	Season	2.19	18.81	0.000
	Agricultural system	3.03	26.05	0.000
	Site × Season	0.05	0.43	0.512
	Site × Agricultural system	0.59	5.10	0.027
	Season × Agricultural system	0.57	4.93	0.030
Prostigmata	Site	0.08	8.10	0.006
	Season	0.68	72.90	0.000
	Agricultural system	0.01	0.90	0.346
	Site × Season	0.08	8.10	0.006
	Site × Agricultural system	0.41	44.10	0.000
	Season × Agricultural system	0.01	0.90	0.346
Spirobolida	Site	0.00	0.12	0.730
	Season	0.36	20.28	0.000
	Agricultural system	0.05	3.00	0.088
	Site × Season	0.00	0.12	0.730
	Site × Agricultural system	0.05	3.00	0.088
	Season × Agricultural system	0.05	3.00	0.088

Arthropods from our field captures were distributed in three classes and eight orders (Table 4). In La Maná, we found that 56% of individuals belonged to order Collembola and 32% to Hymenoptera, while in El Carmen, 61.5% of them were classified into the order Hymenoptera. The least represented group were from the order Spirobolida and Prostigmata, with less than 0.5% values at both sites.

Table 4. Arthropod abundance distributed by taxonomic groups during the rainy and dry seasons in two ecological sites (El Carmen and La Maná) in 2014.

		La Maná				El Carmen			
Class	**Order**	**Rainy Season**	%	**Dry Season**	%	**Rainy Season**	%	**Dry Season**	%
Arachnida	Arachnida	0	0	44	1.9	4	1.4	48	3.3
	Prostigmata	0	0	6	0.3	0	0	12	0.8
Hexapoda	Coleoptera	22	1	58	2.5	17	5.7	40	2.7
	Collembola	1321	67	1038	44.9	47	15.9	98	6.7
	Diptera	40	2	107	4.6	48	16.3	158	10.8
	Hemiptera	40	2	93	4.0	30	10.2	13	1.0
	Hymenoptera	502	26	894	38.6	148	50.2	1063	72.8
	Orthoptera	40	2	67	2.9	1	0.3	22	1.5
Diplopoda	Spirobolida	0	0	7	0.3	0	0	6	0.4
Total		1965	100	2314	100	295	100	1460	100

Irrespectively of the site and season factors, the type of agricultural system factor seems to influence the presence of orders Coleoptera, Collembola, Hemiptera and Orthoptera and project as good indicators of the differences between monocultivar vs. mixed systems. Differences were found between ecosystems: La Maná had more abundance in the quantity of specimens than El Carmen. La Maná is an area with higher relative humidity which indicates that this habitat presents more favorable conditions for the development and conservation of these insect groups, making this agroecosystem more stable and diverse.

Our results on species abundance are in accordance with [19], who affirmed that the collembola play an important functional role in the decomposition processes of dead plant matter, the nutrient cycle, and help in the formation of soil characteristics. As observed in agroecosystems from other latitudes [20,21], Collembola abundance can be related to climatic and edaphic factors, availability of nutrients and biodiversity. Collembola seems to represent the diversity and abundance of species in the agroecosystems, since there was a direct relation between their abundance and the edaphic humidity [22–24]. Thus, in La Maná, the abundance in Collembola indicates that there would be sufficient environmental humidity, even in the dry season.

The largest number of individuals (2686) was obtained in the mixed system from La Maná (Table 5), while the amount of arthropods captured in El Carmen was quite similar in both production systems, and in total, had 40% fold inferior to that of La Maná. The difference between sites can be due to the site characteristics where higher precipitation, temperature, relative humidity and the different soil texture, and a higher presence of soil organic matter measured in La Maná. Furthermore, both the stability of the system (5 years vs. 30 years) and the application of agrochemicals higher in the monocultivar than in mixed systems may account for the observed differences. Our results are in concordance with those of Lavelle et al. [8], who found that arthropod in soils in humid tropics, as El Carmen, are inferior in abundance because they must be adapted to habitats where micro-climate fluctuations can be very strong and usually have compact soils with low oxygen concentration and high brightness, and with few open spaces and low availability and quality of food.

La Maná presented an index of 0.59, indicating that this site has the characteristics suitable for the habitat of different organisms, thus, being a balanced ecosystem. In contrast, in El Carmen an index of 0.49 was obtained, being this ecosystem largely unbalanced and with less favorable conditions for the development of many arthropods (Table 6).

According to data from the Simpson (1-D) diversity indexes, it can be observed that the highest diversity of arthropods is found in soils from mixed farming systems. The agricultural system influences the equity (e ^ H/S). The number of arthropods was higher in the mixed system plots than in the monocultivar ones (Figure 1). On the other hand, agricultural systems and sites influenced the dominance (D) and equity (J) of arthropods. The dominance index was higher in monocultivar systems, while the equity index was higher in mixed systems. To compare different agricultural systems (mixed versus monocultivar), several authors have used the richness of families. Wickramasinghe et al. [19] compared family richness among 24 pairs of mixed and monocultivar farming systems in Great Britain. They observed higher abundance and richness of insect species in mixed systems than in soils with single cultivars.

Table 5. Number of individuals registered in each type of production system at both sampling sites in 2014.

Site	Production System	N° Individuals	Percentage %
La Maná	Mixed	2686	44.5
	Monocultivar	1593	26.4
El Carmen	Mixed	936	15.5
	Monocultivar	819	13.6
	Total	6034	100

Table 6. The Simpson diversity index for each taxonomic group at each site in 2014.

	La Maná		El Carmen	
Order	**N° Individuals**	**Simpson**	**N° Individuals**	**Simpson**
Arachnida	52	0.00015	44	0.00063
Coleoptera	80	0.00035	57	0.00106
Collembola	2359	0.30280	145	0.00689
Díptera	147	0.00118	206	0.01390
Hemíptera	133	0.00096	43	0.00061
Hymenoptera	1396	0.10604	1211	0.48051
Orthoptera	107	0.00062	23	0.00017
Prostigmata	6	0.00000	12	0.00005
Spirobolida	7	0.00000	6	0.00001
Total	4287	0.41209	1747	0.50384
	1-D	0.58791	1-D	0.49616

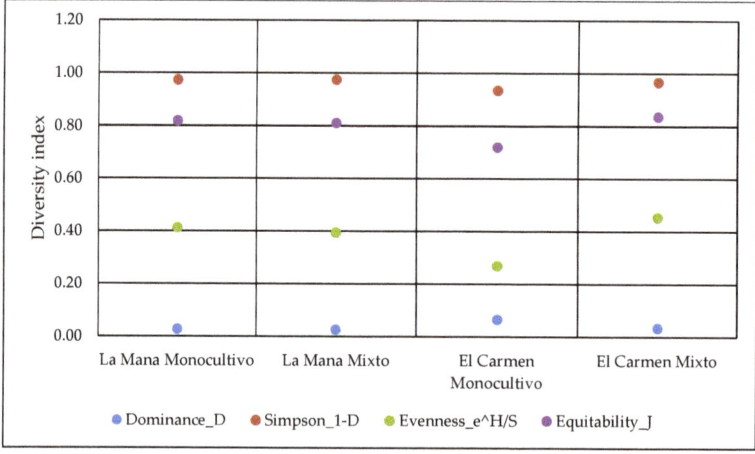

Figure 1. Simpson diversity index (1-D) and its three components; Dominance (D), Evenness (e^H/S) and Equitability (J) in function of the agroecosystem and locality, considering the amount of arthropods found.

According to the Jaccard coefficient, the highest similarity based on arthropod families was found between the mixed and the monocultivar systems at both localities, with 80% (Table 7). At the level of orders, a closer similarity of 100% between monocultivar (La Maná) and mixed (El Carmen) systems was observed. Mixed systems (La Maná) and monocultivar (El Carmen) presented low percentage of similarity (Table 8).

Table 7. Jaccard index of similarity (%) based on different arthropod families between sites and production systems in 2014.

Site/Agricultural System	El Carmen, Mixed	La Maná, Monocultivar	La Maná, Mixed
El Carmen, monocultivar	0.80 (80%)	0.55 (55%)	0.42 (42%)
El Carmen, Mixed	–	0.57 (57%)	0.52 (52%)
La Maná, monocultivar	–	–	0.79 (79%)

Table 8. Jaccard index of similarity (%) based on different arthropod orders between sites and production systems in 2014.

Site/Agricultural System	El Carmen, Mixed	La Maná, Monocultivar	La Maná, Mixed
El Carmen, Monocultivar	0.88 (88%)	0.88 (88%)	0.56 (56%)
El Carmen, Mixed	–	1.00 (100%)	0.88 (88%)
La Maná, Monocultivar	–	–	0.88 (88%)

Among the observed arthropods, there were families categorized as predators or parasitoids of pests. These are highly beneficial for the crops. In banana, parasitoids are important biological control agents that regulate several pests, mainly diverse lepidopteran defoliators [20–23]. Braconidae was one of the most common families with beneficial insects. These parasitoid wasps are recognized as being endo or ectoparasitoids with idiobiont strategy exclusively attacking Lepidoptera, Coleoptera and *Diptera* during different stages of development. The tiny wasps of the Diapriidae family were commonly found in wet microhabitats and shady areas, such as La Maná ecosystem. They are mainly endoparasitoids and primary predators on larvae and pupae of a wide range of insects, especially flies (Diptera) [24,25]. Several studies in different polyculture models have shown that the composition and diversity of plants in or around the production systems have an influence on the number of insects (including parasitoids). The reason is that the diversity of plants attracts natural enemies by offering resources such as micro-habitat and food that, specifically on parasitoids, can be determinant for their species richness, their longevity and their level of parasitism [26,27]. The parasitic Hymenoptera are also good indicators of pesticide impacts, because they are more sensitive to pesticides than most other insects, including their host species [28,29].

Diapriids were the only parasitoids more abundant and with a higher species richness in conventional banana than in the mixture plot [30]. In our study, Braconidae and Diapriids (Table 9) were more abundant in mixed systems with moderate agrochemical inputs than in monocultivar with intensive agrochemical applications in both sites. Taking into consideration all the orders found independent of the agroecosystem, La Maná had three times more beneficial arthropods than El Carmen.

The principal component analysis (PCA) revealed scarce dispersion of the points for sites orders agricultural systems (Figure 2). This could be related to the relatively high number of specific orders favoring the characterization of ecosystems. PCA showed higher dispersion and fewer arthropods in the rainy season than the dry season, where the largest number of arthropods was concentrated. A greater variability among the orders observed in the different localities and agricultural systems were found in both seasons of the year (rainy and dry). In La Maná, Hemiptera and Collembola were closer related to the dry season and the mixed agricultural system, whereas Coleoptera and Orthoptera were more related to dry season and monocultivar production system. In El Carmen site, Hymenoptera, Arachnida and Diptera are more abundant in the dry season and mixed production system and Prostigmata in the dry season and monocultivar agroecosystem. These two axes account for 55.0% of the total variability observed (Figure 2).

Table 9. Number of beneficial arthropods found in the different agroecosystems and sites.

Order	Family	La Maná		El Carmen	
		Mixed	Monocultivar	Mixed	Monocultivar
Coleoptera	Coccinelidae	2	1	4	1
	Escaleridae	0	1	0	0
	Staphylinidae	10	6	1	0
Diptera	Asilidae	5	8	1	2
	Dolichopodidae	28	15	37	6
Hymenoptera	Braconidae	37	23	5	2
	Diapriidae	120	81	36	19
	Encyrtidae	11	11	2	3
	Myridae	5	7	0	0
	Sphecidae	0	2	0	0
Total		218	155	86	33

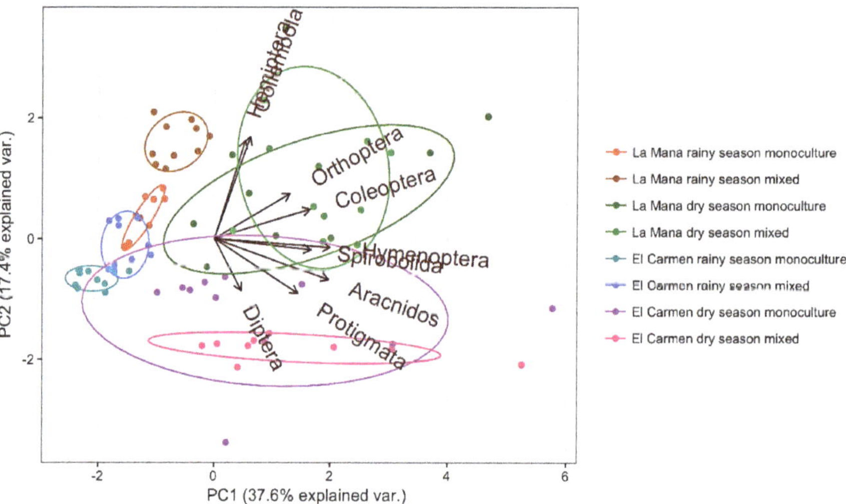

Figure 2. Principal component analysis (biplot) between sites, season, agricultural systems and number of arthropod orders found.

One of the most important aspects derived from these results are that a more diverse system presents a greater diversity of associated arthropods, depending on scale and context. This diversity is closely linked to the climatic conditions of the site, and the diversity of plants in the ecosystem [31], and it seems that the intraspecific diversity used in this study is enough to fulfil the needs of a diverse community of arthropods. Haddad et al. [32] demonstrated that the identity of plant species determines the abundance of arthropods, as well as affects different ecological processes within agricultural agroecosystems. The most relevant processes are biomass accumulation, decomposition rates and floor moisture. Thus, the results of arthropod diversity can be more influenced by the effects of the composition of plants than by the effects of the number of species [32,33]. The study by Rzanny et al. [33] was conducted in a temperate region, where winters induce an annual collapse in the abundance of all taxa. Contrastingly, a tropical climate is relatively stable and supports a relatively less perturbed community. We argue that the relatively stable conditions in the mixed *Musa*-based agroecosystems of the current study (these systems were at least five years old) allowed a stable plant community over years, and the presence of a perennial crop reinforces this stability. Another factor that contributes to the establishment of a relatively stable arthropod community in this mixed Musaceae agroecosystem is the low to moderate application of pesticide treatments.

4. Conclusions

Abundance analysis of the macro edaphic fauna and identification of the beneficial families allowed the determination of the equilibrium level of the Musaceae agroecosystem. The mixed type of production system provides plant diversity, which favors arthropod abundance and permits lower agrochemical application without yield penalties in comparison to the monocultivar. Within Hexapoda, the orders that presented larger populations were Collembola and Hymenoptera, based on the abundance and distribution they presented. The order Hymenoptera dominated in all the treatments, both by its abundance and by its distribution in the studied localities, even in ecosystems with ecological imbalance. Consistent with results from temperate studies, the mixed Musaceae production system was the one with the greatest presence of soil macro fauna, with the order Collembola being the most diverse, which gives us the guideline to say that this order is associated with the quality of the soils.

The management practices in agroecosystems can alter the community structure of pests' natural enemies, which can consequently influence their biocontrol. Since the functional composition of natural enemy communities, rather than taxonomic diversity, drive pest suppression efficiency, it is necessary to employ the functional approach to investigate the impact of management on natural enemies.

Our findings show that intraspecific diversity could be a good option to include in an IPM strategy for small and medium farmers and may help in the design of Musaceae agroecosystems to enhance the ecological regulation of pest management, without putting on the farmer the constraint of management different crops. Further research should explore the effect of combinations of various cultural intraspecific diversity systems with a more detailed study of the arthropod community present, down to at least genus identification in order to better define biodiversity indicators, especially considering that agricultural biodiversity will be essential to cope with the predicted impacts of climate change, and to detect more resilient farm ecosystems.

Author Contributions: Conceptualization, D.V.-A. and C.S.-C., methodology, D.V.-A., C.S.-C., M.L., C.P. and P.D.-S.; formal analysis, D.V.-A. and M.C.-G.; investigation, D.V.-A. and C.S.-C.; writing—original draft preparation, D.V.-A., C.S.-C., M.L., C.P. and P.D.S.; writing—review and editing, D.V.-A. funding acquisition, C.S.-C. All authors have read and agreed to the published version of the manuscript.

Funding: This research was funded by Bioversity International Grant contract number: (I-R_1362).

Acknowledgments: We are very grateful to Bioversity International Rome Italy and Technical State University of Quevedo, Ecuador for their financial and scientific support of this work. I sincerely thank the Universitat Autònoma de Barcelona (UAB), Spain for their contribution in this study.

Conflicts of Interest: The authors declare no conflict of interest.

References

1. Naylor, R.; Ehrlich, P. Natural pest control services and agriculture. In *Nature's Services: Societal Dependence on Natural Ecosystems*; Daily, G.C., Ed.; Island Press: Washington, DC, USA, 1997; pp. 151–174.
2. Quijas, S.; Schmid, B.; Balvanera, P. Plant diversity enhances provision of ecosystem services: A new synthesis. *Basic Appl. Ecol.* **2010**, *11*, 582–593. [CrossRef]
3. Michalko, R.; Košulič, O. The management type used in plum orchards alters the functional community structure of arthropod predators. *Int. J. Pest Manag.* **2020**, *66*, 173–181. [CrossRef]
4. Begon, M.; Harper, J.L.; Townsend, C.R. *Ecology: Individuals, Populations and Communities*, 3rd ed.; Wiley-Blackwell: Oxford, UK, 1996.
5. Tylianakis, J.M.; Tscharntke, T.; Lewis, O.T. Habitat modification alters the structure of tropical host–parasitoid food webs. *Nature* **2007**, *445*, 202–205. [CrossRef]

6. Knops, J.M.H.; Tilman, D.; Haddad, N.M.; Naeem, S.; Mitchell, C.E.; Haarstad, J.; Ritchie, M.E.; Howe, K.M.; Reich, P.B.; Siemann, E.; et al. Effects of plant species richness on invasion dynamics, disease outbreaks, insect abundances and diversity. *Ecol. Lett.* **1999**, *2*, 286–293. [CrossRef]
7. Maleque, M.A.; Maeto, K.; Ishii, H.T. Arthropods as bioindicators of sustainable forest management, with a focus on plantation forests. *Appl. Entomol. Zool.* **2009**, *44*, 1–11. [CrossRef]
8. Lavelle, P.; Blanchart, E.; Martin, A.; Spain, A.V.; Martin, S. Impact of soil fauna on the properties of soils in the humid tropics. *Myth. Sci. Soils Trop.* **1992**, *29*, 157–185.
9. Zerbino, S.; Altier, N.; Moron, A.; Rodriguez, C. Efecto del pastoreo de una pradera natural sobre la macrofauna del suelo. In Proceedings of the Seminario efecto del Pastoreo de una Pradera Natural Sobre la Macrofauna del Suelo, Tacuarembó, Uruguay, 20 September 2007; pp. 1–2.
10. Arroyo, J.; Iturrondobeitia, J.C. Differences in the diversity of oribatid mite communities in forests and agrosystems lands. *Eur. J. Soil Biol.* **2006**, *42*, 259–269. [CrossRef]
11. Carlier, J.; Foure, E.; Gauhl, F.; Jones, D.; Lepoivre, P.; Mourinchon, X.; Pasberg–Gauhl, C.; Romero, R.A. Fungal disease of the foliage. In *Disease of Banana, Abaca and Enset*; Jones, D., Ed.; CABI Publishing: Oxon, UK, 1999; pp. 37–141.
12. Marcillo-Delgado, J.C. Contribución Económica De La Biodiversidad De Musas Spp. A La Sostenibilidad De La Producción Agrícola A Nivel Del Pequeño Productor. Caso El Carmen Y La Maná En El Año 2009. Bachelor's Thesis, Escuela Superior Politecnica Nacional, Quito, Ecuador, 2012.
13. Jarvis, D.I.; Fadda, C.; De Santis, P.; Thompson, J. Damage, diversity and genetic vulnerability: The role of crop genetic diversity in the agricultural production system to reduce pest and disease damage. In Proceedings of the International Symposium, Rabat, Morocco, 15–17 February 2011; Jarvis, D.I., Fadda, C., De Santis, P., Thompson, J., Eds.; Bioversity International: Rome, Italy, 2011; p. 348.
14. Moreno, C.E. *Métodos Para Medir La Biodiversidad*; M&T–Manuales y Tesis SEA: Zaragoza, Spain, 2001.
15. Johnson, N.F.; Triplehorn, C.A. *Borror and DeLong's Introduction to the Study of Insects*, 7th ed.; Cengage Learning: Belmont, CA, USA, 2005.
16. Cerdá, X.; Palacios, R.; Retana, J. Ant community structure in citrus orchards in the mediterranean basin: Impoverishment as a consequence of habitat homogeneity. *Environ. Entomol.* **2009**, *38*, 317–324. [CrossRef]
17. Navarrete, J.; Newton, A. *Biodiversidad, Taxonomía Y Biogeografía De Artrópodos De México: Hacia Una Síntesis De Su Conocimiento*; UNAM Instituto de Biología: CDMX, México, 1996.
18. Chamorro Bello, C. El suelo: Maravilloso teatro de la vida. *Rev. Acad. Colomb. Cienc. Exactas Fis. Nat.* **2001**, *25*, 483–494.
19. Wickramasinghe, L.P.; Harris, S.; Jones, G.; Jennings, N.V. Abundance and species richness of nocturnal insects on organic and conventional farms: Effects of agricultural intensification on bat foraging. *Conserv. Biol.* **2004**, *18*, 1283–1292. [CrossRef]
20. Harrison, J.O. The natural enemies of some banana insect pests in Costa Rica. *J. Econ. Entomol.* **1963**, *56*, 282–285. [CrossRef]
21. Ostmark, H.E. Economic Insect Pests of Bananas. *Ann. Rev. Entomol.* **1974**, *19*, 161–176. [CrossRef]
22. Stephens, C. Ecological upset and recuperation of natural control of insect pests in some Costa Rican banana plantations. *Turrialba* **1984**, *34*, 101–105.
23. Thrupp, L.A. Entrapment and escape from fruitless insecticide use: Lessons from the banana sector of Costa Rica. *Int. J. Environ. Stud.* **1990**, *36*, 173–189. [CrossRef]
24. Fernández, F.; Sharkey, M.J. *Introducción A Los Hymenoptera De La Región Neotropical*; Sociedad Colombiana de Entomología: Bogotá, Colombia, 2006.
25. Quintero, E.M.; López, I.C.; Kondo, T. Manejo integrado de plagas como estrategia para el control de la mosca del botón floral del maracuyá Dasiops inedulis Steyskal (Diptera: Lonchaeidae). *Cienc. Tecnol. Agropecu.* **2012**, *13*, 31–40. [CrossRef]
26. Araj, S.-E.; Wratten, S.; Lister, A.; Buckley, H. Floral diversity, parasitoids and hyperparasitoids—A laboratory approach. *Basic Appl. Ecol.* **2008**, *9*, 588–597. [CrossRef]
27. Letourneau, D.K.; Armbrecht, I.; Rivera, B.S.; Lerma, J.M.; Carmona, E.J.; Daza, M.C.; Escobar, S.; Galindo, V.; Gutiérrez, C.; López, S.D.; et al. Does plant diversity benefit agroecosystems? A synthetic review. *Ecol. Appl.* **2011**, *21*, 9–21. [CrossRef]
28. Plapp, F.W.; Vinson, S.B. Comparative toxicities of some insecticides to the tobacco budworm 1 and its ichneumonid parasite, campoletis sonorensis 23. *Environ. Entomol.* **1977**, *6*, 381–384. [CrossRef]

29. Croft, B. *Arthropod Biological Control Agents and Pesticides*; Wiley: New York, NY, USA, 1990.
30. Matlock, R.B.; de la Cruz, R. An inventory of parasitic Hymenoptera in banana plantations under two pesticide regimes. *Agric. Ecosyst. Environ.* **2002**, *93*, 147–164. [CrossRef]
31. Dassou, A.G.; Dépigny, S.; Canard, E.; Vinatier, F.; Carval, D.; Tixier, P. Contrasting effects of plant diversity across arthropod trophic groups in plantain-based agroecosystems. *Basic Appl. Ecol.* **2016**, *17*, 11–20. [CrossRef]
32. Haddad, N.M.; Tilman, D.; Haarstad, J.; Ritchie, M.; Knops, J.M.H. Contrasting effects of plant richness and composition on insect communities: A field experiment. *Am. Nat.* **2001**, *158*, 17–35. [CrossRef] [PubMed]
33. Rzanny, M.; Kuu, A.; Voigt, W. Bottom–up and top–down forces structuring consumer communities in an experimental grassland. *Oikos* **2013**, *122*, 967–976. [CrossRef]

Article

Intercropping Winter Lupin and Triticale Increases Weed Suppression and Total Yield

Nicolas Carton, Christophe Naudin, Guillaume Piva and Guénaëlle Corre-Hellou *

USC LEVA, INRA, Ecole Supérieure d'Agricultures, Univ. Bretagne Loire, SFR 4207 QUASAV 1, 55 rue Rabelais, 49007 Angers CEDEX, France; nicolas.carton1@gmail.com (N.C.); c.naudin@groupe-esa.com (C.N.); g.piva@groupe-esa.com (G.P.)

* Correspondence: g.hellou@groupe-esa.com

Received: 15 June 2020; Accepted: 21 July 2020; Published: 1 August 2020

Abstract: Lupin (*Lupinus* sp.) produces protein-rich grains, but its adoption in cropping systems suffers from both its low competitive ability against weeds and its high yield variability. Compared with legume sole cropping, grain legume–cereal intercropping benefits include better weed suppression and higher yield and yield stability. However, the potential of enhancing crop competitive ability against weeds in additive winter grain legume–cereal intercrops is not well-known, and this potential in long crop cycles is even less studied. We studied how intercropping with a triticale (×*Triticosecale*) alters weed biomass and productivity of winter white lupin (*Lupinus albus* L.). The experimental setup consisted of eleven sites during a two-year period in western France. In each site-year, winter white lupin sole cropping was compared to winter white lupin-triticale intercropping in an additive sowing design. We found that intercropping reduced weed biomass at lupin flowering by an average of 63%. The rapid growth and high soil N acquisition of triticale compensated for the low competitive ability of lupin against weeds until lupin flowering. Competition from triticale in the intercrop reduced lupin grain yield (−34%), but intercropping produced a higher total grain yield (+37%) than did lupin sole cropping while maintaining the total protein grain yield.

Keywords: intercropping; lupin; triticale; weeds; legumes; nitrogen

---

1. Introduction

In Europe, the livestock sector mostly relies on imported soybean cake as protein-rich feed [1]. Local protein-rich crop products are needed to increase self-sufficiency. Among candidate crops, lupins (*Lupinus albus* (white lupin), *L. angustifolius* (narrow-leafed lupin)) produce seeds that have the highest grain protein content (30–42%) among grain legumes, and these seeds can partly substitute for soybean in ruminant, pig and poultry diets [2]. Lupins can also be an alternative to soybean in food diets including more plant proteins [3]. Lupins also fix significant amounts of atmospheric nitrogen (N_2) with an average fixation rate of 75% [4]. Like other grain legumes, lupins can provide farming systems with additional services by contributing to crop diversification and reducing synthetic nitrogen (N) fertilizer requirement in crop rotations. However, lupins are not widely cultivated in the European Union (EU) (approximately 120,000 ha in 2014; [5]) because of the high yield variability of this crop [6]. This variability is presumably associated with its high susceptibility to biotic and abiotic stresses. High weed infestation levels are usually reported [7,8]; these infestations are likely due to the slow ground cover and the long cropping season of the crop, especially for winter white lupin.

Compared with sole cropping, intercropping of grain legumes with cereals is a cropping strategy that can increase yield and improve yield stability, especially under low-input conditions [9–11]. This phenomenon generally results from the improved use of abiotic resources (light interception and use of both soil mineral N and atmospheric N_2). Intercropping can also reduce insect pests [12], diseases [13] and weeds [14,15].

Farmers often cite weeds as the main challenge in grain legumes. Intercropping a grain legume with a cereal can reduce weed growth. For instance, in spring pea, intercrops with barley suppress more weed biomass than do sole pea crops [15,16].

While pea– and faba bean–cereal intercropping has been the subject of numerous studies, white lupin–cereal intercropping for grain harvest is an innovative practice that has received little attention from academic researchers. Among grain legumes, lupin exhibits both the most variable yield [6] and the least competitive ability against weeds [7,17], meaning that intercropping winter white lupin for grain may have a high potential of development.

The objective of the practice described in this study is to produce protein-rich grains and the originality is that lupin is the main crop and the cereal is a companion crop that is also expected to produce grain ("harvested companion crop"). Intercropping lupin with a cereal could promote lupin cropping if the practice can reliably circumvent the two main shortcomings of lupin sole cropping by increasing competitive ability against weeds and securing grain and protein production.

The combination of intercropped species that exhibit contrasting traits could increase both the use of available resources and the competitive ability of the mixture against weeds. Compared with legume sole cropping, intercropping two species supposedly results in a higher competitive ability, especially at the beginning of the crop cycle, due to the contrasting traits of both species. In lupin–cereal intercrops, we expect both lupin early growth and soil N acquisition to be low and cereal growth and N acquisition to compensate for the low early competitive ability of lupin against weeds. In a multisite study in Europe, cereal competitive ability for soil mineral N was decisive regarding the higher weed suppression in organic spring pea–barley intercrops than in sole-cropped pea [16]. Species interactions can vary over time, especially during long cycle crops. To better understand the ultimate performances of legume–cereal intercrops, the systematic description of the relative dominance of each species before the period of maximum growth and maximum N_2 fixation rate of the legume would be useful because the benefits of intercropping for resource use start in the early growth phase [18] and because early dominance can shape the interactions in the second half of the growth cycle.

The aim of the present study was to compare the weed suppression and yield performance between winter white lupin-triticale intercropping and lupin sole cropping. The original aspect of this work is to study the effect of the addition of triticale by analyzing two phases: from sowing to lupin flowering and from lupin flowering to maturity. Moreover, the interest of this study lies in the fact that a range of contrasting growth conditions was used which will help to understand the conditions needed to guarantee the success of this practice. This was achieved by comparing winter white lupin–triticale intercropping and lupin sole cropping throughout a set of eleven experiments during a two-year period in western France.

2. Materials and Methods

2.1. Field Sites

Field experiments were carried out in the 2014/15 and 2015/16 growing seasons in western France for a total of eleven site-years (see details in Table 1). The 20-year average annual rainfall in the area is 718 mm, and average annual air temperature is 12.5 °C. The weather patterns of the two study years deviated similarly from the 20-year average. Specifically, the main deviation from the average data involved the October-February air temperatures, which were 8.9 °C (2014/15) and 9.6 °C (2015/16) averaged over the study sites, whereas the 20-year average was 8.2 °C (Figure 1). At each site, two winter white lupin cropping strategies were compared: lupin sole cropping and lupin–winter triticale intercropping.

Table 1. Site details of the experimental fields.

Year	2015						2016				
Site	**A**	**B**	**C**	**D**	**E**	**F**	**G**	**H**	**I**	**J**	**K**
Location	47.40 N, 1.32 W	47.87 N, 0.20 E	47.51 N, 1.48 W	47.06 N, 1.31 W	47.46 N, 1.23 W	47.64 N, 1.51 W	47.53 N, 1.03 W	47.47 N, 0.40 W	47.39 N, 1.33 W	47.51 N, 1.48 W	47.78 N, 1.45 W
Experimental design	strips	blocks	strips	blocks	strips	strips	strips	blocks	strips	strips	blocks
Soil texture	Loam	sand	silt loam	sandy loam	loam	silt loam	loam	sand	loam	silt loam	silt loam
Sowing date (day/month)	27/09	25/09	13/09	26/09	5/10	2/10	30/09	1/10	25/09	23/09	29/09
Varieties: Lupin Triticale	Lumen Ragtac	Clovis Kaulos	Clovis Ragtac	Clovis Ragtac	Lumen Ragtac	Lumen Ragtac	Lumen -Clovis (50% mix) Ragtac	Magnus Vuka	Orus Ragtac	Lumen Ragtac	Magnus Ruminac
Sowing density (kernels·m^{-2}): Lupin Triticale	 28 75	 25 75	 30 70	 25 70	 27 72	 31 70	 25 87	 30 75	 30 60	 25 75	 25 75
Triticale sowing	close to lupin rows	same rows	same rows	alternating rows	alternating rows	alternating rows	same rows	alternating rows	close to lupin rows	same rows	alternating rows
Preceding crop	winter wheat	winter wheat	winter wheat	winter barley	winter wheat	winter wheat	winter wheat	rapeseed	winter wheat	winter wheat	forage maize
Weed control C: chemical M: mechanical (number of operations)	C (1) M (1)	C (1)	C (1)	C (2)	C (2)	C (1)	C (1)	0	C (1)	C (1)	C (2)
Lupin plant density after winter in the sole crop (pl·m^2)	17	n.d.	40	n.d.	27	24	14	29	26	22	n.d.
Lupin plant density after winter in the intercrop (pl·m^2)	20	n.d.	21	n.d.	24	41	14	26	24	17	n.d.
Available N at sowing (0–90 cm) (kg·ha^{-1})	81	n.d.	119	n.d.	67	206	121	95	104	124	n.d.

n.d.: not determined.

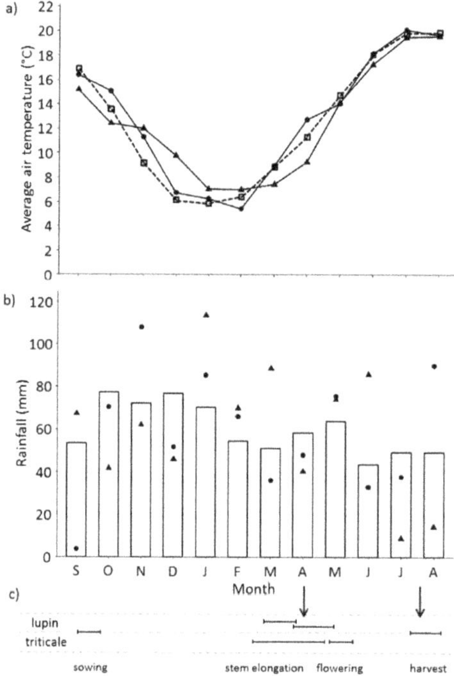

Figure 1. (**a**) Monthly average air temperature, (**b**) monthly rainfall, averaged over the study sites in 2015 (dots) and 2016 (triangles) as well as the 20-year average (temperature: open squares, rainfall: bars) and (**c**) average timing of key events of the crop cycle. Arrows represent the two harvests: at lupin main stem flowering and at lupin maturity.

The eleven sites were seven lupin-triticale intercropped farm fields (minimum 1 ha) on real farms including a wide strip (minimum 10 × 100 m) of sole-cropped lupin and four microplot experiments in a randomized block design with four replicates, including sole-cropped winter white lupin and lupin-triticale intercrop (individual plots ranging from 3 × 10 m to 6 × 17 m). The interest of these two sources of data was to involve different actors (farmers, advisers, researchers) and investigate the effect of intercropping in a wide range of situations. All fields were managed with conventional farming practices. Decisions of agricultural practices including cultivars; sowing date; preceding crop; row width; intercrop spatial arrangement; and pest, disease and weed control were made by both farmers and experiment managers and varied among sites (Table 1). However, the sole-cropped lupin and intercrop were managed identically, with the exception of the site F, in which the intercrop seedbed preparation included less soil tillage than did the lupin sole crop and 50 kg N·ha^{-1} was added only to the intercrop in April. The other sites were managed without N fertilization both in sole and intercrops. Lupin row spacing ranged between 12 and 75 cm (average: 34 cm). In all intercrops, the two species were sown on the same day or within one day. The lupin cultivars were chosen from the dwarf determinate branched cultivars that are typically cultivated in western France. Triticale cultivars were chosen for their late maturity, which is desirable for the simultaneous maturity of both species in the intercrop. Tested lupin cultivars do not show major differences except for the greater plant height of cv. Magnus. Tested triticale cultivars do not show major differences; all have a medium plant height. The lupin seeds were inoculated with *Bradyrhizobium lupini* in accordance with commonly recommended practices. The sole-cropped lupin was sown at an average of 27 kernels·m^{-2} (Table 1; SD = 2.4 kernels·m^{-2}). Lupin in the lupin-triticale intercrop was sown at the same density as lupin in the sole crop; the average triticale sowing density was 73 kernels·m^{-2} (SD = 6.1 m^{-2}), corresponding to an additive design in which lupin (L) was sown at the recommended density, and triticale (T) was sown at 30% of the sole crop recommended sowing density (L100:T30). All sites were rain fed and received no supplemental irrigation. All sites except H received a chemical control with only one application before emergence for sites A, B, C, D, F, G, I, J and with two applications, before emergence and during the winter for sites E and K. One site (A) also received a mechanical weeding in spring. These chemical or mechanical operations were similar in intercrops and sole crops.

2.2. Measurements, Sampling and Analysis

At all sites, the aboveground parts of lupin, triticale and weeds were hand-harvested twice throughout the crop cycle: at lupin main-stem flowering (April) and at lupin maturity (July until beginning of August). In the seven real farms, six plots (20 × 30 m) were defined randomly: three in the intercropped strips and three in the sole-cropped lupin strips. In each plot, the plants were harvested in three randomly defined subplots that covered 1 m × 2 lupin or lupin + triticale rows, and the values were averaged across subplots. In the four randomized block design experiments, plot size ranged between 1.7 × 10 m to 4.5 × 20 m and the plants were harvested in each plot in a randomly defined subplot (minimum area: 0.3 m^2; maximum area: 1.05 m^2). In the same plots later used for biomass sampling, all weeds were identified. The aboveground dry matter (DM) was determined after oven drying at 70 °C for 48 h until constant weight. At harvest, the grain and straw were threshed and then weighed. For N content measurements on aboveground biomass of lupin and triticale at lupin flowering, and on lupin and triticale grain and straw at maturity and aboveground weed biomass at flowering and maturity, the samples were pooled each across blocks and ground (120-mm mesh netting; "Pulverisette 19" universal cutting mill, "Laborette 27" sampler, and "Pulverisette 14" variable speed rotor mill; Fritsch, Idaroberstein, Germany). The total N concentration and ^{15}N:^{14}N ratio measurements were performed using a CHN analyser (EA3000; Euro Vector, Milan, Italy) and a mass spectrometer (IsoPrime; Elementar, Hanau, Germany). The mineral soil N content of representative soil samples from a 0–90-cm depth at sowing was measured via segmented flow analysis (Skalar Analytical B.V., Breda, Netherlands), which enables the determination of nitrate and ammonium contents by extraction with KCl [19]. At eight sites (A, C, E, F, G, H, I, J), crop plant density after emergence and lupin density after winter were recorded, and he mineral soil N content was measured after winter (0–90 cm). Protein content was determined by N content multiplied by 6.25.

2.3. Calculations

Weed reduction (WR) was assessed to characterize the ability of the intercrop to suppress weeds compared to the lupin sole crop. The index was determined according to the following equation:

$$WR = 100 \times ((\text{weed DM in the lupin sole crop} - \text{weed DM in the intercrop })/ (\text{weed DM in the lupin sple crop})) \quad (1)$$

The percentage of accumulated N derived from the air (%Ndfa) in lupin was determined on the two sampling dates using the ^{15}N natural abundance method [20]. Triticale served as the non-fixing reference in the calculation. The following equation was used:

$$\%Ndfa = 100 \times ((\ \delta^{15}N_{legume} - \delta^{15}N_{reference}) - (\beta fix - \delta^{15}N_{reference})) \quad (2)$$

where $\delta^{15}N_{legume}$ and $\delta^{15}N_{reference}$ are the natural ^{15}N enrichment values of the legume and triticale, respectively. The β-values for lupin ("βfix") were derived from the minimum values attained by $\delta^{15}N$ at all the sites: −0.88 at flowering (site H) and −1.03 at maturity (site D).

The normality and homoscedasticity of model residuals were tested using Shapiro's and Levene's tests, respectively ($\alpha = 0.05$). For the across-sites statistics, per site means of the data were used and the differences between sole-cropped lupin and the intercrop were assessed by Student's paired T-tests ($\alpha = 0.05$) except when non-normality was detected for model residuals. In those cases, Wilcoxon's signed-rank test ($\alpha = 0.05$) was used. Linear regressions using model II (Reduced Major Axis) were computed to assess relationships between variables. The absence of outliers in the data was assessed with Grubbs' test of model residuals [21]. For the per site statistics, per block data were used and the differences between sole-cropped lupin and the intercrop on each site were assessed by Student's *T*-tests ($\alpha = 0.05$), using the pooled variance estimate calculated using all sites. The Benjamini and Hochberg method was used to control the false discovery rate, i.e., the expected proportion of false discoveries amongst the rejected hypotheses [22]. Individual per site *T*-tests were used for lupin grain yield and crop total grain yield, where the global model was not applicable because of variance heterogeneity. All statistical analyses were performed using R software [23] version 3.3.2.

3. Results

3.1. Weed Suppression

The treatments were compared under various situations of weed infestation and growing conditions. In the lupin sole crop, at maturity, weed biomass ranged from 0 g·m^{-2} (site K) to 567 g·m^{-2} (site G; Table 2). The sites differed also in weed communities (Table 3).

Crop biomass of the different treatments at maturity also varied widely across sites, from 109 g·m^{-2} to 1238 g·m^{-2} in the lupin sole crop and from 416 g·m^{-2} to 1850 g·m^{-2} in the lupin-triticale intercrop. The variability of crop and weed growth was higher among sites than between the two years; therefore, the year effect was not isolated in the analyses.

The weed biomass at lupin flowering was lower in the intercrop (on average 38 g·m^{-2}; Table 4) than in the lupin sole crop (average of 100 g·m^{-2}).

The weed reduction (WR) reached an average of 63%. The difference in weed biomass between the lupin sole crop and the intercrop was higher with higher levels of weed biomass (Figure 2a).

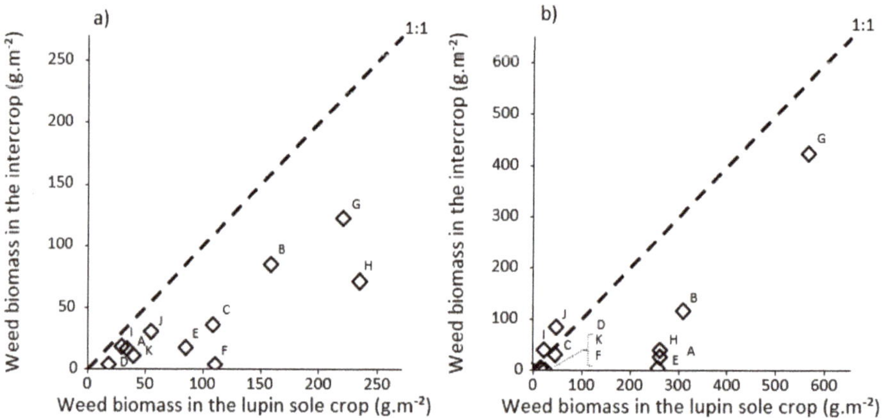

Figure 2. (**a**) Weed biomass in the intercrop against weed biomass in the lupin sole crop at lupin flowering and (**b**) at lupin maturity. Letters next to points identify study sites. The dashed lines represent the theoretical situations with equal weed biomass in both cropping strategies.

Table 2. Weed biomass (dry matter), crop biomass (dry matter), lupin grain dry yield and total grain dry yield in lupin sole crop (SC) and lupin-triticale intercrop (IC) on the 11 sites.

		Weed Biomass (g. m²)								Crop Biomass at Lupin Flowering (g·m²)						Lupin Grain Yield (g·m²)			Total Grain Yield (g·m²)		
		At Lupin Flowering				at Maturity				Lupin			Crop Total								
Site	Treatment	Mean	SD	Global *t*-test	WR (%)	Mean	SD	Global *t*-test	WR (%)	Mean	SD	Global *t*-test	Mean	SD	Global *t*-test	Mean	SD	Per Site *t*-test	Mean	SD	Per Site *t*-test
A	Lupin SC	34	25	n.s.	53	264	131	***	90	74	33	n.s.	74	33	***	33	21	n.s.	33	21	*
	IC	16	20			26	17			51	5		424	93		53	13		248	39	
B	Lupin SC	160	34	*	47	311	97	***	63	80	53	n.s.	80	53	***	236	117	n.s.	236	117	n.s.
	IC	85	48			117	26			122	56		302	133		315	161		462	163	
C	Lupin SC	109	37	n.s.	67	47	58	n.s.	37	528	58	***	528	59	***	271	80	n.s.	271	80	n.s.
	IC	35	22			30	11			196	57		852	34		118	6		228	111	
D	Lupin SC	19	18	n.s.	81	17	30	n.s.	79	333	129	***	333	129	***	370	82	n.s.	370	82	n.s.
	IC	4	5			4	6			206	88		898	302		227	82		523	62	
E	Lupin SC	86	50	n.s.	79	259	86	***	99	117	40	n.s.	117	40	***	381	102	*	381	102	n.s.
	IC	18	10			3	2			77	17		501	103		123	31		608	30	
F	Lupin SC	111	67	*	97	267	24	n.s.	100	243	75	*** [1]	243	75	***	445	21	*	445	21	n.s.
	IC	3	4			0	0			399	25		1070	126		401	8		628	72	
G	Lupin SC	221	51	*	45	567	90	**	25	59	32	n.s.	59	32	***	75	43	n.s.	75	43	n.s.
	IC	122	47			424	49			47	22		386	13		40	8		68	10	
H	Lupin SC	235	52	***	70	263	75	***	85	154	34	n.s.	154	34	***	218	86	n.s.	218	86	*
	IC	71	42			40	29			111	21		552	167		183	43		399	55	
I	Lupin SC	29	25	n.s.	37	23	28	n.s.	−70	325	47	***	325	47	n.s.	259	9	n.s.	259	9	n.s.
	IC	19	13			40	11			160	19		430	46		208	54		285	37	
J	Lupin SC	55	26	n.s.	45	48	42	n.s.	−73	355	70	***	355	70	**	301	53	*	301	53	n.s.
	IC	31	20			83	59			113	31		550	62		181	25		278	40	
K	Lupin SC	40	74	n.s.	73	0	0	n.s.	n.a.	445	55	***	445	55	***	663	154	*	663	154	n.s.
	IC	11	5			0	0			300	76		999	112		299	42		621	57	

The significance levels of comparisons between IC and lupin SC were assessed with a *t*-test using the pooled variance estimate calculated using all sites ("global *t*-test") or individual per site *t*-tests where the global model was not applicable. n.s.: not significant, n.a.: not applicable; [1] on site F, lupin biomass at flowering was significantly higher in the intercrop than in the sole crop. ***, **, indicate significant differences among species at $p < 0.001$, $p < 0.05$ respectively.

Table 3. List of weed species (species present with more than one plant per m^2 in at least one third of the subplots at lupin flowering) in lupin sole crops (SC) and intercrops (IC).

Sites	SC/IC	List of Weed Species
A	SC	*Atriplex patula, Epilobium tetragonum, Polygonum aviculare*
	IC	*Attiplex patula, Epilobium tetragonum*
B	SC	*Erodium cicutarium, Fallopia convolvulus, Juncus bufonius, Poa annua, Senecio vulgaris, Viola arvensis, Conyza sumatrensis*
	IC	*Erodium cicutarium, Fallopia convolvulus, Juncus bufonius, Poa annua, Senecio vulgaris, Viola arvensis*
C	SC	*Hypericum perforatum, Poa annua, Atriplex patula, Epilobium tetragonum, Polygonum aviculare, Ranunculus sardous, Senecio vulgaris*
	IC	*Poa annua, Phleum pratense*
D	SC	*Hypericum perforatum, Polygonum aviculare, Stellaria media, Atriplex patula, Chenopodium album, Conyza sumatrensis, Epilobium tetragonum, Poa annua, Portulaca oleracea, Ranunculus sardous*
	IC	*Hypericum perforatum, Poa annua, Ranunculus sardous*
E	SC	*Juncus bufonius, Lysimachia arvensis, Epilobium tetragonum, Hypericum perforatum, Poa annua, Ranunculus*
	IC	*Raphanus raphanistrum, Epilobium tetragonum, Geranium dissectum, Ranunculus sardous, Senecio vulgaris*
F	SC	*Hypericum perforatum, Juncus bufonius, Geranium dissectum, Epilobium tetragonum, Fallopia convolvulus, Tripleurospermum inodorum, Poa annua, Polygonum aviculare, Ranunculus sardous, Senecio vulgaris*
	IC	*Hypericum perforatum, Juncus bufonius, Tripleurospermum inodorum, Poa annua, Polygonum aviculare, Ranunculus sardous*
G	SC	*Bromus mollis, Tripleurospermum inodorum, Poa annua, Arabidopsis thaliana, Daucus carota, Fumaria officinalis, Rumex crispus*
	IC	*Dactylis glomerata, Tripleurospermum inodorum, Poa annua, Aphanes arvensis, Rumex crispus*
H	SC	*Poa annua, Juncus bufonius, Senecio vulgare, Arabidopsis thaliana, Capsella bursa-pastoris, Conyza sumatrensis, Tripleurospermum inodorum, Sonchus asper*
	IC	*Poa annua, Juncus bufonius, Senecio vulgare, Arabidopsis thaliana, Capsella bursa-pastoris, Tripleurospermum inodorum*
I	SC	*Juncus bufonius, Poa annua, Polygonum aviculare, Stellaria media, Veronica hederifolia*
	IC	*Juncus bufonius, Poa annua, Polygonum aviculare, Daucus carota, Hypericum perforatum, Ranunculus sardous*
J	SC	*Daucus carota, Poa annua*
	IC	*Daucus carota, Poa trivialis*
K	SC	*Elytrigia repens*
	IC	*Elytrigia repens*

Table 4. Crop and weed biomass production and soil N acquisition at the 11 sites in lupin sole crops (SCs) and lupin–triticale intercrops (ICs) during two periods, from sowing to lupin flowering, and from lupin flowering to maturity.

		From Sowing until Lupin Flowering		From Lupin Flowering until Maturity	
		Crops	**Weeds**	**Crops**	**Weeds**
Biomass production ($g \cdot m^{-2}$)	Lupin SC	247 (162)	100 (76)	526 (285)	66 (138)
	IC total	633 (270)	38 (38)	494 (263)	32 (93)
	Lupin in IC	162 (109)		307 (180)	
	Triticale in IC	471 (181)		187 (201)	
comparisons	Lupin SC-IC total	$t_{10} = 6.3$ ***	$t_{10} = -4.2$ **	$t_{10} = 0.4$ (n.s.)	$t_{10} = -1.1$ (n.s.)
	Lupin SC - Lupin in IC	$t_{10} = -2.1$ (n.s.)		$t_{10} = -3.3$ **	
	Lupin in IC - Triticale in IC	$t_{10} = -7.9$ ***		$t_{10} = 1.4$ (n.s.)	
Soil N uptake ($g \cdot m^{-2}$)	Lupin SC	2.5 (1.6)	1.9 (1.0)	6.7 (6.9)	0.4 (2.0)
	IC total	7.0 (3.2)	0.7 (0.6)	3.5 (3.6)	0.5 (1.6)
	Lupin in IC	0.9 (0.9)		3.0 (3.2)	
	Triticale in IC	6.2 (2.7)		0.5 (2.5)	
Comparisons	Lupin SC - IC total	$t_{10} = 7.4$ ***	$t_{10} = -5.5$ ***	V = 6 *	$t_{10} = 0.3$ (n.s.)
	Lupin SC - Lupin in IC	$t_{10} = -4.8$ ***		V = 5 **	
	Lupin in IC - Triticale in IC	$t_{10} = -7.4$ ***		$t_{10} = 1.9$ (n.s.)	

The significance levels of comparisons were assessed with *T*-tests except where *V*, the test statistic of Wilcoxon's signed-rank test, is given. n.d.: not determined; n.s.: not significant. All values are the means (SDs) of plant aboveground dry matter and soil N uptake, $n = 11$. ***, **, indicate significant differences among species at $p < 0.01$, $p < 0.05$ respectively.

A significant effect of intercrop on weed biomass was observed on four sites (B, F, G, H) (Table 2). These sites had a high weed biomass (higher than 110 $g \cdot m^2$).

The weed biomass at maturity was also significantly lower in the intercrop (on average 70 $g \cdot m^{-2}$) than in the lupin sole crop (on average 166 $g \cdot m^{-2}$). However, the WR was lower (average of 43%) at crop maturity than at flowering. Weed reduction in the intercrop compared to sole-cropped lupin occurred mainly from sowing until lupin flowering, and to a lesser extent from lupin flowering until crop maturity. The variability across sites of WR at maturity (coefficient of variation (CV) of 151%) was higher than WR at flowering. (CV of 30%). The average WR at maturity was only 15% for the six sites that had the lowest weed biomass in the sole crop and weed reduction was not statistically significant for these sites. However, the WR was highly significant at maturity and reached 60% for the five sites on which weed biomass in the sole crop surpassed 200 $g \cdot m^{-2}$ (Table 2).

Triticale produced more biomass than did lupin in the intercrop from sowing until lupin flowering (Table 4). The addition of triticale systematically significantly increased total crop biomass at lupin flowering except on site I (average of +387 $g \cdot m^{-2}$, i.e., +157% DM, $t_{10} = 6.3$, $p = 8$ E^{-5}, minimum: +32% (site I), maximum: +549% (site G), Figure 3. At lupin flowering, the WR was linearly correlated with the crop total biomass gain allowed by the integration of triticale (Figure 4).

Intercropping allowed a median crop biomass of 551 $g \cdot m^{-2}$, and the weed biomass was maintained at less than 50 $g \cdot m^{-2}$ on more than 60% of the sites, even the sites with the lowest crop biomass (Figure 5). At half of the sites, the biomass of sole-cropped lupin at flowering was less than 243 $g \cdot m^{-2}$. At this level of lupin biomass, weed growth was below 50 $g \cdot m^{-2}$ at a probability of less than 10% (Figure 5). At sites with a higher lupin biomass (higher than 243 $g \cdot m^{-2}$), weed biomass at lupin flowering could be maintained under 50 $g \cdot m^{-2}$ for 65% of those sites.

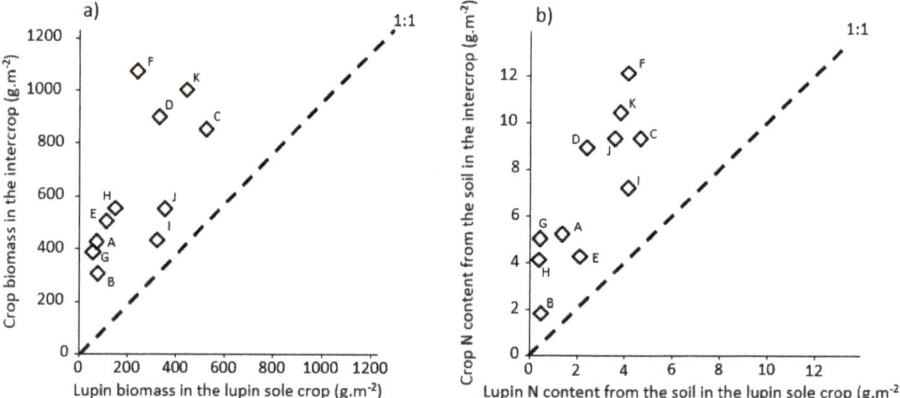

Figure 3. (**a**) Crop biomass in the intercrop against lupin biomass in the lupin sole crop at lupin flowering. (**b**) Crop N content from the soil in the intercrop against lupin N content from the soil in the lupin sole crop at lupin flowering. Letters next to points identify study sites. The dashed lines represent the theoretical situations with equal values in both cropping strategies.

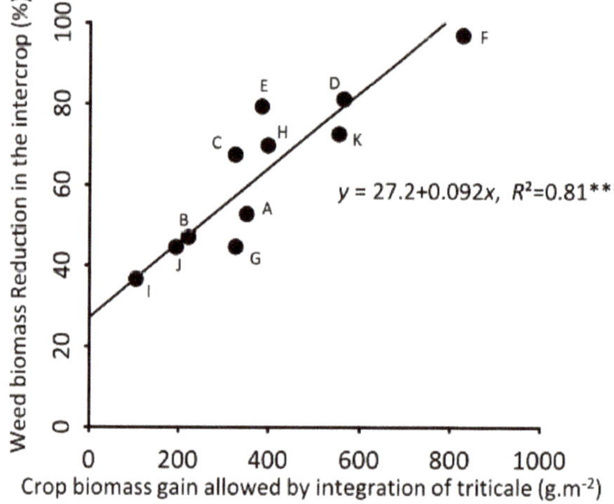

Figure 4. Correlation between weed biomass reduction in the intercrop compared to the lupin sole crop (WR) (%) and crop biomass gain allowed by addition of triticale at lupin flowering. Letters next to points identify study sites. *** indicate that the correlation is significant at $p < 0.001$.

From lupin flowering to maturity, the intercrop and the lupin sole crop produced similar amounts of DM. In the intercrop, the lupin biomass increased at a higher rate than did triticale biomass (+189% vs. +40%, respectively, Table 4). The variability of WR at maturity across sites was not explained by increases in crop or lupin biomass (no correlation, $p = 0.3$ and $p = 0.4$, respectively). However, the weed growth after lupin flowering was maintained at less than 60 g·m^{-2} when the crop biomass at lupin flowering attained the threshold value of 400 g·m^{-2} (Figure 6). This crop biomass value was attained in the intercrop at nine of eleven sites and in the lupin sole crop at two of eleven sites. The weed growth between lupin flowering and maturity was negative at some sites because of low weed growth and the decomposition of weed biomass in the end of the crop cycle.

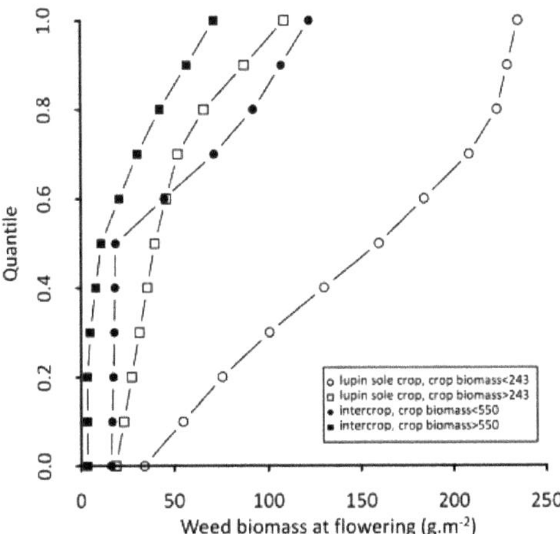

Figure 5. Distribution of weed biomass at lupin flowering across the 11 sites for different crop biomasses in the intercrop and in the lupin sole crop. Lupin sole crop and intercrop were separated in two pools according to median crop biomass so that for each cropping strategy, the sites with lowest crop biomass are represented with circles and the sites with highest crop biomass are represented with squares. Median crop biomass in the lupin sole crop: 243 g·m^{-2}, median crop biomass in the intercrop: 550 g·m^{-2}.

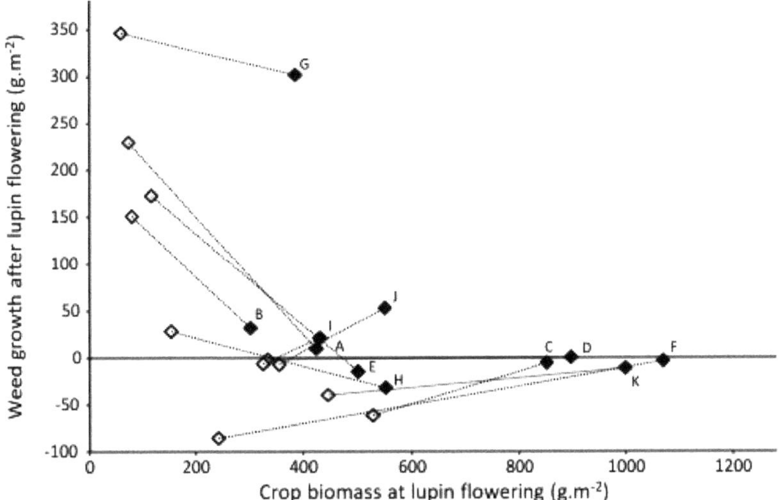

Figure 6. Weed growth after flowering against crop biomass at lupin flowering. Open symbols refer to lupin sole crops and full symbols refer to intercrops. Letters next to points identify study sites.

The mean proportion of triticale in the intercrop biomass was 75% at lupin flowering (CV: 14%) This proportion decreased to 59% at maturity and with a higher variability (CV = 28%). The addition of triticale reduced the proportion of weeds in total plant biomass. At flowering, the weed biomass represented 33% of the canopy in pure lupin and 7% in the intercrop. At maturity, the weed biomass represented 24% of the canopy in pure lupin and 8% in the intercrop.

The addition of triticale reduced weed biomass but also the diversity of weed species compared to lupin sole crop (Figure 7a). However, the proportion of mono/dicotyledonous species in the total number of weed species was not greatly modified except in site C (Figure 7b).

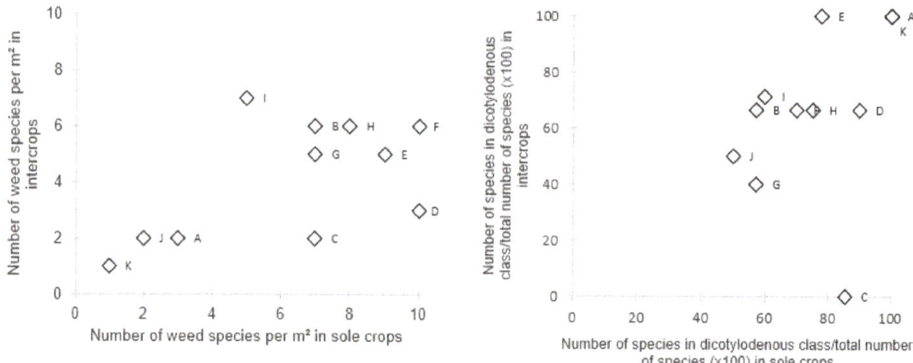

Figure 7. (**a**) Number of weed species in intercrops against number of weed species in lupin sole crops. (**b**) The percentage of dicotyledonous species in the total number of weed species in intercrops against the percentage in lupin sole crop. The dashed lines represent the theoretical situations with equal values in both cropping strategies.

3.2. Soil N Acquisition by Crops and Weeds before and after Flowering

Soil N acquisition was very low in lupin sole crop at the beginning of the crop cycle. It reached only 2.5 g·m^{-2} at the beginning of lupin flowering. Crop mineral soil N acquisition until lupin flowering was enhanced by the addition of triticale: the average crop mineral soil N acquisition gain was +4.5 g·m^{-2}, i.e., +181%, Table 4). In the intercrop, triticale acquired 88% of the crop mineral soil N acquired by the mixture until flowering. Triticale biomass was on average 3 times higher than lupin biomass and triticale soil mineral N acquisition was on average 8 times higher than lupin soil mineral N acquisition until flowering. Lupin acquired on average 5 mg of soil N per g of crop biomass produced, whereas triticale acquired 13 mg of soil N per g of crop biomass.

Weeds had a high ability to acquire soil mineral N: they acquired on average 19 mg of soil N per g of weed biomass at the beginning of lupin flowering. However, weed mineral soil N acquisition was reduced in the intercrop (−63% on average, Table 4) in comparison with weeds observed in the lupin sole crop. The lupin %Ndfa at the beginning of lupin flowering was significantly higher in the intercrop (84%, SD = 12) than in the lupin sole crop (66%, SD = 14, $t_{10} = 3.0$, $p = 0.01$; Table 5). The intercrop acquired more mineral soil N than did the lupin sole crop despite lupin depending less on mineral soil N in the intercrop.

At the end of winter, the integration of triticale had a tendency to reduce the mineral soil N content (53 kg·ha^{-1} in the sole crop and 43 kg·ha^{-1} in the intercrop, $t_7 = -2.2$, $p = 0.06$; Table 5), showing that an effect on the available N occurred rather early in the cropping season.

Total crop mineral soil N acquisition until lupin flowering varied less in the intercrop than in the lupin sole crop (Table 4). The CV for sole-cropped lupin was 64%, whereas the CV for the intercrop was 45%.

Table 5. Soil mineral N content and lupin %Ndfa (percentage of N accumulated in aboveground parts derived from N_2 fixation).

Site	Treatment	Soil Mineral N Content in the End of Winter (kg·ha^{-1})	Lupin %Ndfa at Flowering	Lupin %Ndfa at Maturity
A	Lupin SC	83	45	56
	IC	78	95	69
B	Lupin SC	18	77	79
	IC	n.d.	85	84
C	Lupin SC	72	75	69
	IC	45	81	74
D	Lupin SC	32	78	81
	IC	n.d.	95	100
E	Lupin SC	72	46	66
	IC	37	98	99
F	Lupin SC	86	49	62
	IC	38	80	75
G	Lupin SC	31	78	73
	IC	28	79	71
H	Lupin SC	25	89	71
	IC	25	94	76
I	Lupin SC	39	60	47
	IC	50	56	44
J	Lupin SC	61	68	24
	IC	43	74	37
K	Lupin SC	67	65	50
	IC	n.d.	85	70
mean	Lupin SC	53	66	62
	IC	43	84	73
SD	Lupin SC	25	15	17
	IC	16	12	19
comparison		$t_7 = -2.2$ n.s.	$t_{10} = 3.0$ *	$t_{10} = 3.4$ **

The significance levels of comparisons were assessed with *T*-tests. n.d.: not determined; n.s.: not significant. **, *, indicate significant differences among species at $p < 0.05$, $p < 0.01$, respectively.

In the lupin sole crop, only 27% of soil N was acquired before flowering, whereas in the intercrop, 67% of soil N was acquired before flowering. Thus, from lupin flowering until maturity, the lupin sole crop acquired more mineral soil N than did the intercrop (Table 4). As observed at lupin flowering, the lupin %Ndfa at maturity was significantly higher in the intercrop (73%, SD = 19) than in the lupin sole crop (62%, SD = 17, $t_{10} = 3.4$, $p = 6$ E^{-3}; Table 5).

3.3. Grain and Productivity Yield of the Lupin Sole and Intercrop

In the lupin sole crop, the mean grain yield was 296 g·m^{-2} (minimum: 33 (site A), maximum: 663 (site K), Table 2), and the mean protein yield was 104 g·m^{-2}. The mean lupin yield was 34% lower in the intercrop than in the sole crop, but the effect of the addition of triticale on lupin yield differed among sites (Figure 8). Lupin yield was significantly lower in the intercrop than in the sole crop on four sites (Sites E, F, J and K) (Table 2). At sites A and B, two sites with low-to-medium lupin yields, the lupin grain yield tended to be higher in the intercrop than in the lupin sole crop. The lupin protein concentration did not significantly differ between the lupin sole crop and the intercrop (35.7% and 35.2%, respectively, with a CV of 17% in the sole crop and 15% in the intercrop; data not shown).

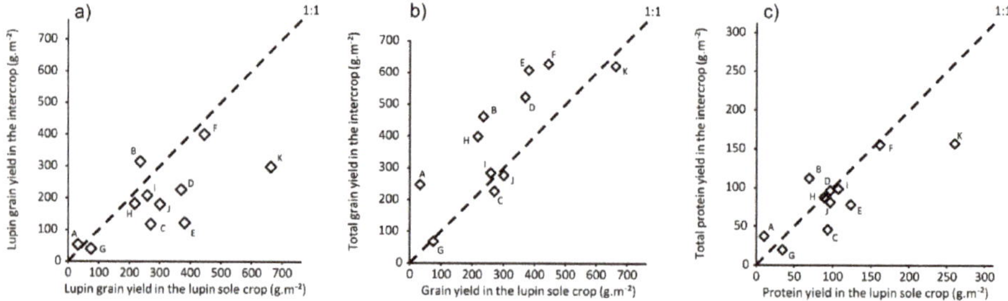

Figure 8. (**a**) Lupin grain yield in the intercrop against lupin grain yield in the lupin sole crop. (**b**) Total grain yield in the intercrop against lupin grain yield in the lupin sole crop. (**c**) Total protein yield in the intercrop against protein grain yield in the lupin sole crop. Letters next to points identify study sites. The dashed lines represent the theoretical situations with equal values in both cropping strategies.

Triticale produced on average 201 g·m^{-2} grain (minimum: 28 (site G), maximum: 485 (site E), Table 2), i.e., a similar yield as that of lupin in the intercrop (195 g·m^{-2} on average, $t_{10} = 0.12$, $p = 0.9$). Triticale grain had a mean protein concentration of 10.0% (data not shown). When considering total grain production, the intercrop produced more grain than did the lupin sole crop on average over all sites (+37%, mean: 395 g·m^{-2}, minimum: 68 (site G), maximum: 628 g·m^{-2} (site F)). At site A and site H, the lupin yield did not differ between the intercrop and the sole crop, but the total intercrop yield was significantly higher than the lupin sole crop yield (Table 2). At site A, the lupin grain yield was less than 100 g·m^{-2} in the sole crop and the addition of triticale allowed a total production of 248 g·m^{-2} grain (Table 2). At the four sites where the lupin yield was significantly lower in the intercrop than in the lupin sole crop, the intercrop total yield did not differ from the lupin sole crop yield (Table 2). The triticale proportion in intercrop grain biomass ranged from 27% to 80% (mean 49.5%).

On average, the total protein yield of the intercrop did not significantly differ from that of the lupin sole crop, but differences across sites were recorded (Figure 8). The reduction in protein production due to intercropping was highest at the site with the highest lupin yield in the sole crop (site K, −103 g·m^{-2} protein), as triticale protein production could not compensate for the reduction of protein-rich lupin grain production. However, on site B, the intercrop produced 43 g·m^{-2} more protein than did the lupin sole crop (Figure 8). In the intercrop, lupin represented on average 49% of grain yield and 78% of protein yield. The total grain and protein production values varied less across the eleven site-years in the intercrop than in the lupin sole crop (Table 2).

4. Discussion

Our results showed that the addition of triticale has a great ability to reduce weed biomass in lupin crops, especially when weed pressure is high, while maintaining protein yield. Weed biomass reduction can be explained by the increased crop biomass and mineral soil N acquisition especially at the beginning of the crop cycle through the addition of triticale. Moreover, crop biomass, mineral soil N acquisition, grain yield and protein yield were more stable in intercrops.

4.1. Weed Reduction Allowed by the Addition of Triticale and Underlying Processes

Weed reduction values at lupin flowering were consistently greater than 37% across a wide range of practices, pedo-climatic conditions and weed growth potentials (Figure 2, Table 2), demonstrating that the addition of triticale at 30% of its recommended density in the sole crop is effective at reducing weed growth. This finding is in line with previous studies on other intercropping systems. Focusing on intercrops that have a short growing season, Corre-Hellou et al. [16] studied spring pea–barley additive intercrops in which barley was sown at 150 kernels·m^{-2} (100:50) in five countries in Europe and obtained a mean WR of 55% at the beginning of pea flowering. Using an additive intercrop design consisting of spring pea and oat (60 kernels·m^{-2}, 100:20), Gronle et al. [24] reported WR values of 14% and 27% at the beginning and end of pea flowering, respectively.

Weed growth in winter white lupin seems to utilize an available ecological niche related to the low growth rate of lupin until flowering. The addition of triticale in lupin occupied this niche; this occupation strongly and systematically increased crop weed suppression before flowering. Intercropping allowed attainment of crop biomass levels that ensure high weed control; this high crop biomass was rarely observed in lupin sole crops and is consistent with the finding that triticale can particularly compensate for the low competitive ability of a legume crop that produces low levels of crop biomass [24]. The less competitive a legume sole crop is, the more the addition of a cereal facilitates weed suppression.

In the lupin sole crop, mineral soil N not used by lupin was taken up by weeds, whereas in the intercrop, triticale acquired a large amount of mineral soil N to the detriment of weeds (Table 4). Cereals have a higher soil N requirement than legumes, and this demand is often associated with rapid root growth and a dense root system [9,25–28]. Integrating a cereal into a legume crop can result in the use of mineral soil N to produce cereal grain instead of weed biomass [15]. Lupin acquired much less soil mineral N in the intercrop than in the lupin sole crop due to the combined effects of higher %Ndfa and lower biomass, mitigating the effects of triticale soil mineral N acquisition. Hauggaard-Nielsen et al. [29] also reported that grain legumes accumulated less soil N when intercropped than was expected from sole crop acquisition.

Despite not being measured, competition for light is also assumed to be an important mechanism in the higher competitive ability of the intercrop than sole crop, as observed in other intercropping systems [30] due to complementary traits for leaf area distribution in the canopy and an increase in spatial homogeneity [31,32]. The spatial homogeneity may depend on sowing patterns of intercrops. Sowing the triticale in alternating rows in an additive design may decrease early heterogeneity of crop ground cover by filling the wide inter-row space in the lupin sole crop [33]. In our study, in the five sites (sites D, E, F, H, K) with highest WR (higher than 70%) at lupin flowering, intercrops were sown with triticale and lupin on alternating rows (Tables 1 and 2).

Lupin and triticale differ in their growth dynamics, and two contrasting periods were studied in the long crop cycle; the limit was the time of lupin flowering. Until lupin flowering, lupin had a low biomass production, which favored weeds in the lupin sole crop because the weeds could develop virtually without crop competition during autumn and winter and could gain an initial advantage over lupin during the first growing period (Tables 2 and 4). Triticale had a high biomass production during the initial growth stages (Tables 2 and 4). As reported in spring barley and pea [26,34], the maximum growth rate of the cereal occurred before that of the grain legume. Beginning at lupin flowering, the lupin growth rate strongly increased, whereas the triticale growth rate strongly decreased (Table 4). The clearly offset period of maximum growth rate and opposite growth patterns between lupin and triticale may allow temporal complementarity of resource use. In our study, compared with winter legume sole cropping, intercropping with triticale reduced weed growth before flowering but not after flowering (Table 4).

Complementarity of resource use in time and space in intercropping may be not the unique mechanism explaining weed suppression. Allelopathy effects can also contribute to weed suppression [35], but this mechanism was not investigated in our study and would require specific experiments. A better understanding on the mechanisms behind weed control and other benefits of intercropping systems can guide the design of various species combinations with traits that maximize positive or minimize negative interactions and reach expected services [36,37].

Weed species complexes encountered in the field network may have interacted with the effects of triticale. Nevertheless, the effect of intercropping seems effective for a wide range of weed compositions. Some weed species taller than triticale might limit the intercropping effect, but such situations were rare (only observed punctually during the crop cycle with *Erodium cicutarium* (site B), *Triplospermum inodorum* (site G), *Dactylis glomerata* (site G), and *Poa trivialis* (site J)). Our results indicated a reduction of the number of weed species in intercropping but without a modification of the repartition of the species in monocotyledonous and dicotyledonous classes. Nevertheless, these results need to be confirmed with a wider range of situations and with additional data (plant density and biomass per weed species) to investigate more in depth the effect of intercropping on the structure of the weed community in relation to weed and crop functional traits.

4.2. Productivity of Intercropping Compared to Lupin Sole Cropping

This study confirmed the high potential of lupin to yield large amounts of grain and protein both in sole cropping and in intercropping systems. Although triticale reduced the lupin growth and yield, triticale grain production increased average total grain production by 37%. The mean lupin yield reduction by triticale (34%) was lower than that obtained by Hauggaard-Nielsen et al. [29] using a 50:50 substitutive design with narrow-leafed lupin and spring barley during a three-year period, during which intercropping reduced the lupin grain yield by 62%. It is likely that the lupin 100:triticale 30 additive design used in our study better maintains lupin yield than does a balanced substitutive design and still allows satisfactory cereal production. In our study, lupin compensated for the reduced soil N availability and fulfilled its N requirements by increasing the proportion of N derived from fixation; this mechanism is in agreement with the results of numerous studies [9,10,26,38,39]. However, the lupin biomass decreased, hence the reduced total N amount in lupin. The sites H and A showed promising results: despite triticale proportions in the intercrop biomass being higher than 80%, lupin yield was little decreased at site H (−16%) and even increased at site A (+60%). The reasons for these results are not clear, but they may have been in part favored by the sowing design, in which lupin and triticale were not sown on the same rows [40,41]. However, site A showed low lupin yields in both the sole crop and intercrop. A minimum of lupin plant density is important to reach satisfactory lupin yield and weed control levels: on the site with the lowest lupin density after winter (14 plants·m^{-2} in the lupin sole crop and in the intercrop, site G), lupin biomass at flowering and lupin grain yield in the intercrop were the lowest of all sites and weed biomass at maturity were by far the highest of all sites in the sole crop and in the intercrop.

Willey [42] reported that the growth of species sown in intercrops at the same density as that of sole crops is always less than the growth the sole crop. This phenomenon shows that full complementarity between intercropped species cannot be achieved. Lupin-triticale intercropping is a system in which the crop producing favored yield lets the companion crop have an initial advantage. This phenomenon is not ideal but seems inevitable when the main species exhibits slow early growth, as observed in grain legumes. In our study, this effect has possibly been further enhanced by the particularly mild autumn and winter during both study years. A lower temperature during crop establishment and early growth would probably have mitigated the cereal growth and dominance in the intercrop because it would have delayed the beginning of cereal maximal growth phase and reduced tillering, whereas lupin maximal growth phase and branching takes place later and would not have been affected.

Grain yield variability across a wide range of situations was lower when considering total intercrop yield rather than lupin sole crop yield. This result is consistent with previous results on both spring intercrops [9] and winter intercrops [43] but contrasts with those of Hauggaard-Nielsen et al. [29], who reported no yield stability differences between narrow-leafed (spring) lupin-barley intercrops and narrow-leafed lupin sole crops. The higher yield stability measured in this study needs to be assessed in long-term studies. The level of yield variability remained high in the intercrop; however, here, we mostly characterized between-site variability, whereas farmers may be more interested in ways to increase inter-annual stability. If for lupin, intercropping proves to be an efficient way to secure yield, this could be a convincing argument for some farmers who could decide to replace lupin sole crop with lupin–cereal intercrop or start to grow lupin using the intercropping strategy.

4.3. Perspectives for the Use of Lupin-Based Intercrops

Effective weed control combined with the maintenance of lupin yield in the intercrop occurred for instance on site F where the combination of increased total crop biomass and a high proportion of lupin in the crop biomass occurred. In situations where sole-cropped lupin can perform well (limited biotic and abiotic stresses under a favorable climate, the absence of water logging and the use of pesticides and herbicides, e.g., site K in our study), adding a cereal will very likely reduce protein yields. However, intercropping has a high potential for lupin growth in suboptimal conditions that are becoming increasingly frequent and unpredictable due to climate change. Intercropping should also be promoted as part of an integrated agronomic strategy in combination with other agronomic measures such as crop rotation, cover cropping and mechanical weeding to limit or forfeit the use of herbicides. Adding cereals in grain legume crops seems to allow maintaining protein productivity while keeping weed biomass within acceptable thresholds without or with a low use of herbicides. This additive intercropping design should therefore be promoted as a strategy to facilitate production of grain legumes following the need to reduce the use of herbicides, in the same way as other (mostly substitutive) intercropping designs have been promoted as strategies to increase total grain production and cereal protein concentration in low-input systems [44,45].

Although the triticale cultivars and density levels used in this study showed good performances, further adjustment of cereal species or cultivar choice or density fine-tuning is needed, as competition from the companion crop on lupin must be reduced. We hypothesize that on two sites (F and H), the alternating row design played a role in allowing the competition of triticale against lupin to be lower than that at other sites that had the same level of triticale proportion in the intercrop biomass. Specific experiments are needed to compare different spatial arrangements. Triticale cultivars or other cereal species with a shorter height after stem elongation may be favorable to maintain lupin yield. Selecting lupin cultivars for traits best adapted to intercropping with cereals could further increase the benefits of this cropping strategy [46].

Most farmers and experimenters managing experimental fields had no previous experience in lupin intercropping, suggesting that large room for optimization of field choice and management practices exists and that higher performances of the intercrop can be expected. In our field network, sole-cropped lupin management was not optimal since practices adapted to the intercrop were applied to both cropping strategies. Specifically, in lupin sole crops, a post-emergence herbicide treatment was typically applied in conventional fields at the time of the study and it has not been used here in eight of the eleven sites. The potential of sole-cropped lupin may have been underestimated in this study.

5. Conclusions

Comparing the intercrop and the sole crop in the context of the transition to low-input crop management strategies is increasingly needed as solutions for chemical weeding are becoming scarce. In this context, we showed that the lupin-triticale intercrop is a relevant option. Because a moderate lupin yield reduction can lead to a high protein yield loss, intercropping lupin with triticale does not seem to potentially perform better than sole cropping lupin regarding protein productivity on an area basis. At a broader scale, intercropping could allow an increase in lupin cropping area via increased lupin adoption by farmers due to increased weed suppression and secured total productivity. In this case, lupin intercropped with cereals could significantly contribute to the production of protein-rich grains in Europe.

Author Contributions: Conceptualization, N.C., G.P., C.N. and G.C.-H.; methodology, N.C., G.P., C.N. and G.C.-H.; validation, N.C., G.P., C.N. and G.C.-H.; formal analysis, N.C.; investigation, N.C.; writing—original draft preparation, N.C., G.P., C.N. and G.C.-H.; writing—review and editing, N.C., G.P., C.N. and G.C.-H.; visualization, N.C.; supervision, G.C.-H.; project administration, G.P. and G.C.-H.; funding acquisition, G.P. and G.C.-H. All authors have read and agreed to the published version of the manuscript.

Funding: This research was funded by EAFRD and Regional Council (Brittany) within the project PROGRAILIVE carried out by the association Pôle Agronomique de l'Ouest, and by the French National Research Agency (ANR) within the project LEGITIMES, grant number ANR-13-AGRO-0004. The APC was funded by Ecole Supérieure d'Agricultures, Angers.

Acknowledgments: We gratefully acknowledge all farmers and experimenters involved and the staff members at LEVA (Légumineuses, Ecophysiologie Végétale, Agroécologie) and FNAMS (Fédération Nationale des Agriculteurs Multiplicateurs de Semences) for their excellent technical assistance. We are most grateful to François Boissinot, Céline Bourlet and Martine Mauline for coordination of the study sites and to Erik Steen Jensen for valuable comments on an earlier version of the manuscript. We thank PLATIN' (Plateau d'Isotopie de Normandie) core facility for all element and isotope analysis used in this study.

Conflicts of Interest: The authors declare no conflict of interest.

References

1. De Visser, C.L.; Schreuder, R.; Stoddard, F. The EU's dependency on soya bean import for the animal feed industry and potential for EU produced alternatives. *OCL* **2014**, *21*, D407. [CrossRef]
2. Van Barneveld, R.J. Understanding the nutritional chemistry of lupin (*Lupinus* spp.) seed to improve livestock production efficiency. *Nutr. Res. Rev.* **1999**, *12*, 203–230. [CrossRef] [PubMed]
3. Lucas, M.M.; Stoddard, F.; Annicchiarico, P.; Frias, J.; Martinez-Villaluenga, C.; Sussmann, D.; Duranti, M.; Seger, A.; Zander, P.; Pueyo, J. The future of lupin as a protein crop in Europe. *Front. Plant Sci.* **2015**, *6*, 705. [CrossRef] [PubMed]
4. Herridge, D.F.; Peoples, M.B.; Boddey, R.M. Global inputs of biological nitrogen fixation in agricultural systems. *Plant Soil* **2008**, *311*, 1–18. [CrossRef]
5. FAO. FAOSTAT. (Ed Food and Agriculture Organization of the United Nations). 2017. Available online: www.Fao.org/faostat/en/# (accessed on 3 August 2017).
6. Cernay, C.; Ben-Ari, T.; Pelzer, E.; Meynard, J.M.; Makowski, D. Estimating variability in grain legume yields across Europe and the Americas. *Sci. Rep.* **2015**, *5*, 11171. [CrossRef]
7. Putnam, D.H.; Wright, J.; Field, L.A.; Ayisi, K.K. Seed yield and water-use efficiency of white lupin as influenced by irrigation, row spacing, and weeds. *Agron. J.* **1992**, *84*, 557–563. [CrossRef]
8. Folgart, A.; Price, A.; Van Santen, E.; Wehtje, G. Organic weed control in white lupin (*Lupinus albus*, L.). *Renew. Agr. Food Syst.* **2011**, *26*, 193–199. [CrossRef]
9. Jensen, E.S. Grain yield, symbiotic N_2 fixation and interspecific competition for inorganic N in pea-barley intercrops. *Plant Soil* **1996**, *182*, 25–38. [CrossRef]
10. Hauggaard-Nielsen, H.; Gooding, M.; Ambus, P.; Corre-Hellou, G.; Crozat, Y.; Dahlmann, C.; Dibet, A.; von Fragstein, P.; Pristeri, A.; Monti, M.; et al. Pea–barley intercropping for efficient symbiotic N_2-fixation, soil N acquisition and use of other nutrients in European organic cropping systems. *Field Crop. Res.* **2009**, *113*, 64–71. [CrossRef]
11. Bedoussac, L.; Journet, E.P.; Hauggaard-Nielsen, H.; Naudin, C.; Corre-Hellou, G.; Jensen, E.S.; Prieur, L.; Justes, E. Ecological principles underlying the increase of productivity achieved by cereal-grain legume intercrops in organic farming. A review. *Agron. Sustain. Dev.* **2015**, *35*, 911–935. [CrossRef]
12. Risch, S.J. Intercropping as cultural pest control: Prospects and limitations. *Environ. Manag.* **1983**, *7*, 9–14. [CrossRef]
13. Boudreau, M.A. Diseases in Intercropping Systems. *Annu. Rev. Phytopathol.* **2013**, *51*, 499–519. [CrossRef] [PubMed]
14. Liebman, M.; Dyck, E. Crop rotation and intercropping strategies for weed management. *Ecol. Appl.* **1993**, *3*, 92–122. [CrossRef] [PubMed]
15. Hauggaard-Nielsen, H.; Ambus, P.; Jensen, E.S. Interspecific competition, N use and interference with weeds in pea-barley intercropping. *Field Crop. Res.* **2001**, *70*, 101–109. [CrossRef]
16. Corre-Hellou, G.; Dibet, A.; Hauggaard-Nielsen, H.; Crozat, Y.; Gooding, M.; Ambus, P.; Dahlmann, C.; von Fragstein, P.; Pristeri, A.; Monti, M.; et al. The competitive ability of pea-barley intercrops against weeds and the interactions with crop productivity and soil N availability. *Field Crop. Res.* **2011**, *122*, 264–272. [CrossRef]
17. Bohm, H.; Aulrich, K. Effects of row distances and seed densities on yield and quality of blue lupin (L. Augustifolius) in organic farming. In *Proceedings of the 13th International Lupin Conference, Poznan, Poland*; Naganowska, B., Kachlicki, P., Wolko, B., Eds.; International Lupin Association: Temuco, Chile, 2011.
18. Carton, N.; Naudin, C.; Piva, G.; Corre-Hellou, G. Variability of traits associated with early N acquisition in lupin and early complementarity in lupin-triticale mixed stands. *Aob. Plants* **2018**, *10*, ply001. [CrossRef]

19. ISO. *Soil quality—Determination of Nitrate, Nitrite and Ammonium in Field-Moist Soils by Extraction with Potassium Chloride Solution—Part 2: Automated Method with Segmented Flow Analysis*; International Organization for Standardization: Geneva, Switzerland, 2005.
20. Amarger, N.; Mariotti, A.; Mariotti, F.; Durr, J.C.; Bourguignon, C.; Lagacherie, B. Estimate of symbiotically fixed nitrogen in field grown soybeans using variations in 15N Natural abundance. *Plant Soil* **1979**, *52*, 269–280. [CrossRef]
21. Komsta, L. Outliers: Tests for Outliers. Available online: http://www.komsta.net/ (accessed on 24 January 2011).
22. Benjamini, Y.; Hochberg, Y. Controlling the false discovery rate: A practical and powerful approach to multiple testing. *J. R. Stat. Soc. Ser. B (Methodol.)* **1995**, *57*, 289–300. [CrossRef]
23. R Core Team. *R: A Language and Environment for Statistical Computing*; R Foundation for Statistical Computing: Vienna, Austria, 2013; Available online: http://www.R-project.org/ (accessed on 2 September 2016).
24. Gronle, A.; Böhm, H.; Heß, J. Effect of intercropping winter peas of differing leaf type and time of flowering on annual weed infestation in deep and shallow ploughed soils and on pea pests. *Landbauforschung Volkenrode.* **2014**, *64*, 31–44. [CrossRef]
25. Hauggaard-Nielsen, H.; Ambus, P.; Jensen, E.S. Temporal and spatial distribution of roots and competition for nitrogen in pea-barley intercrops—A field study employing 32P technique. *Plant Soil* **2001**, *236*, 63–74. [CrossRef]
26. Andersen, M.K.; Hauggaard-Nielsen, H.; Ambus, P.; Jensen, E.S. Biomass production, symbiotic nitrogen fixation and inorganic N use in dual and tri-component annual intercrops. *Plant Soil* **2004**, *266*, 273–287. [CrossRef]
27. Corre-Hellou, G.; Crozat, Y. Assessment of Root System Dynamics of Species Grown in Mixtures under Field Conditions using Herbicide Injection and 15N Natural Abundance Methods: A Case Study with Pea, Barley and Mustard. *Plant Soil* **2005**, *276*, 177–192. [CrossRef]
28. Corre-Hellou, G.; Brisson, N.; Launay, M.; Fustec, J.; Crozat, Y. Effect of root depth penetration on soil nitrogen competitive interactions and dry matter production in pea–barley intercrops given different soil nitrogen supplies. *Field Crop. Res.* **2007**, *103*, 76–85. [CrossRef]
29. Hauggaard-Nielsen, H.; Jørnsgaard, B.; Kinane, J.; Jensen, E.S. Grain legume–cereal intercropping: The practical application of diversity, competition and facilitation in arable and organic cropping systems. *Renew. Agric. Food Syst.* **2008**, *23*, 3–12. [CrossRef]
30. Abraham, C.; Singh, S. Weed management in sorghum-legume intercropping systems. *J. Agric. Sci.* **1984**, *103*, 103–115. [CrossRef]
31. Schwinning, S.; Weiner, J. Mechanisms determining the degree of size asymmetry in competition among plants. *Oecologia* **1998**, *113*, 447–455. [CrossRef]
32. Weiner, J.; Griepentrog, H.W.; Kristensen, L. Suppression of weeds by spring wheat *Triticum aestivum* increases with crop density and spatial uniformity. *J. Appl. Ecol.* **2001**, *38*, 784–790. [CrossRef]
33. Olsen, J.M.; Griepentrog, H.W.; Nielsen, J.; Weiner, J. How important are crop spatial pattern and density for weed suppression by spring wheat? *Weed Sci.* **2012**, *60*, 501–509. [CrossRef]
34. Bellostas, N.; Hauggaard-Nielsen, H.; Andersen, M.K.; Jensen, E.S. Early interference dynamics in intercrops of pea, barley and oilseed rape. *Biol. Agric. Hortic.* **2003**, *21*, 337–348. [CrossRef]
35. Reiss, A.; Fomsgaard, I.S.; Mathiassen, S.K.; Kudsk, P. Weed suppressive traits of winter cereals: Allelopathy and competition. *Biochem. Syst. Ecol.* **2018**, *76*, 35–41. [CrossRef]
36. Brooker, R.W.; Bennett, A.E.; Cong, W.F.; Daniell, T.J.; George, T.S.; Hallett, P.D.; Hawes, C.; Iannetta, P.P.M.; Jones, H.G.; Karley, A.J.; et al. Improving intercropping: A synthesis of research in agronomy, plant physiology and ecology. *New Phytol.* **2015**, *206*, 107–117. [CrossRef] [PubMed]
37. Damour, G.; Dorel, M.; Quoc, H.T.; Meynard, C.; Risède, J.M. A trait-based characterization of cover plants to assess their potential to provide a set of ecological services in banana cropping systems. *Eur. J. Agron.* **2014**, *52*, 218–228. [CrossRef]
38. Corre-Hellou, G.; Fustec, J.; Crozat, Y. Interspecific competition for soil N and its interaction with N_2 fixation, leaf expansion and crop growth in pea-barley intercrops. *Plant Soil* **2006**, *282*, 195–208. [CrossRef]
39. Naudin, C.; Corre-Hellou, G.; Pineau, S.; Crozat, Y.; Jeuffroy, M.H. The effect of various dynamics of N availability on winter pea-wheat intercrops: Crop growth, N partitioning and symbiotic N_2 fixation. *Field Crop. Res.* **2010**, *119*, 2–11. [CrossRef]

40. Yunusa, I. Effects of planting density and plant arrangement pattern on growth and yields of maize (*Zea mays*, L.) and soya bean (*Glycine max* (L.) Merr.) grown in mixtures. *J. Agric. Sci.* **1989**, *112*, 1–8. [CrossRef]
41. Kristensen, L.; Olsen, J.; Weiner, J. Crop density, sowing pattern, and nitrogen fertilization effects on weed suppression and yield in spring wheat. *Weed Sci.* **2008**, *56*, 97–102. [CrossRef]
42. Willey, R. Intercropping—its importance and research needs: Part 1. Competition and yield advantages. *Field Crop Abstr.* **1979**, *32*, 1–10.
43. Naudin, C.; Aveline, A.; Corre-Hellou, G.; Dibet, A.; Jeuffroy, M.H.; Crozat, Y. Agronomic analysis of the performance of spring and winter cereal-legume intercrops in organic farming. *J. Agric. Sci. Technol.* **2009**, *3*, 17–28.
44. Gooding, M.J.; Kasyanova, E.; Ruske, R.; Hauggaard-Nielsen, H.; Jensen, E.S.; Dahlmann, C.; von Fragstein, P.; Dibet, A.; Corre-Hellou, G.; Crozat, Y.; et al. Intercropping with pulses to concentrate nitrogen and sulphur in wheat. *J. Agric. Sci.* **2007**, *145*, 469–479. [CrossRef]
45. Bedoussac, L.; Justes, E. The efficiency of a durum wheat-winter pea intercrop to improve yield and wheat grain protein concentration depends on N availability during early growth. *Plant Soil* **2010**, *330*, 19–35. [CrossRef]
46. Hauggaard-Nielsen, H.; Jensen, E.S. Evaluating pea and barley cultivars for complementarity in intercropping at different levels of soil N availability. *Field Crop. Res.* **2001**, *72*, 185–186. [CrossRef]

Article

Approach for Image-Based Semantic Segmentation of Canopy Cover in Pea–Oat Intercropping

Sebastian Munz [1,*] and David Reiser [2,*]

1 Institute of Crop Science, University of Hohenheim, 70599 Stuttgart, Germany
2 Institute of Agricultural Engineering, University of Hohenheim, 70599 Stuttgart, Germany
* Correspondence: s.munz@uni-hohenheim.de (S.M.); david.reiser@uni-hohenheim.de (D.R.); Tel.: +49-711-459-22359 (S.M.); +49-711-459-24552 (D.R.)

Received: 2 July 2020; Accepted: 11 August 2020; Published: 13 August 2020

Abstract: Intercropping systems of cereals and legumes have the potential to produce high yields in a more sustainable way compared to sole cropping systems. Their agronomic optimization remains a challenging task given the numerous management options and the complexity of interactions between the crops. Efficient methods for analyzing the influence of different management options are needed. The canopy cover of each crop in the intercropping system is a good determinant for light competition, thus influencing crop growth and weed suppression. Therefore, this study evaluated the feasibility to estimate canopy cover within an intercropping system of pea and oat based on semantic segmentation using a convolutional neural network. The network was trained with images from three datasets during early growth stages comprising canopy covers between 4% and 52%. Only images of sole crops were used for training and then applied to images of the intercropping system. The results showed that the networks trained on a single growth stage performed best for their corresponding dataset. Combining the data from all three growth stages increased the robustness of the overall detection, but decreased the accuracy of some of the single dataset result. The accuracy of the estimated canopy cover of intercropped species was similar to sole crops and satisfying to analyze light competition. Further research is needed to address different growth stages of plants to decrease the effort for retraining the networks.

Keywords: convolutional neural network; light competition; transfer learning; growth stages; mixed cropping

---

1. Introduction

Intercropping systems comprise two or more crop species grown on the same field with overlapping growth periods [1]. It is widely practiced in developing countries under resource-limited conditions and recently gained more interest in European countries as well, especially in organic agriculture [2,3]. In particular, intercropping of cereals and legumes is very common and provides various advantages mainly due to the complementary use of nitrogen [4,5]. Compared to growing a sole crop, intercropping can increase resource use efficiency (e.g., nitrogen and water), total productivity, yield stability, and protein concentration in the cereal grain [6]. In addition, external inputs can be decreased such as synthetic fertilizers and herbicides due strong weed suppression by the cereal [2,6].

In addition to these numerous advantages, the agronomic optimization remains a challenging task given the large number of possible crop and cultivar combinations, spatial and temporal arrangement, and management [7,8]. The pathway to the implementation of intercropping into agricultural systems follows a combination of academic research and on-farm experimentation [3]. To cope with the complexity of intercropping systems and their adoption by farmers, efficient methods are needed that enable (i) to analyze the interactions between crop species efficiently (large area in short time) and (ii) to allow an easy implementation without the need of sophisticated equipment.

The interaction between crop species for light capture (light competition) has a large impact on both crop growth and weed suppression, especially during early growth stages. The canopy cover, i.e., the proportion of soil that is covered by the plant, is a good determinant for light interception and hence in intercropping systems for light competition.

For quantifying the canopy cover, different methods were used, e.g., visual estimation, light measurements above and below the canopy, or image analysis [9]. Visual estimation is time consuming and prone to subjectivity. Light measurements do not allow the differentiation between plant species. Whereas, image analysis with semantic segmentation could provide a precise estimation of canopy cover while being objective, able to differentiate between plant species, and efficient.

However, plant species in intercropping systems can overlap even early after emergence, which complicates this task. This is similar for close to crop weed detection, which is mandatory for autonomous site-specific weeding with robots or automated implements. However, for weeding applications, the position of the plants is sufficient. For applications like harvesting, phenotyping, plant health evaluation, and plant monitoring, semantic segmentation of image data is mandatory, like it is for canopy cover detection in intercropping systems [10].

Typical methods for the image-based canopy cover estimation are index-based [11], feature-based [12], and learning-based methods [13]. Several different learning-based methods where evaluated in the agricultural context for image identification, like random forest classifiers, support vector machines, or convolutional neural network (CNN) structures [10,14,15]. During the last years, especially CNN structures have gained great popularity and have been used for a wide range of applications in research. Most published research focused on weed and crop detection [13]. Moreover, research was conducted to segment fruits, flowers, pests, and plant diseases using CNNs [16–18]. For gaining additional information about the crop and environment, some research focused on segmentation of three-dimensional point clouds. Especially for crop detection and phenotyping, this additional dimension could supply more accurate results of plant position and shape [19].

The image analysis of intercropping systems with CNNs was hardly investigated. The only research conducted, to the best knowledge of the authors, is by Mortensen et al. using a modified version of VGG16 deep neural network to separate oil radish, barley, and weed. They reached pixel accuracies of 79% and an intersection over union (IoU) of 66% [20]. For image-based semantic segmentation of intercropping of cereals and legumes, no research is known to the authors.

Related research focusing on crop-weed detection reached different accuracies for semantic plant detection dependent on the application and setup. For instance, Lottes et al. reached a mean average precision (mAP) between 40.1% and 69.7% for segmenting sugar beet and weed pixels in artificially illuminated images of three different datasets [21]. Dutta et al. reached an mAP of 77.6% using pre-trained CNNs in close range images for weed classification [22]. Challenges arise for all networks when transferring them to other environments or when the growth stage of a plant (i.e., size and structure) changes, implying a high effort for training individual networks for each situation or retraining an existing network with new data according to the changes before good results can be expected [23]. Transfer learning between fruit flowers (apple, peach, and pear) worked quite well [17]; however, the transfer between different plant species and growth stages resulted in a large decrease in accuracy [21,24].

The mentioned related research used data augmentation to extend the input data and to limit the number of labeled images. Mostly geometric data augmentation (e.g., flipping, mirroring, and cropping) was applied, but also the additional integration of index-based features and edge detection was used [15]. It was shown by Taylor and Nitschke that data augmentation can result in an increase of accuracy of over 14% [25]. However, in the agricultural domain, we deal with unstructured objects in unstructured environments increasing the variability for CNNs to an infinite number of possibilities [26]. Therefore, data augmentation seems to be a good start to achieve a better transferability of trained networks [27].

This study presents the first step—by evaluating the feasibility of using image-based semantic segmentation for estimating the canopy cover in intercropping systems—towards an efficient, field-applicable method with the aim to optimize the agronomic management of intercropping systems. The evaluation focuses on the transferability of networks (i) trained with just one single growth stage of a crop to analyze different growth stages, and (ii) trained with images of single crops to differentiate between these crops grown in an intercropping system. For this study, we selected two important crops in intercropping systems in Europe—pea and oat. The images were taken under normal outdoor conditions assuring an easy field application.

2. Materials and Methods

2.1. Field Experiment and Image Acquisition

For this study, data acquisition was conducted within a field experiment on pea–oat intercropping conducted at the experimental station 'Ihinger Hof' of the University of Hohenheim (48°44′ N, 8°55′ E, 478 m above sea level) in Southwest Germany in 2019. The field size was 80 m × 36 m and had a slope of around 10 m height difference. On 28 March, pea cv. Respect and oat cv. Troll (both from IG Pflanzenzucht, Ismaning, Germany) were sown in strips of 2 m width along 80 m of length. The crops were sown both as single and mixed crops with six replicates (Figure 1). Row distance was 12 cm. The sowing density (seeds m^{-2}) was 80 and 60 for pea and 320 and 160 for oat, in single and mixed crops, respectively.

Figure 1. Overview of the field experiment. The first three strips from right to left are oat, pea–oat intercrop, and pea.

Images were taken on three dates—25 April, 2 May, and 16 May—to capture the temporal dynamics in canopy cover during early growth stages. In the following, the three dates will be denoted as low, intermediate, and high cover, respectively. The canopy covers of pea and oat varied between 3.8% and 51.8% across dates and cropping systems (Table 1). The weed cover was comparably low (0.4–2.0%) across all dates, crops, and cropping systems. Pea and oat were in the growth stages (BBCH, [28]) between 12–32 and 12–21, respectively. Besides variability in plant size and structure, overlapping of plants, and weed pressure, differences in illumination conditions (sunny–cloudy) and soil reflectance (dry–wet) occurred.

Table 1. Mean values of the canopy cover of pea, oat, and weed in sole crops and intercrops for the three datasets.

Date	Name of Dataset	Cover in Sole Crops [%]			Cover in Intercrops [%]		
		Pea	Oat	Weed	Pea	Oat	Weed
25 April	Low cover	7.0	14.1	0.4	3.8	7.9	0.6
2 May	Intermediate cover	10.1	20.7	1.2	4.8	13.0	1.4
16 May	High cover	17.6	51.8	1.9	11.4	30.3	2.0

The images were acquired with a D3100 equipped with an AF-S DX NIKKOR 18–55 mm 1:3, 5-5, 6G VR lens (Nikon Corporation, Tokio, Japan) at a distance between 0.5 m and 1 m to capture at least three crop rows in each image. These distances are applicable for moving vehicles like tractors or robots, to automate the image acquisition. The shutter speed was adapted to given illumination conditions and ranged between 1/160 and 1/1250 s with a shorter exposure time under very bright conditions. All the other settings were kept constant (ISO 400, F/8). The image size was 3072 × 4608 pixels with a resolution of 240 dpi. The images had a spatial resolution of 3–6 pixel/mm, depending on the acquisition distance. The images were taken hand-held, i.e., horizontal leveling, and therefore, distance did vary to a certain extent between images. The images were captured along the 80 m of each of the 18 strips, which resulted in a total of 300–400 single images per date. In Figure 2, three exemplary images of the sole and intercrops are shown for the three acquisition dates with different canopy cover, denoted as low, intermediate, and high cover dataset.

(a) (b) (c)

Figure 2. Exemplary images of the three datasets used for the evaluation. (**a**) Low cover dataset (25 April), (**b**) intermediate cover dataset (2 May), and (**c**) high cover dataset (16 May). From top to bottom: pea, oat, and pea–oat intercrops.

2.2. Image Processing and CNN Architecture

For subsequent analysis, first, the software imageJ was used to cut out a section of 2600 × 2600 pixels in the center of each image before processing [29]. Next, these images were analyzed by a semantic CNN to generate the different pixel classes. The output images were afterwards filtered with a Matlab script (Matlab R2019a, The MathWorks Inc., Natick, MA, USA) to get rid of small single objects in binary image results, which corresponded to weeds and other outliers. Therefore, all objects in the binary image with a pixel number smaller than 2400 pixels were removed from the image.

For the semantic segmentation of the acquired images, the online platform of the company Wolution GmbH & Co. KG (Planegg, Germany) was used. They supply an easy and accessible online interface, which could be used to upload, label, and train images with different machine learning algorithms. In our case, we used a semantic CNN based on a deep neural network, similar as described by Havaei et al. [30]. The CNN architecture exploits local and contextual features, and therefore, is able to model the local details and the global context at the same time. The CNN applies a series of layers, constructed from convolutions and activation functions (in this case a rectified linear unit), to the image. In a CNN, each successive layer results in a more abstract feature map, providing more global (contextual) features. The final layer applies a softmax activation function, which results in normalized class probabilities for each pixel in the image. The CNN is trained with the Adam optimizer [31] until the validation loss converges. For this purpose, the dataset is split into 80% training data and 20% validation data.

The Adam optimizer was trained with a learning rate of 10^{-4}. The neural network training was performed with randomly cropped image patches of size 64 × 64 pixels. Each image patch was randomly mirrored and rotated to enlarge the dataset. One batch (for stochastic gradient descent) included 128 image patches, randomly selected from the training images. After each 1000th batch (1000 × 128 image patches), a validation step was performed to check the training error. For validation, 1000 batches from the validation images were randomly selected and evaluated. For the current datasets, we found that the training error converged after about 20 epochs of 1000 batches. In total, the training process took about three hours per dataset.

2.3. CNN Training

The training was done with manually pixel-wise labeled images of pea and oat. The training was performed only with images of single plants of pea and oat. Intercropping images were not used for training. For improving the results, the tendrils and the leaves of pea were labeled into two individual classes. Existing weeds in the images were not labeled, and therefore, were part of the soil/background class. This was possible without a high error rate as the datasets showed a low weed pressure (see Table 1).

In total, five different networks were trained and evaluated. The first network just used training data from the low cover dataset (LC), the second just from the intermediate cover dataset (IC), and the third just from the high cover dataset (HC). The fourth network combined training data from LC and IC (LC + IC). The fifth network combined all three datasets LC, IC, and HC (LC + IC + HC) for training. Examples of labeled images for the three different networks LC, IC, and HC are shown in Figure 3.

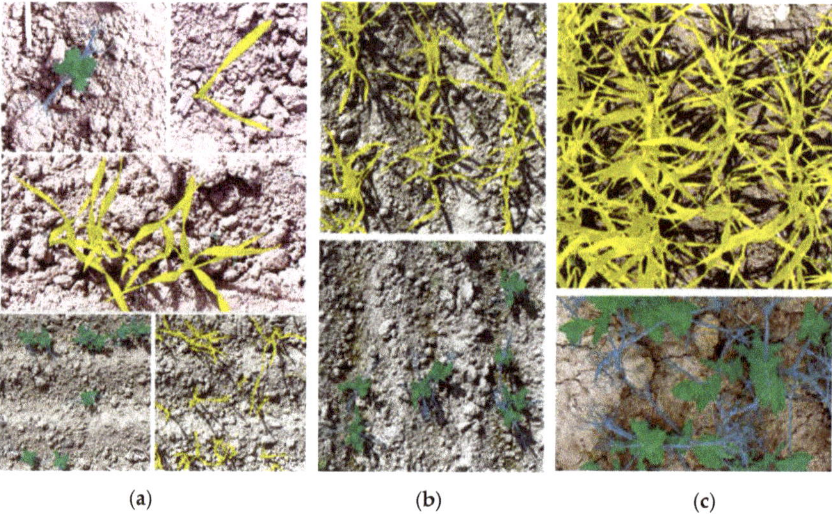

(a) (b) (c)

Figure 3. Examples of labeled images for training of the three networks: (**a**) low cover (LC), (**b**) intermediate cover (IC), and (**c**) high cover (HC). Three classes were labeled: pea leaves (green), pea tendrils (blue), and oat plants (yellow).

The reason behind this training procedure is that labeling at early growth stages is much less time consuming. It can be partly automated using index-based segmentation as plants overlap only to a small extent. In addition, labeling sole crops is much easier and faster than to differentiate and individually label species in a mixed canopy of an intercropping system.

Therefore, the network input data was ordered from low to high effort for creating the training input. Additionally, the evaluation of the light competition within an intercropping system has to start at an early growth stage with low cover. If we could increase the accuracy of a good working network with just a few additional training images, this could facilitate development time of highly accurate networks for different growth stages.

The idea was to first check the performance of the networks based on each dataset individually, and how the combination of training data influenced the results and robustness. In Table 2, the differently trained networks and their number of input images and plants, and the pixels per class are given in detail. The comparability between the networks individually trained on a specific dataset was assured by a similar number of pea and oat pixels contained in the training images.

Table 2. Description of the training input for the five networks.

	Low Cover (LC)	Intermediate Cover (IC)	High Cover (HC)	LC + IC	LC + IC + HC
No. of images	42	7	7	49	56
No. of pea plants	102	40	39	142	181
No. of oat plants	158	98	64	256	320
Pea pixels	3.628.537	2.518.875	2.129.794	6.147.412	8.277.206
Oat pixels	3.473.450	3.456.233	4.020.441	6.929.683	10.950.124
Soil pixels	87.489.759	41.344.890	5.614.212	128.834.649	134.448.861

2.4. Evaluation

For evaluation, 15 images (each 2600 × 2600 pixels) of each dataset, which were not part of the training data, were randomly selected. From each dataset (low, intermediate, high), five images were selected from each sole crop and the intercrop. Each individual image was divided into three different classes: soil, oat, and pea. The two classes of pea used for training (leaves and tendrils) were combined to compare to ground truth. The results of the CNNs were evaluated with the DPA-Software, which includes a pixel-wise comparison between ground truth image and result [11]. The transferability of the CNNs was evaluated for

(1) The different datasets (low, intermediate, and high cover) by analyzing all three datasets with each of the five trained networks including both sole and intercrops;

(2) The intercrops specifically, by comparing the results achieved for the sole and the intercrops.

For analyzing the accuracy of the networks, the True Positives (*TP*), False Negatives (*FN*), and False Positives (*FP*) for each single class were evaluated and the Precision and Recall were calculated:

$$Precision = \frac{TP}{TP + FP}, \tag{1}$$

$$Recall = \frac{TP}{TP + FN}. \tag{2}$$

Additionally, we calculated the intersection over union (*IoU*) according to

$$IoU = \frac{|A \cap B|}{|A \cup B|}, \tag{3}$$

where *A* corresponds to the quantity of ground truth pixels and *B* to the quantity of result pixels of each class.

3. Results and Discussion

First, the performance of the networks over single and intercrop images was tested and evaluated. In all three tested datasets, the classes of oat, pea, and soil pixels were detected at high rates. The networks in general detected oat more reliable than pea in the images. The individually trained networks (LC, IC, and HC) performed best for their corresponding dataset with an average precision of 91% (88–95%) and 75% (64–83%), a recall of 84% (81–89%) and 74% (65–83%), and an Intersection over Union (IoU) of 78% (73–81%) and 60% (48–68%) for oat and pea, respectively (Table 3). The network trained on all datasets (LC + IC + HC) showed almost equal performance and even slightly increased the performance when applied on the intermediate and high cover datasets.

The transfer of the networks to other datasets, especially for the HC-trained network, showed a strong decrease in the mean Intersection over Union (mIoU). The transfer of LC onto the intermediate cover dataset and IC onto the low cover dataset yielded in a mIoU higher than 69%. Whereas, the transfer of LC and IC onto the high cover dataset and the HC onto the other two datasets showed a strong decrease of the mIoU with values between 40% and 50%.

An example for the performance of the three networks individually trained with sole crop images from the three datasets on the intermediate cover dataset is shown in the Figure 4.

Table 3. Intersection over union (IoU, %), precision (%), and recall (%) for pea, oat, and soil for the three evaluation datasets (low cover—LC, intermediate cover—IC, and high cover—HC) each analyzed by the five different networks. Individually trained networks in bold.

Dataset	Network	IoU [%]				Precision [%]			Recall [%]		
		mIoU	Soil	Pea	Oat	Soil	Pea	Oat	Soil	Pea	Oat
Low Cover	**LC**	**81.1**	**96.7**	**67.7**	**79.1**	**97.9**	**78.7**	**95.1**	**98.7**	**82.9**	**82.5**
	IC	69.1	94.1	55.8	57.5	95.8	65.1	97.9	98.2	79.6	58.2
	HC	41.4	85.6	30.2	8.4	91.9	31.9	98.0	92.6	85.3	8.4
	LC + IC	77.7	96.1	60.7	76.4	97.3	78.3	93.9	98.8	73.0	80.3
	LC + IC + HC	81.1	96.6	67.9	78.7	97.9	77.4	95.5	98.7	84.7	81.8
Int. Cover	LC	75.5	93.5	55.6	77.3	96.8	81.8	82.5	96.5	63.5	92.4
	IC	**80.3**	**94.9**	**64.5**	**81.4**	**97.1**	**82.5**	**90.2**	**97.7**	**74.7**	**89.3**
	HC	46.0	80.7	31.8	25.5	89.3	33.6	88.6	89.4	85.2	26.3
	LC + IC	75.8	93.6	55.9	77.8	96.6	77.7	85.8	96.9	66.6	89.3
	LC + IC + HC	81.4	95.1	66.5	82.5	97.7	81.6	88.3	97.3	78.3	92.6
High Cover	LC	50.3	64.2	35.8	50.7	76.8	40.7	87.9	79.7	75.2	54.5
	IC	40.1	59.5	33.0	27.9	69.1	37.8	95.1	81.1	71.9	28.3
	HC	**66.3**	**78.6**	**47.5**	**72.9**	**86.5**	**64.3**	**88.1**	**89.6**	**64.5**	**80.9**
	LC + IC	41.3	58.1	30.1	35.8	70.4	34.1	90.9	76.9	71.7	37.1
	LC + IC + HC	67.5	78.1	51.1	73.3	88.3	62.3	86.2	87.1	74.1	83.0

Figure 4. Examples for images from the intermediate cover dataset analyzed by the three individually trained networks. From top to bottom: pea, oat, and pea–oat intercrops (yellow: oat, green: pea, blue: pea tendrils, white: pea combined).

These results indicate that the transferability across different growth stages (respectively, degree of canopy cover) is challenging. Therefore, the need for retraining the network for a new dataset seems the only option to optimize the performance. The difference between the accuracy of oat and pea pixels could be a result of the increased complexity of the plant and the overall cover of the two species in the images. Especially in intercropping, the cover of pea was less than half of oat (Table 1). Additionally, the tendrils of pea where hard to detect, especially when their share in cover increased during growth and more tendrils with a small diameter were present. Interestingly, the mIoU dropped considerably for the later growth stage, where overlapping of plants increased and the canopy cover was higher. The main reason for this was the lower quality in detection of the intercrops as shown in Table 4.

Table 4. Intersection over union (IoU, %) of pea and oat in sole crops and intercrops for the three evaluation datasets (low, intermediate, and high cover) each analyzed by the individually trained network (LC, IC, HC) and the network trained on all datasets (LC + IC + HC).

Dataset	Network	Sole Crops		Intercrops	
		Pea IoU [%]	Oat IoU [%]	Pea IoU [%]	Oat IoU [%]
Low Cover	LC	81.7	83.1	60.2	73.7
	LC + IC + HC	82.2	84.3	58.3	70.4
Int. Cover	IC	76.5	86.6	52.6	79.6
	LC + IC + HC	76.6	89.7	56.8	79.9
High Cover	HC	74.4	82.1	25.6	63.3
	LC + IC + HC	72.3	82.7	38.5	64.2

The sole crops were detected well across all datasets with an IoU between 72% and 90%. The transfer of the networks trained on sole crop images onto the intercrops showed a good performance for intercropped oat for the first two datasets (IoU: 70–80%). However, the IoU of intercropped pea decreased considerably and for the high cover dataset, the network performed poorly for the intercrops and especially pea.

The largest error was associated with the small tips of the oat plants, which were detected as pea tendrils. The center of the plants resulted in another typical zone of errors. A reason might be shading, which created a different coloration for the center compared with the rest of the plant.

Reasons for this behavior could be the comparable low training input, the change of color, and the different shape of the plants in the high cover dataset as both species reached the next main growth stage (pea: stem elongation, oat: tillering). This leads to the point that future CNN architectures for applications in agriculture should address different growth stages in the networks. To gain a network with the given method that fits all growth stages seems challenging. However, future architectures could address the special needs for extracting invariant features for agricultural plants under different growth stages. The better performance of the LC + IC + HC network on intercropped pea for the high cover dataset indicated that better results might be obtained with more training images of pea. A few example images of this network are shown in the following Figure 5.

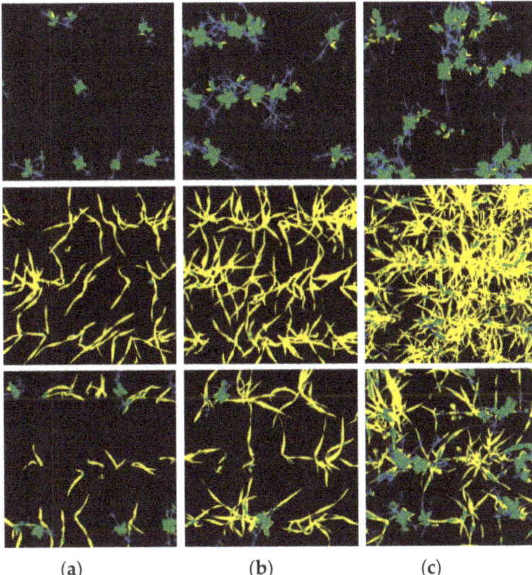

Figure 5. Examples for images from the three datasets: (**a**) low, (**b**) intermediate, and (**c**) high cover analyzed by the network trained on all datasets (LC + IC + HC). From top to bottom: pea, oat, and pea–oat intercrops (yellow: oat, green: pea, blue: pea tendrils).

For the general use in agriculture, the absolute precision of plant pixels is not mandatory. As todays machines for fertilization and weeding are not plant-specific, mean estimates over the field or a specific site are sufficient. Therefore, the mean crop cover over an area of interest is enough to estimate the light competition between the intercrops for a given agronomic practice (e.g., sowing density of each intercrop) or site. With this in mind, the two best performing networks for each dataset were selected and the mean average cover ($\overline{m}$) evaluated by the CNN was compared to the ground truth on an absolute (Δa) and relative (Δr) scale (Table 5).

Table 5. Mean estimated canopy cover ($\overline{m}$) of pea and oat in sole crops and intercrops and absolute (Δa) and relative (Δr) difference to ground truth for the three evaluation datasets (low, intermediate, and high cover).

Dataset	Network	Sole Crops						Intercrops					
		Pea Cover [%]			Oat Cover [%]			Pea Cover [%]			Oat Cover [%]		
		$\overline{m}$	Δa	Δr	$\overline{m}$	Δa	Δr	$\overline{m}$	Δa	Δr	$\overline{m}$	Δa	Δr
Low Cover	LC + IC + HC	6.3	−0.7	−10.0	12.6	−1.5	−10.6	4.7	0.9	23.7	6.2	−1.7	−21.5
	LC	6.2	−0.8	−11.4	12.4	−1.7	−12.1	4.3	0.5	13.2	6.5	−1.4	−17.7
Int. Cover	LC + IC + HC	9.5	−0.6	−5.9	20.6	−0.1	−0.5	4.3	−0.5	−10.4	13.2	0.2	1.5
	IC	9.3	−0.8	−7.9	19.2	−1.5	−7.2	3.6	−1.2	−25.0	13.0	0.0	0.0
High Cover	LC + IC + HC	17.0	−0.6	−3.4	50.5	−1.3	−2.5	13.8	2.4	21.1	28.1	−2.2	−7.3
	HC	16.5	−1.1	−6.3	46.5	−5.3	−10.2	9.3	−2.1	−18.4	29.1	−1.2	−4.0
Mean ($\overline{m}$)	LC + IC + HC		−0.6	−6.4		−1.0	−4.5		0.9	11.5		−1.2	−9.1
	single net		−0.9	−8.5		−2.8	−9.8		−0.9	−10.1		−0.9	−7.2

The results showed that the absolute cover was estimated quite accurately. The maximum absolute difference to ground truth was −1.1% for pea and −5.3% for oat with a relative difference not exceeding 12.1%. The estimated canopy cover of the intercrops were on absolute scale in the same magnitude as the sole crops. However, given their lower canopy cover, relative differences were higher reaching a maximum of 25%. Interestingly, a lower IoU as shown for the high cover dataset (Tables 3 and 4) does not necessarily result in a lower accuracy of estimated canopy cover.

Compared to the existing state of the art work in the field, the trained networks did perform quite well. The reached mIoU between 66% and 81% for the networks is in the range of published results as mentioned in the introduction. This study confirmed that transferring a CNN to another dataset resulted in a considerable decrease in IoU. This corresponds to the results of Lottes et al. and Bosilj et al. [21,23].

Shorten and Khoshgoftaar highlighted in their review that there are no existing augmentation techniques that can correct a dataset that has a very poor diversity with respect to the test data [27]. However, especially in an agricultural domain, this is the most challenging point, as diversity of the possible real test environment is huge. Therefore, future research has to address the generalization and transferability of networks in the agricultural domain, as we deal with unstructured objects in unstructured environments [25]. Data augmentation techniques like geometric transformations, color space transformations, kernel filters, mixing images, random erasing, feature space augmentation, adversarial training, generative adversarial network-based augmentation, neural style transfer, and meta-learning schemes could help to gain better transferability of networks in the future.

4. Conclusions

The results of this study showed that it is feasible to estimate the canopy cover in intercropping systems with a satisfying accuracy based on sole crop training data. However, the transferability of trained networks onto other datasets—than the one used for training—has to be improved in future research to reduce the effort for retraining the networks for new situations. In a next step, the network will also be trained with another dataset having a higher weed pressure to estimate weed cover separately (not as a part of the soil/background class). Besides the use of the estimated canopy cover to analyze light competition between intercrops and identify promising management practices, a combination of the results with site-specific management would open new possibilities to dynamically influence the interactions between crop species to maximize yield and weed suppression.

Author Contributions: Conceptualization, S.M. and D.R.; formal analysis, S.M. and D.R.; funding acquisition, S.M.; methodology, S.M. and D.R.; project administration, S.M.; software, D.R.; validation, S.M. and D.R.; visualization, S.M. and D.R.; writing—original draft, S.M. and D.R.; writing—review & editing, S.M. and D.R. All authors have read and agreed to the published version of the manuscript.

Funding: This research was funded by the European Union's Horizon 2020 Programme for Research & Innovation under grant agreement n°727217 (ReMIX: Redesigning European cropping systems based on species MIXtures).

Acknowledgments: We thank Jonas Boysen for helping with preparing the ground truth data, Alexander Stana and Nils Lüling for proofreading, and Julian Zachmann and the technical staff of the research station for managing the field experiment. Additionally, we thank Dominik Wuttke and Moritz Münchmeyer for their help and support while preparing this manuscript.

Conflicts of Interest: The authors declare no conflict of interest.

References

1. Vandermeer, J. *The Ecology of Intercropping*, 1st ed.; Cambridge University Press: Cambridge, UK, 1989; ISBN 0521345928.
2. Jensen, E.S.; Bedoussac, L.; Carlsson, G.; Journet, E.; Justes, E.; Hauggaard-Nielsen, H. Enhancing Yields in Organic Crop Production by Eco-Functional Intensification. *Sustain. Agric. Res.* **2015**, *4*, 42–50. [CrossRef]
3. Verret, V.; Pelzer, E.; Bedoussac, L.; Jeu, M. Tracking on-farm innovative practices to support crop mixture design: The case of annual mixtures including a legume crop. *Eur. J. Agron.* **2020**, *115*. [CrossRef]
4. Hauggaard-Nielsen, H.; Gooding, M.; Ambus, P.; Corre-Hellou, G.; Crozat, Y.; Dahlmann, C.; Dibet, A. Pea-Barley intercropping for efficient symbiotic N_2-fixation, soil N acquisition and use of other nutrients in European organic cropping systems. *Field Crop. Res.* **2009**, *113*, 64–71. [CrossRef]
5. Jensen, E.S. Intercropping of grain legumes and cereals improves the use of soil N resources and reduces the requirement for synthetic fertilizer N: A global-scale analysis. *Agron. Sustain. Dev.* **2020**, *40*, 1–9. [CrossRef]
6. Bedoussac, L.; Journet, E.; Hauggaard-Nielsen, H.; Naudin, C.; Corre-Hellou, G.; Jensen, E.S. Ecological principles underlying the increase of productivity achieved by cereal-grain legume intercrops in organic farming. A review. *Agron. Sustain. Dev.* **2015**, *35*, 911–935. [CrossRef]
7. Gaudio, N.; Escobar-Gutiérrez, A.J.; Casadebaig, P.; Evers, J.B.; Gérard, F.; Louarn, G.; Colbach, N.; Munz, S.; Launay, M.; Marrou, H.; et al. Current knowledge and future research opportunities for modeling annual crop mixtures. A review. *Agron. Sustain. Dev.* **2019**, *39*, 1–20. [CrossRef]
8. Munz, S.; Graeff-Hönninger, S.; Lizaso, J.I.; Chen, Q.; Claupein, W. Modeling light availability for a subordinate crop within a strip-intercropping system. *Field Crop. Res.* **2014**, *155*, 77–89. [CrossRef]
9. Shepherd, M.J.; Lindsey, L.E.; Lindsey, A.J. Soybean Canopy Cover Measured with Canopeo Compared with Light Interception. *Agric. Environ. Lett.* **2018**, *3*, 1–3. [CrossRef]
10. Kusumam, K.; Kranjík, T.; Pearson, S.; Cielniak, G.; Duckett, T. Can You Pick a Broccoli? 3D-Vision Based Detection and Localisation of Broccoli Heads in the Field. In Proceedings of the 2016 IEEE International Conference on Intelligent Robots and Systems (IROS), Daejeon, Korea, 9–14 October 2016; pp. 646–651.
11. Riehle, D.; Reiser, D.; Griepentrog, H.W. Robust index-based semantic plant/background segmentation for RGB-images. *Comput. Electron. Agric.* **2020**, *169*, 105201. [CrossRef]
12. Åstrand, B.; Baerveldt, A.J. An agricultural mobile robot with vision-based perception for mechanical weed control. *Auton. Robots* **2002**, *13*, 21–35. [CrossRef]
13. Kamilaris, A.; Prenafeta-Boldú, F.X. Deep learning in agriculture: A survey. *Comput. Electron. Agric.* **2018**, *147*, 70–90. [CrossRef]
14. Lottes, P.; Hoeferlin, M.; Sander, S.; Stachniss, C. An Effective Classification System for Separating Sugar Beets and Weeds for Precision Farming Applications. In Proceedings of the IEEE International Conference on Robotics and Automation (ICRA), Stockholm, Sweden, 16–21 May 2016; pp. 5157–5163.

Water-soluble carbohydrates are also important for DF because leaf growth is related to a decrease in WSC concentration [5,6]. When carbohydrate synthesis is greater than its utilization, WSCs are temporarily stored in the base of leaf sheaths with a lower proportion stored in leaf blades [7]. Water-soluble carbohydrates are an immediate energy source for plant growth after defoliation; therefore, the quantity of stored WSC pre-defoliation is associated directly with the speed of pasture regrowth [8].

In pastures, the amount of stored WSCs is associated with the plant's phenological stage [9]. *Lolium perenne* uses WSCs as an energy source immediately following defoliation until approximately one leaf has expanded completely and can generate enough energy via photosynthesis to cover maintenance requirements and tissue development [10,11]. Once the third leaf has emerged, it can be assumed that the plant has accumulated enough WSC reserves to tolerate subsequent defoliation.

Defoliation frequency influences the mobilization and storage of WSCs in grass pastures. Short intervals between defoliation do not allow plants to recover enough WSCs, reducing pasture persistence, whereas longer intervals permit plants to synthesize enough carbohydrates to support post-defoliation regrowth [9,12,13]. The season also influences WSC recovery. Most studies have been conducted during spring or in controlled environments because it is known that the concentration of WSCs in grass increases during daylight as a result of the positive balance between photosynthesis and respiration [10,14]. In C_3 species, the accumulation and mobilization of stored WSCs, predominantly fructans, are of great importance in the synthesis of new tissues during regrowth [15]. Hence, the concentration of WSCs in plants varies through the day and from season to season [16,17], playing an important role in the balance between photosynthesis, carbon contributions, and C requirements for plant growth and development [18].

Some studies have concluded that *L. perenne* has a higher concentration of carbohydrates than *B. valdivianus*. The concentration of WSCs in species such as *L. perenne* and *B. valdivianus* increases in the afternoon because of increased photosynthetic activity in the presence of full light [19,20]. In fall, post-defoliation regrowth is slow compared to spring because the temperature drops and luminosity is lower [9]. However, fall is an important season because, along with spring, it provides the most forage during the year. Therefore, the two objectives of this study were (1) to examine the dynamics of WSC use and the recovery of leaf sheaths and blades of *L. perenne* and *B. valdivianus* subjected to two DFs; (2) to evaluate how DF influences regrowth and accumulated herbage mass (AHM) during fall.

2. Materials and Methods

2.1. Site Description

This study was carried out at the Austral University of Chile (UACh) in the Austral Agriculture Research Station (39°46′ S, 73°13′ W, 12 m a.s.l.) in Valdivia, southern Chile, during the fall of 2017, from 20 March to 23 June. The precipitation during the study period (March to June) was 677 mm with a mean temperature of 11 °C (Figure 1), and the annual precipitation and mean temperature at the experimental site was 2210 mm and 11.7 °C, respectively. Climate information for the experiment was collected daily via a meteorological station (Agromet-Inia) located 5 m from the study site. The topography was flat with a slope of less than 2%. The soil corresponded to a Duric Hapludand Andisol, in the Valdivia series [21], with a pH of 5.4, organic matter level of 14.6%, 17.4 mg kg^{-1} P_{olsen}, and 4.9% aluminum saturation (Soil Lab, Institute of Agricultural Engineering and Soils, Faculty of Agriculture, Universidad Austral de Chile, Valdivia, Chile). The fertilizer application was 150 kg ha^{-1} year^{-1} of nitrogen, 44 kg ha^{-1} year^{-1} of phosphorous, and 50 kg ha^{-1} year^{-1} of potassium. The soil was suitable for perennial species.

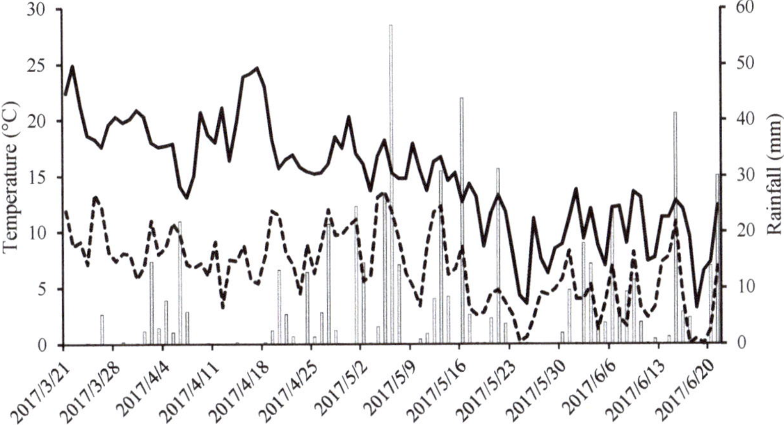

Figure 1. Daily maximum (solid line) and minimum (dashed line) temperature and rainfall (bars) for Valdivia, Chile, during the fall of 2017.

2.2. Experimental Design

The experiment included 12 plots of 15 m^2 (3 m wide × 5 m long) randomly distributed into three blocks. In March 2015, six plots were sown with 30 kg ha^{-1} *L. perenne* cv. Alto and six with 45 kg ha^{-1} *B. valdivianus* Phil. Each plot was defoliated with one of the following fixed DFs based on AGDDs: DF1 = 135 AGDDs that corresponded to the 1.5 leaf stage (LS) for *L. perenne* and 1.7 for *B. valdivianus*; DF2 = 270 AGDDs that corresponded to the 3 and 3.4 LS for *L. perenne* and *B. valdivianus*, respectively, using a rotary mower equipped with a collection bag and with a residual height of 5 cm left.

Each block comprised four plots, and each plot corresponded to the interaction between species (*L. perenne* and *B. valdivianus*) and DFs (135 and 270 AGDDs). The interaction between these two factors was considered the treatment (TTm). All plots were defoliated according to the assigned DF during the plot conditioning period from October 2016 to March 2017. Sampling was conducted from 21 March to 21 June 2017.

The following equation was used to calculate accumulated thermal time:

$$\text{AGDD} = \sum [(T_{\text{max}} + T_{\text{min}})/2] - T_{\text{base}}, \tag{1}$$

where T_{max} is the daily maximum air temperature, T_{min} is the daily minimum air temperature, and T_{base} is the temperature below which the observed process does not occur. A base temperature of 5 °C was used for both species, and the temperature accumulated from day 1 to day *n* [22,23]. The air temperature was measured at 150 cm high above the ground level, and it was taken from the meteorological station close to the field experiment.

2.3. Evaluated Variables and Sampling

Three cores (90 cm in diameter and 10 cm deep) were collected from each plot at 08:00 every third day, until each plot reached the assigned AGDDs. Each core corresponded to a sample composed of soil with root, leaf blades, and leaf sheaths which were stored on ice to avoid WSC losses until they were transported to the Animal Nutrition Laboratory of the UACh, where they were washed to remove soil residue. The root was discarded, and the grass blades were separated from the sheaths.

The blade and sheath materials were measured separately for green weight and then dried in a forced-air oven at 65 °C for 72 h to measure dry matter (DM) content and dry weight. Dried samples were ground to 1 mm using a Wiley mill (Model Digital ED-5, Thomas Scientific, Swedesboro, NJ, USA) and stored in plastic containers at ambient temperature for subsequent WSC analysis.

Water-soluble carbohydrate concentration was determined in the blade and sheath by near infrared reflectance spectroscopy (NIRS) with a FOSS-NIRSystems Model 6500 and Version 4.4 FOSS-ISIScam software and equations developed in the UACh Animal Nutrition Laboratory. The equation for WSC concentration was calibrated using the anthrone method described by Yemm and Willis [24], where WSCs are expressed as glucose and determined using spectroscopy based on the blue–green color created when carbohydrates are heated with anthrone in sulfuric acid. The standard errors of cross-validation and R^2 for WSCs were 6.99 and 0.96, respectively.

The AHM was estimated every six days using a rising plate meter (Ashgrove Plate Meter, Hamilton, New Zealand). Each value was calculated using the average height of 10 samples per plot. To estimate production per hectare (kg DM ha^{-1}), an equation specific to fall pastures in southern Chile was used [25]:

$$Y = 120X + 350\ (r^2 = 0.74), \tag{2}$$

where Y = AHM in kg DM ha^{-1}, and X = compressed average height.

When each plot accumulated the assigned AGDDs, it was defoliated again for further sampling. Therefore, during the fall, plots with 135 AGDDs were defoliated twice, and plots with 270 AGDDs were defoliated once. The growth was calculated from AHM at the defoliation time minus the residue that was measured in the past defoliation and calculated at 1200 kg DM ha^{-1}, and the growth of 135 AGDDs was adding up the growth for both periods.

2.4. Statistical Analysis

Response variables included WSC concentration (g kg^{-1}) in blades and sheaths and AHM production (kg DM ha^{-1}). Data were analyzed using the MIXED procedure of SAS (SAS Institute, V9.0, 2008) in a complete randomized block design with a factorial arrangement of treatments (two species and two DF), where AGDDs and species were fixed effects, and the field block and their interaction were random effects. Sampling time was included as a repeated measure in the model. The covariance structure [26] was based on the probability test and the Akaike information criterion, test according to (a) no structure, (b) composite symmetry (CS), (c) heterogeneity of compound symmetry (HCS), (d) Toeplitz (TOEP), and (e) Toeplitz heterogeneity (HTOEP). Prior to analysis, all data were checked for normality and homogeneity of variances. When there were significant differences ($p < 0.05$), a multiple mean comparison test (LSMEANS) was performed with the PDIFF command.

3. Results

3.1. Effect of Defoliation Frequency on WSC Concentration in Leaf Sheaths and Blades

Lolium perenne L. and *B. valdivianus* defoliated after 135 AGDDs had similar WSC levels ($p > 0.05$, Figure 2a,b) in both the blades and sheaths. When the DF was extended to 270 AGDDs, significant differences ($p < 0.05$, Figure 3a,b) were found between the two species. This change became apparent after 195 AGDDs, at which point *L. perenne* regrowth was able to accumulate approximately 33% more WSCs than *B. valdivianus*. This difference remained constant between the two species from 195 to 270 AGDDs.

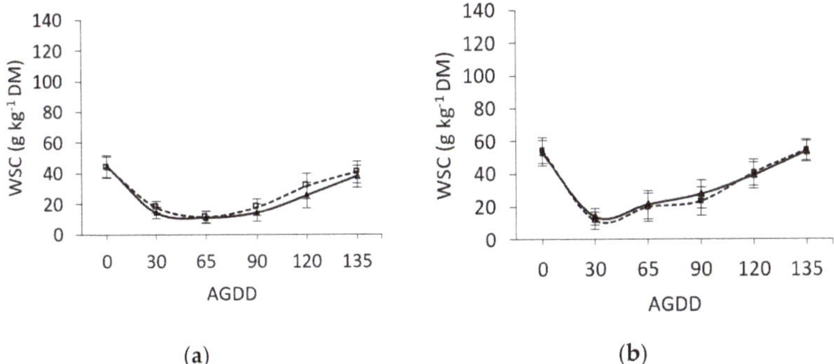

Figure 2. Mean of two growth cycles for water-soluble carbohydrate (WSC) concentration (g kg^{-1} dry matter (DM)) in leaf blades (**a**) and leaf sheaths (**b**) for *L. perenne* (–□–) and *B. valdivianus* Phil. (-△-) plots defoliated at 135 accumulated growing degree days (AGDDs). Bars indicate standard error.

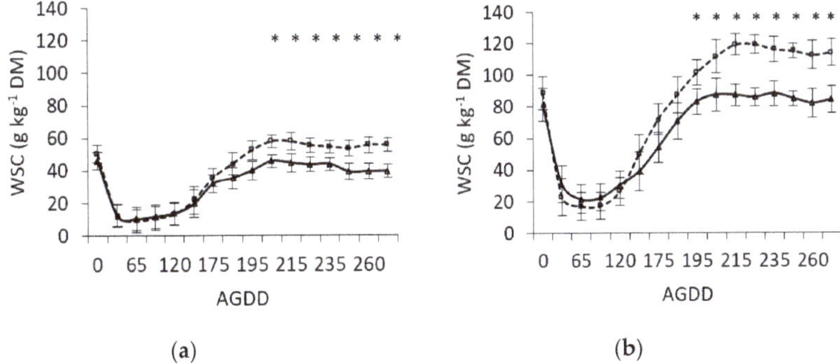

Figure 3. Mean water-soluble carbohydrate (WSC) concentration (g kg^{-1} DM) post-defoliation in leaf blades (**a**) and leaf sheaths (**b**) for *L. perenne* (–□–) and *B. valdivianus* Phil. (-△-) plots defoliated at 270 accumulated growing degree days (AGDDs). * Indicates significant differences ($p < 0.05$) between species. Bars indicate standard error.

It was observed that in both *L. perenne* and *B. valdivianus*, the concentration of WSCs in the blades and sheaths significantly decreased following defoliation until 30 AGDDs ($p < 0.05$, Figures 2 and 3), (**a**) and (**b**) when plants began to accumulate WSCs. This process was similar for both blades and sheaths.

After both DFs, leaf blade WSC concentration was reduced by approximately 80% in *L. perenne* and 74% in *B. valdivianus*. For both species, plants began to recover WSCs after 65 to 90 AGDDs post-defoliation when defoliated every 135 AGDDs. For plants defoliated every 270 AGDDs, the recovery of WSCs was initiated after 90 to 120 AGDDs had been reached. On the other hand, the recovery rate of WSCs in leaf sheath was different between species defoliated at 270 AGDDs. The *L. perenne* recovered faster than the *B. valdivianus* using 1.5 AGDDs per 1 g of WSC, whereas *B. valdivianus* needed more TT (2.3 AGDDs) to accumulate the same amount of WSCs. When the species were defoliated at 135 AGDDs, the recovery rate was the same for *L. perenne* and *B. valdivianus* with 2.5 AGDDs per 1 g of WSCs.

3.2. Water-soluble Carbohydrates and Pasture Regrowth

After the initial defoliation (day 0) and during the subsequent growth cycle of 135 AGDDs DF, there were no significant differences in WSC levels between species for both leaf blades and sheaths

($p > 0.05$, Figure 4), except at the end of the experiment, when the leaf blade WSC concentration of both species had not fully recovered. In contrast, leaf sheaths recovered WSCs to the pre-defoliation levels measured on day 0 ($p < 0.05$, Figure 4). Mean concentrations of WSC were 25% and 28% greater in leaf sheaths than in leaf blades for *L. perenne* and *B. valdivianus*, respectively, when defoliation occurred every 135 AGDDs (Figure 4a; $p < 0.05$).

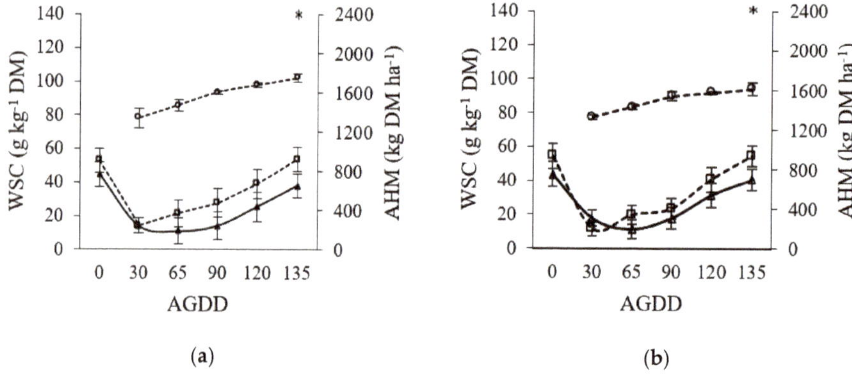

Figure 4. Mean of two growth cycle for water soluble carbohydrates (WSCs, g kg^{-1} DM) and accumulated herbage mass (AHM, kg DM ha^{-1}) in *Bromus valdivianus* Phil. (**a**) and *Lolium perenne* L. (**b**) in: leaf blades (-△-) and leaf sheath (–□–) defoliated at 135 accumulated growing degree days (AGDDs) and AHM (–o–). * Indicates significant differences ($p < 0.05$) between leaf sheaths and blades for the same species. Bars indicate the standard error.

When plants were defoliated after 270 AGDDs (Figure 5), the WSC concentration was significantly greater for leaf sheaths than for leaf blades at day 0 and after 135 AGDDs ($p < 0.05$, Figure 5). Between 30 and 120 AGDDs, the WSC concentration was similar for both leaf portions (Figure 5). The same tendency was observed in pasture plots defoliated every 135 AGDDs, where the WSC concentration reduced from day 0 to 30 AGDDs (Figures 2–4).

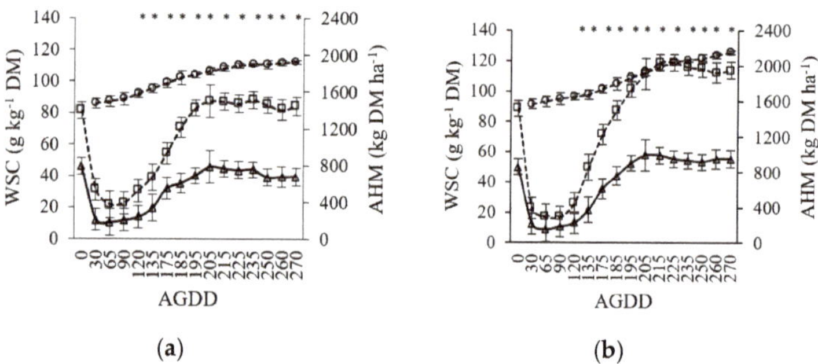

Figure 5. Mean water soluble carbohydrates (WSCs, g kg^{-1} DM) and accumulated herbage mass (AHM, kg DM ha^{-1}) in *Bromus valdivianus* Phil. (**a**) and *Lolium perenne* L. (**b**) in: leaf blade (-△-) and leaf sheath (–□–) defoliated at 270 accumulated growing degree days (AGDDs) and AHM (–o–). * Indicates significant differences ($p < 0.05$) between leaf sheaths and blades for the same species. Bars indicate the standard error.

In *L. perenne* plots defoliated every 270 AGDDs, the WSC concentration was reduced by 85% compared with the pre-defoliation levels (day 0) and did not recover until approximately 215 AGDDs, when plants reached the WSC concentration used for regrowth. In *B. valdivianus* plots defoliated with the same DF, the WSC concentration levels decreased by 75% post-defoliation and required 205 AGDDs to completely recover the WSC reserves used during regrowth (Figure 5a,b).

Similar patterns in the movement of WSCs throughout the regrowth cycle were observed in both *L. perenne* and *B. valdivianus* plots defoliated after 135 and 270 AGDD. Plants in both DFs utilized WSC immediately after defoliation, but the replenishment of WSC levels before the next defoliation was slower at 270 AGDDs than at 135 AGDDs (Figures 4 and 5).

Plots defoliated at 270 AGDDs had time to accumulate greater concentrations of WSC compared to those defoliated at 135 AGDDs. In both forage species and at both DFs, the concentration of WSC was higher in leaf sheaths than in leaf blades during the entire regrowth cycle. When considering the combined effect of DF and species, the highest concentration of WSCs corresponded to leaf sheaths of *L. perenne* plots defoliated at 270 AGDDs, followed by sheaths of *B. valdivianus* defoliated at the same frequency.

The water-soluble carbohydrate concentration in leaf sheaths did not significantly differ between species when defoliated at 135 AGDDs ($p > 0.05$, Table 1). The highest leaf blade WSC concentration levels were found in the interaction among *L. perenne* plots defoliated at 270 AGDDs, but there were no other significant differences in leaf blade WSC concentration level interactions ($p > 0.05$, Table 1). A significant interaction between species and DF was observed for both blades and sheaths. In *B. valdivianus*, WSC in blades was similar in both DF, whereas for *L. perenne*, a greater WSC concentration was observed for DF 270 compared with 135. In the case of sheaths, the interaction showed a greater increase from DF 135 to DF 270 for *L. perenne* (+58 g kg^{-1} DM) compared with *B. valdivianus* (+31 g kg^{-1} DM). The same trend was obtained for WSCs per hectare, where *L. perenne* produced 110 kg more than *B. valdivianus*, both defoliated at 270 AGDDs, while *L. perenne* and *B. valdivianus*, which were defoliated at 135 AGDDs, did not show a significant difference.

Table 1. Mean water-soluble carbohydrate (WSC) content (kg kg^{-1} DM) in leaf sheaths and leaf blades and production per hectare (kg WSC ha^{-1}) for *Bromus valdivianus* Phil. and *Lolium perenne* L. defoliated at 135 and 270 accumulated growing degree days (AGDDs).

Species	AGDDs	Blade	Sheath	Hectare
B. valdivianus	135	37.95 ^{b}	53.70 ^{c}	193 ^{c}
	270	39.63 ^{b}	84.28 ^{b}	238 ^{b}
L. perenne	135	40.72 ^{b}	54.65 ^{c}	192 ^{c}
	270	55.61 ^{a}	113.93 ^{a}	348 ^{a}
SEM		1.481	1.923	36.7
p-Value		0.001	0.001	0.001

Means within a column with different superscripts differ ($p < 0.05$); SEM = standard error of the mean.

3.3. Defoliation Frequency and Growth

The increase in growth followed the same trend as WSC replenishment levels. As shown in Figures 4 and 5, growth began to increase after 30 AGDDs in both DF treatments. In contrast, WSCs did not begin to recover until 90 AGDDs in plots defoliated at 135 AGDDs (Figure 4) and 45 AGDDs in plots defoliated at 270 AGDDs (Figure 5). Table 2 shows that the average value of the interaction between DF and species for growth was not significant ($p > 0.05$, Table 2), illustrating that similar growth occurs in fall grazing with intervals of 135 and 270 AGDDs. However, *B. valdivianus* defoliated at 135 AGDDs produced 52 (6%) kg DM ha^{-1} more than *L. perenne* defoliated at 270 AGDDs; that in one hectare is not significant but in several hectares it could be significative. In addition, pastures defoliated to 135 AGDDs have a better nutritive value in terms of crude protein and metabolizable energy than defoliated pastures at 270 AGDDs.

Table 2. Mean growth (kg DM ha^{-1}) of *Bromus valdivianus* Phil. and *Lolium perenne* L. defoliated after 135 and 270 accumulated growing degree days (AGDDs).

Species	AGDDs	Growth
B. valdivianus	135	906 ^{x}
	270	723
L. perenne	135	816 ^{x}
	270	854
SEM		17.319
p-Value		0.415

^{x} Accumulated value of two growth cycles each one at 135 AGDDs; SEM = standard error of the mean.

4. Discussion

The present field study highlights the effect of DF, determined by AGDDs, on the utilization, concentration levels, and recovery of WSC concentrations in *L. perenne* and *B. valdivianus* pastures during fall. More frequent defoliation (135 AGDDs) reduced the levels of WSCs in both leaf sheaths and blades, but reserves were replaced faster. When defoliated with a lower frequency (270 AGDDs), WSC levels were higher but recovery was slower.

4.1. Defoliation Frequency and WSCs

During most of the growth cycle, the WSC level and recovery rate werer greater in leaf sheaths than in leaf blades, supporting previous studies that report the sheath as the main WSC storage component in grass species [7]. Moraes et al. [27] evaluated 24 *poa* species and reported that the WSC concentration was greater in sheaths than in blades, with the exception of *Echinolaena inflexa*. The storage of WSCs in leaf sheaths is probably a survival mechanism given that the reserves are located close to the growth points and allow for efficient use. If WSCs were predominately stored in the leaf blades, they would be removed by grazing activity and the plant would lose this resource for regrowth in the subsequent cycle [28].

In most of terrestrial plants, the use of WSCs begins immediately after a plant is defoliated and after recovery process the reserves are restored to the original concentration [29]. In the present study, plants defoliated with the lower DF of 270 AGDDs recovered the original concentration (between 100 to 120 g kg^{-1} DM) after approximately 195 AGDDs. Similar results were reported by Donaghy and Fulkerson [5], who measured 125 g kg^{-1} DM of WSC for *L. perenne* pastures defoliated at the three-leaf stage with 5 cm of residue. Turner et al. [6] reported that *L. perenne, B. willdenowwii* Kunth., and *Dactylis glomerata* L. recovered WSCs after 2.5 to 3 leaves had appeared. Many studies have shown that AGDDs are directly related to the phyllochron and the leaf stage [30]. The new leaf appearance in *L. perenne* takes 88 AGDDs, while for *B. valdivianus* it is only 77 AGDDs [4]. After 195 AGDDs, pasture grass has between two and three leaves per tiller and some 80% of WSC reserves are restored, which supports recovery after the next grazing cycle [31].

Berone et al. [32] evaluated the impact of DF on WSC concentrations in winter pastures of *L. perenne* and *B. stamineus* in Argentina using leaf stage as a grazing criterion. They reported that the WSC concentration was greater in plants defoliated with less frequency (5 leaf stage) than in plants defoliated more frequently (3 leaf stage), and the WSC mean values were 25 g kg^{-1} DM for *B. stamineus* and 15 g kg^{-1} DM for *L. perenne*. The response to different DFs is consistent with the trend measured in the present study; however, they reported lower WSC values. This may be due to the fact that the two studies were conducted under different seasons and climate conditions.

The measurements of WSC concentration following defoliation at intervals of 135 and 270 AGDDs in *L. perenne* and *B. valdivianus* indicated that WSC recovery was initiated at the same time in both leaf sheaths and blades, with a consistently greater concentration in the sheaths. This supports the conclusion that, regardless of the DF or species, leaf sheaths are the main organ for WSC reserves [10].

The levels of WSC in sheaths of *L. perenne* defoliated at 135 and 270 AGDDs measured in the present study (54.65 and 113.93 g kg^{-1} DM, respectively) were lower than values reported by Loaiza et al. [33] with more than 150 g kg^{-1} DM for *L. perenne* plants defoliated at the three-leaf stage during fall. This difference between WSC levels could be explained by N fertilization, as N supply induces sucrose cleavage to release hexoses capable of supporting regrowth during fall growing conditions [15]. The season is a main factor affecting pasture growth, because weather changes modify the physiological stages of each species (water requirements, soil type, water absorption, and transpiration); therefore, the season determines how WSCs are stored [34,35]. In spring and summer, plants maintain greater concentrations of WSCs as well as DM production because of the increased photosynthesis activity from higher temperatures and luminosity [7]. Water-soluble carbohydrates are only accumulated when the synthesis of carbohydrates exceeds their use, which generally occurs when plants have sufficient photosynthetically active leaf blades expanded to support the energy requirements to continue the growth cycle [36,37].

4.2. Defoliation Frequency and AHM

The higher AHM with defoliation every 270 AGDDs compared to 135 AGDDs demonstrates that there is a relationship with WSC recovery, given that higher WSC concentrations were measured with higher AHM (20% and 27% more than 135 AGDDs, respectively). This is supported by both Donaghy and Fulkerson [5] and Lee et al. [38], who reported that there is a positive linear relationship between these variables (r^2 = 0.52, leaf blade (g kg^{-1} DM) = 1.04 + 0.99 WSCs). This behavior is replicated in the majority of forage species, corroborating the close relationship between WSC concentration and AHM [39,40].

In addition to their role in pasture regrowth, WSCs are also important at the nutritional level, in particular the portion stored in the leaf blade, since this is the plant part consumed by grazing animals. Cajarville et al. [9] measured WSC concentrations of 39.1 g kg^{-1} DM in *Festuca arundinacea* pastures in fall when harvested in the morning and also found that there was a positive relationship between in vitro gas production and WSC concentration. Beltran [41] observed that *L. perenne* pastures had higher WSC concentrations in the afternoon compared to the morning (57 g and 82 g kg^{-1} DM, respectively), but this did not significantly affect milk production. The average WSC concentration for the two species measured in the present study were 40 g and 100 g kg^{-1} DM when defoliated at 135 and 270 AGDDs, respectively.

5. Conclusions

The present study confirmed that the leaf sheath was the principal storage organ for WSC reserves, having higher concentrations than leaf blades in *L. perenne* and *B. valdivianus* fall pastures. Water-soluble carbohydrates are easily accessible energy sources that support plant physiological requirements immediately after defoliation. Approximately 80% of total WSC was used during the regrowth process before WSC storage recommenced. Defoliation frequency affected WSC concentration, with longer intervals between defoliation (270 AGDDs) being preferred, because the plants could recover 99% of WSC reserves and could tolerate another grazing event better. Defoliation with greater frequency (135 AGDDs) diminished the synthesis and storage of WSCs in *L. perenne* and *B. valdivianus* and led to slower regrowth. We encourage further research under other seasons and with others defoliation frequencies.

Author Contributions: Conceptualization: I.C., O.B., M.A., J.P.K. and I.L. Data curation: I.C. and M.A.; Formal analysis: I.C., J.P.K. and I.L. Funding acquisition: O.B.; Investigation, I.C., O.B. and M.A.; Methodology: I.C., O.B. and M.A.; Resources: O.B.; Supervision: M.A., J.P.K. and I.L. Writing—Original draft, I.C.; Writing—Review and editing, O.B. All authors have read and agreed to the published version of the manuscript.

Funding: This research was funded by The National Fund for Scientific and Technological Development, Chile. Project Fondecyt 1180767.

Acknowledgments: Special thanks to Felipe Bozo, Ignacio Beltran, and Katira Navarro who participated in the data collection. The first author obtained a CONICYT-PCHA Doctorado Nacional 2015 Scholarship towards his PhD program.

Conflicts of Interest: The authors declare no conflict of interest.

References

1. Clerget, B.; Dingkuhn, M.; Goze, E.; Rattunde, H.F.; Ney, B. Variability of phyllochron, plastochron and rate of increase in height in photoperiod-sensitive sorghum varieties. *Ann. Bot.* **2008**, *101*, 579–594. [CrossRef]
2. Chaves, G.G.; Cargnelutti Filho, A.; Alves, B.M.; Lavezo, A.; Wartha, C.A.; Uliana, D.B.; Pezzini, R.V.; Kleinpaul, J.A.; Neu, I.M.M. Phyllochron and leaf appearance rate in oat. *Bragantia* **2017**, *76*, 73–81. [CrossRef]
3. Wilhelm, W.; McMaster, G.S. Symposium on the phyllochron: Importance of the phyllochron in studying development and growth in grasses. *Crop Sci.* **1995**, *35*, 1–3. [CrossRef]
4. Calvache, I.; Balocchi, O.; Alonso, M.; Keim, J.P.; López, I.F. Thermal Time as a Parameter to Determine Optimal Defoliation Frequency of Perennial Ryegrass (*Lolium perenne* L.) and Pasture Brome (*Bromus valdivianus* Phil.). *Agronomy* **2020**, *10*, 620. [CrossRef]
5. Donaghy, D.; Fulkerson, W. Priority for allocation of water-soluble carbohydrate reserves during regrowth of *Lolium perenne*. *Grass Forage Sci.* **1998**, *53*, 211–218. [CrossRef]
6. Turner, L.; Donaghy, D.; Lane, P.; Rawnsley, R. Effect of defoliation management, based on leaf stage, on perennial ryegrass (*Lolium perenne* L.), prairie grass (*Bromus willdenowii* Kunth.) and cocksfoot (*Dactylis glomerata* L.) under dryland conditions. 1. Regrowth, tillering and water-soluble carbohydrate concentration. *Grass Forage Sci.* **2006**, *61*, 164–174.
7. Pelletier, S.; Tremblay, G.F.; Bélanger, G.; Bertrand, A.; Castonguay, Y.; Pageau, D.; Drapeau, R. Forage nonstructural carbohydrates and nutritive value as affected by time of cutting and species. *Agron. J.* **2010**, *102*, 1388–1398. [CrossRef]
8. Smith, K.; Simpson, R.; Culvenor, R.; Humphreys, M.O.; Prud'Homme, M.; Oram, R. The effects of ploidy and a phenotype conferring a high water-soluble carbohydrate concentration on carbohydrate accumulation, nutritive value and morphology of perennial ryegrass (*Lolium perenne* L.). *J. Agric. Sci.* **2001**, *136*, 65–74. [CrossRef]
9. Cajarville, C.; Britos, A.; Errandonea, N.; Gutiérrez, L.; Cozzolino, D.; Repetto, J.L. Diurnal changes in water-soluble carbohydrate concentration in lucerne and tall fescue in autumn and the effects on in vitro fermentation. *N. Z. J. Agric. Res.* **2015**, *58*, 281–291. [CrossRef]
10. Donaghy, D.J.; Turner, L.R.; Adamczewski, K.A. Effect of defoliation management on water-soluble carbohydrate energy reserves, dry matter yields, and herbage quality of tall fescue. *Agron. J.* **2008**, *100*, 122. [CrossRef]
11. Zhao, D.; MacKown, C.T.; Starks, P.J.; Kindiger, B.K. Interspecies variation of forage nutritive value and nonstructural carbohydrates in perennial cool-season grasses. *Agron. J.* **2008**, *100*, 837–844. [CrossRef]
12. Savitch, L.V.; Gray, G.R.; Huner, N.P. Feedback-limited photosynthesis and regulation of sucrose-starch accumulation during cold acclimation and low-temperature stress in a spring and winter wheat. *Planta* **1997**, *201*, 18–26. [CrossRef]
13. Xue, G.P.; McIntyre, C.L.; Jenkins, C.L.; Glassop, D.; van Herwaarden, A.F.; Shorter, R. Molecular dissection of variation in carbohydrate metabolism related to water-soluble carbohydrate accumulation in stems of wheat. *Plant Physiol.* **2008**, *146*, 441–454. [CrossRef]
14. Ciavarella, T.A.; Simpson, R.J.; Dove, H.; Leury, B.J.; Sims, I.M. Diurnal changes in the concentration of water soluble carbohydrates in *Phalaris aquatica* L. pasture in spring, and the effect of short-term shading. *Aust. J. Agric. Res.* **2000**, *51*, 749–756. [CrossRef]
15. Liu, Q.; Rasmussen, S.; Johnson, L.J.; Xue, H.; Parsons, A.J.; Jones, C.S. Molecular Mechanisms Regulating Carbohydrate Metabolism During *Lolium perenne* Regrowth Vary in Response to Nitrogen and Gibberellin Supply. *J. Plant Growth Regul.* **2020**, *39*, 1342–1355. [CrossRef]
16. Cairns, A.J.; Begley, P.; Sims, I.M. The structure of starch from seeds and leaves of the fructan-accumulating ryegrass, *Lolium temulentum* L. *J. Plant Physiol.* **2002**, *159*, 221–230. [CrossRef]
17. Smith, A.M.; Zeeman, S.C.; Thorneycroft, D.; Smith, S.M. Starch mobilization in leaves. *J. Exp. Bot.* **2003**, *54*, 577–583. [CrossRef] [PubMed]

18. Amiard, V.; Morvan-Bertrand, A.; Cliquet, J.B.; Billard, J.P.; Huault, C.; Sandström, J.P.; Prud'homme, M.P. Carbohydrate and amino acid composition in phloem sap of *Lolium perenne* L. before and after defoliation. *Can. J. Bot.* **2004**, *82*, 1594–1601. [CrossRef]
19. Delagarde, R.; Peyraud, J.; Delaby, L.; Faverdin, P. Vertical distribution of biomass, chemical composition and pepsin—cellulase digestibility in a perennial ryegrass sward: Interaction with month of year, regrowth age and time of day. *Anim. Feed Sci. Technol.* **2000**, *84*, 49–68. [CrossRef]
20. Griggs, T.C.; MacAdam, J.W.; Mayland, H.F.; Burns, J.C. Nonstructural carbohydrate and digestibility patterns in orchardgrass swards during daily defoliation sequences initiated in evening and morning. *Crop Sci.* **2005**, *45*, 1295–1304. [CrossRef]
21. CIREN. Centro de Información de Recursos Naturales. *Agrological Study X Region, Descriptions of Soils, Materials and Symbols*; Publicación CIREN N° 123: Santiago, Chile, 2003.
22. Biligetu, B.; Coulman, B. Responses of Three Bromegrass (Bromus) Species to Defoliation under Different Growth Conditions. *Int. J. Agron.* **2010**, *20*, 1–5. [CrossRef]
23. McMaster, G.S.; Wilhelm, W. Growing degree-days: One equation, two interpretations. *Agric. For. Meteorol.* **1997**, *87*, 291–300. [CrossRef]
24. Yemm, E.; Willis, A. The estimation of carbohydrates in plant extracts by anthrone. *Biochem. J.* **1954**, *57*, 508. [CrossRef] [PubMed]
25. Canseco, C.; Demanet, R.; Balocchi, O.; Parga, J.; Anwandter, V.; Abarzúa, A.; Teuber, N.; Lopetegui, J. Determinación de la disponibilidad de materia seca de praderas en pastoreo. *Manejo Del Pastor. Imprenta Am.* **2007**, *10*, 23–50.
26. Littell, R.; Henry, P.; Ammerman, C. Statistical analysis of repeated measures data using SAS procedures. *J. Anim. Sci.* **1998**, *76*, 1216–1231. [CrossRef] [PubMed]
27. Moraes, M.G.; Chatterton, N.J.; Harrison, P.A.; Filgueiras, T.S.; Figueiredo-Ribeiro, R.C.L. Diversity of non-structural carbohydrates in grasses (Poaceae) from Brazil. *Grass Forage Sci.* **2012**, *68*, 165–177. [CrossRef]
28. White, L.M. Carbohydrate reserves of grasses: A review. *J. Range Manag.* **1973**, *26*, 13–18. [CrossRef]
29. Martínez-Vilalta, J.; Sala, A.; Asensio, D.; Galiano, L.; Hoch, G.; Palacio, S.; Piper, F.I.; Lloret, F. Dynamics of non-structural carbohydrates in terrestrial plants: A global synthesis. *Ecol. Monogr.* **2016**, *86*, 495–516. [CrossRef]
30. Gramig, G.G.; Stoltenberg, D.E. Leaf appearance base temperature and phyllochron for common grass and broadleaf weed species. *Weed Technol.* **2017**, *21*, 249–254. [CrossRef]
31. Pommerrenig, B.; Ludewig, F.; Cvetkovic, J.; Trentmann, O.; Klemens, P.A.W.; Neuhaus, H.E. In concert: Orchestrated changes in carbohydrate homeostasis are critical for plant abiotic stress tolerance. *Plant Cell Physiol.* **2018**, *59*, 1290–1299. [CrossRef]
32. Berone, G.D.; Lattanzi, F.A.; Agnusdei, M.G.; Bertolotti, N. Growth of individual tillers and tillering rate of Lolium perenne and Bromus stamineus subjected to two defoliation frequencies in winter in Argentina. *Grass Forage Sci.* **2008**, *63*, 504–512. [CrossRef]
33. Loaiza, P.A.; Balocchi, O.; Bertrand, A. Carbohydrate and crude protein fractions in perennial ryegrass as affected by defoliation frequency and nitrogen application rate. *Grass Forage Sci.* **2017**, *72*, 556–567. [CrossRef]
34. Taiz, L.; Zeiger, E. *Plant Physiology and Development*, 6th ed.; Sinauer Associates: Sunderland, MA, USA, 2015.
35. Zeeman, S.C.; Smith, S.M.; Smith, A.M. The diurnal metabolism of leaf starch. *Biochem. J.* **2007**, *401*, 13–28. [CrossRef]
36. Bahuguna, R.N.; Jagadish, K.S.V. Temperature regulation of plant phenological development. *Environ. Exp. Bot.* **2015**, *111*, 83–90. [CrossRef]
37. McClung, C.R.; Davis, S.J. Ambient thermometers in plants: From physiological outputs towards mechanisms of thermal sensing. *Curr. Biol.* **2010**, *20*, R1086–R1092. [CrossRef]
38. Lee, J.M.; Donaghy, D.J.; Roche, J.R. Effect of defoliation severity on regrowth and nutritive value of perennial ryegrass dominant swards. *Agron. J.* **2008**, *100*, 308–314. [CrossRef]
39. Herz, K.; Dietz, S.; Haider, S.; Jandt, U.; Scheel, D.; Bruelheide, H. Predicting individual plant performance in grasslands. *Ecol. Evol.* **2017**, *7*, 8958–8965. [CrossRef]
40. Hoogmoed, M.; Sadras, V.O. The importance of water-soluble carbohydrates in the theoretical framework for nitrogen dilution in shoot biomass of wheat. *Field Crops Res.* **2016**, *193*, 196–200. [CrossRef]

41. Beltran, I.E.; Morales, A.; Balocchi, O.; Pulido, P. Interaction between herbage mass and time of herbage allocation modify milk production, grazing behaviour and nitrogen partitioning of dairy cows. *Anim. Prod. Sci.* **2019**, *59*, 1837–1846. [CrossRef]

Publisher's Note: MDPI stays neutral with regard to jurisdictional claims in published maps and institutional affiliations.

Article

Crop Rotation Enhances Agricultural Sustainability: From an Empirical Evaluation of Eco-Economic Benefits in Rice Production

Dun-Chun He [1], Yan-Li Ma [1,2], Zhuan-Zhuan Li [1,2], Chang-Sui Zhong [3], Zhao-Bang Cheng [2] and Jiasui Zhan [4,*]

1 Institute of Eco-technological Economics, School of Economics and Trade, Fujian Jiangxia University, Fuzhou 350108, China; hedc@fjjxu.edu.cn (D.-C.H.); mayl@fafu.edu.cn (Y.-L.M.); lzz@fafu.edu.cn (Z.-Z.L.)
2 Institute of Plant Protection, Jiangsu Academy of Agricultural Sciences, Nanjing 210014, China; czb69jaas@jaas.ac.cn
3 Agro-Tech Extension and Service Station of Fuqing City, Fuzhou 350300, China; zhengrr@fafu.edu.cn
4 Department of Forest Mycology and Plant Pathology, Swedish University of Agricultural Sciences, 75007 Uppsala, Sweden
* Correspondence: jiasui.zhan@slu.se

Abstract: Cropping systems greatly impact the productivity and resilience of agricultural ecosystems. However, we often lack an understanding of the quantitative interactions among social, economic and ecological components in each of the systems, especially with regard to crop rotation. Current production systems cannot guarantee both high profits in the short term and social and ecological benefits in the long term. This study combined statistic and economic models to evaluate the comprehensive effects of cropping systems on rice production using data collected from experimental fields between 2017 and 2018. The results showed that increasing agricultural diversity through rotations, particularly potato–rice rotation (PR), significantly increased the social, economic and ecological benefits of rice production. Yields, profits, profit margins, weighted dimensionless values of soil chemical and physical (SCP) and heavy metal (SHM) traits, benefits and externalities generated by PR and other rotations were generally higher than successive rice cropping. This suggests that agricultural diversity through rotations, particularly PR rotation, is worth implementing due to its overall benefits generated in rice production. However, due to various nutrient residues from preceding crops, fertilizer application should be rationalized to improve the resource and investment efficiency. Furthermore, we internalized the externalities (hidden ecological and social benefits/costs) generated by each of the rotation systems and proposed ways of incenting farmers to adopt crop rotation approaches for sustainable rice production.

Keywords: agricultural sustainability; crop rotation; rice; eco-economic benefit; externality

Citation: He, D.-C.; Ma, Y.-L.; Li, Z.-Z.; Zhong, C.-S.; Cheng, Z.-B.; Zhan, J. Crop Rotation Enhances Agricultural Sustainability: From an Empirical Evaluation of Eco-Economic Benefits in Rice Production. *Agriculture* **2021**, *11*, 91. https://doi.org/10.3390/agriculture11020091

Received: 20 December 2020
Accepted: 19 January 2021
Published: 21 January 2021

1. Introduction

The human population is expected to exceed 9.7 billion by 2050, requiring a substantial increase in agricultural production capacity to secure global food supplies [1] which, on the other hand, are threatened by climate change, environmental pollution and drained natural resources such as water and fossil energy [2,3]. Climate change, environmental pollution and natural resource constraints are also expected to have negative impacts on the productivity and quality of crops. Current agricultural production systems heavily rely on high inputs of natural resources, particularly irrigation water, fertilizers and pesticides. For example, in Samsun, Turkey, the annual energy consumption for wheat production is 35,737 MJ/ha [4]. Up to 15 fungicide sprays are executed annually to control potato late blight in Northern and Western Europe [5] and more than 20 fungicide sprays are applied to control rose mildew in some parts of the world [6]. In apple, more than 12 fungicide applications usually take place each season to control scab caused by *Venturia inaequalis* [7], even though a recent result indicated that only five applications could achieve the similar control purpose [8]. High resource inputs may usually increase gross production of crops but many of them may not generate positive net returns due to the gain in production being over-weighted by excess inputs. Furthermore, when yield is the primary goal of farmers, which is always the case for cereal productions, little attention will be paid to the direct and indirect effects of the production process on society and ecology such as sustainability of food safety, soil quality and ecological resilience generated by high chemical residues which pollute soils and rivers, demolish biodiversity and poison humans and other animals, etc. Therefore, conservation agriculture as a tool for sustainable development is essential so that natural resources can be used in a rational and economical manner for social and ecological sustainability [9]. Recent concern for the sustainability of agriculture and associated natural environments has led to renewed interest in practices that seek to increase production while improving soil health and ecological resilience [10–13] through crop diversification and comprehensive evaluation of the social, economic and ecological impacts of producing systems [14] by internalizing the externalities, i.e., the hidden benefits and costs are not reflected in marketing prices, associated with primary production systems. Agricultural diversification is referred to the reallocation of some farming resources such as lands, equipment and labor to other social or natural services and can be achieved by multiple paths such as changing cropping systems, modifying productive goals and switching to non-farming activities at spatial and/or temporal scales [15]. Among them, crop rotation has been thought to be a promising agricultural practice which could regenerate balanced biotic and abiotic interactions, supporting a synergistic service to both society and nature by enhancing key elements of biodiversity, increasing resource efficiency, reducing pest epidemics and stabilizing the function of ecosystem production over time [16–18]. For example, it has been shown that crop rotation can eliminate soil-borne pathogens, pests and weed reservoirs that cannot be effectively controlled by pesticides and improve soil quality such as nutrition status and physical structure [19]. However, the benefits associated with crop rotation are rarely evaluated by a comprehensive evaluation of social, economic and ecological impacts of crop rotation generated from field data.

Rice, as one of the main nutrient supplies of the world, is especially important in less developed Asian countries. Soil pollution and ecological deterioration associated with current agricultural production systems greatly threaten sustainable rice production [19]. For example, the projected increase in pore-water arsenite—the more toxic form of arsenic—may cause up to a 39% rice grain reduction compared to current soil arsenic concentrations [20]. Even though higher yields do not always result in better economic benefits [21], the problem of overreliance on chemical inputs while targeting to maximize yield [9] is particularly serious in the rice production system of China. It has been reported that Chinese rice cultivation occupies 20% of global production acreage but consumes 26.7% of chemical nitrogen fertilizers (over 180 kg/ha). Only 20–30% of the fertilizers applied are taken by rice crop [22].

Paddy-upland rotations have received particular attention in the era of agricultural diversification and have been adopted by some regions in Southern China [19] where the available irrigation system allows the farmer's practice to be successfully executed. Although better pay-offs were reported when farmers rotated rice with other crops rather than successively growing rice, how this practice may impact other components of production such as ecology and energy conservation and what the best crop to rotate is are hardly understood. Many governments in the world have focused on increasing the total rice cultivation area through the provision of subsidies to reduce production costs. These economic incentives ensure the steady increase in rice production and encourage farmers to invest more in machinery for rice production, able to handle large acreages. However, their production ambitions and economic returns are not always synchronous, generating concerns of production sustainability. The disagreement results from that fact that net income of rice (as with other crops) production depends not only on immediate, direct factors such as yield, price, subsidy and expenses but also on future, indirect factors such as soil health and ecological resilience. A lack of comprehensive assessment of the synergies and trade-offs generated by the short-term and long-term interactions between direct and indirect benefits and economic and ecological benefits has resulted in a poor equilibrium among efficiency, cost, profit and sustainability of production. With regards to the research on the rice cropping system, scientists have focused on fundamental questions such as its links with soil chemical and physical properties or applied issues such as technology development rather than social and ecological economics analyses. Particularly, the externalities of rotations for rice production have rarely been quantitatively studied based on data generated from field experiments but are necessary to ensure sustainable rice production to feed the growing global population [23].

In the current study, data generated from fields with different rice cropping systems over two consecutive years were evaluated in parallel with economic, social and ecological effects in order to develop a more profitable, effective and eco-friendly rice production strategy. The specific goals of the study were to (1) determine the differences in the factors responsible for the economic, social and ecological benefits of rice production within different cropping systems; (2) evaluate the pros and cons of rice production among different cropping systems and develop a practice model of rice production in main rice cultivation areas such as Southern China; (3) quantify the externalities of rice production associated with different cropping systems and make recommendations to policy-makers to increase the sustainability of rice production.

2. Materials and Methods

2.1. Experimental Site

The experimental site (25°33′20.67″ N, 119°25′36.93″ E) was located at the field trial Station of Fujian Agriculture and Forestry University in Jiangjing town, Fuqing city, Fujian province, China. This site has a humid subtropical monsoon climate with mean annual rainfall of 1050~1500 mm and an effective accumulated temperature of 6000–6600 °C, with an average daily temperature of 20–25 °C during growing season. The experimental fields were well equipped with an irrigation system and were either in fallow or planted with watermelon, potato or rice before this study according to the experimental requirements described in the next sections.

2.2. Experimental Design and Crop Management

The experiments were conducted between March and August in 2017 and 2018. Each of the field experiments contained treatments including two rice cultivars—Yiyou 673, provided by the Rice Institute, Fujian Academy of Agricultural Sciences, in Fuzhou, and Fulong 3831, provided by the Longyan Institute of Agricultural Science of Longyan City—and four cropping systems (2 $\times$ 4), and they were laid out in a completely randomized block design with three replicates (a total of 24 experimental units). The two rice cultivars are similar in many agronomic characteristics such as plant height and maturity and have been widely grown in this region for many years. The four cropping systems were successive rice cropping (RR), fallow followed by rice (FR), potato and rice rotation (PR) and watermelon and rice rotation (WR). Each of the experimental units was 0.2 ha in size and was separated from the others by ~50-cm furrows to prevent water and nutrient flows among units.

The rice seeds were sown in seedling trays in late March. Immediately after sowing, the seedling trays were mulched with white plastic films to maintain temperature and moisture while allowing sunlight to transmit. Experimental fields were prepared by ploughing twice with a power tiller, a harrow and a leveler. An ammonium bicarbonate nitrogen fertilizer (N $\geq$ 17.1%) (Anhui Liuguo Chemical Co., Ltd., Tongling, China) was applied as a base fertilizer at 450 kg/ha before transplantation according to the theoretical calculation of rice N demand, local average rice yield and the estimated N content in the soil of the experimental fields. Rice seedlings at the stage of 3–4 leaves were transplanted mechanically (Shanghai Kubota Co., Ltd., Shanghai, China) at a density of 165,000–180,000 hills/ha. A compound fertilizer (N:P:K = 16:16:16, total nutrient $\geq$ 48%, 150 kg/ha) and a urea fertilizer

(N ≥ 46.4%, 75 kg/ha) (Anhui Liuguo Chemical Co., Ltd., Tongling, China) were applied at the beginning of the tillering stage. Water, diseases, pests and weeds were managed according to field conditions. The rice was mechanically harvested in August. The rice straws were returned to the fields after grain thrashing.

2.3. *Traits Measurement and Parameters Estimates*

Identical sampling protocols were used for all treatments of the experiments conducted in the two years. Five sample sites were selected from each experimental unit using a stratified strategy with one site in the center of the unit and two sites each in the ends of the unit. Soil samples (0−15 cm depth) were collected using a tube auger from the five sampling sites in each experimental unit and were thoroughly mixed to form a composite sample for physical and chemical characterizations [24]. Soil pH, organic matter (SOM), available N, available P and available K were measured by a pH meter, the acidified dichromate method, the alkali hydrolysis and diffusion method, the Olsen method and the atomic absorption spectrophotometry, respectively, using a slurry of 1:2.5 soil/water (*v*/*v*) as previously described [25–27]. Concentrations of lead (Pb), mercury (Hg), chromium (Cr), cadmium (Cd), copper (Cu) and zinc (Zn) in the soil samples were determined using graphite furnace atomic absorption and flame atomic absorption [28–30]. Straw biomass and grain yield were also determined from the five sampling sites during harvesting and then converted to total production in each of the experimental units using the total areas measured from the five sites (20 m^2). Grain production was quantified with all crops in each of the experimental units.

The rice marketing price, governmental subsidy and total production cost associated with farmland rent, consumable materials (e.g., fertilizers, pesticides, seeds, plastic tray and film) and labor (sowing, ploughing, transplanting, fertilizing, managing and harvesting, etc.) were calculated by farm gate price, actual government support and expenses and mechanical devaluation was estimated. To obtain the direct information needed for the calculation, a direct survey involving face-to-face interviews with farmers was conducted as described previously [31]. The survey was conducted with a total of 25 farmers across the five towns of the city. Accordingly, the costs of farmland rent, material and labor in seeding, ploughing, transplanting, fertilization, plant protection (diseases, insects and weeds), harvesting and other miscellaneous expenses were set to 692, 265, 230, 230, 138, 138, 127 and 81 USD per hectare, respectively, with a total cost of 1904 USD per hectare. The annual governmental subsidy for rice cropping was 230 USD per hectare.

Harvest index (HI), revenue (R), profit (NP), profit margin (PM), weighted dimensionless values of soil chemical and physical (SCP) and heavy metal (SHM) traits were calculated using the following formulas [24]:

$$\mathrm{HI} = \mathrm{G}/(\mathrm{G} + \mathrm{DS}) \quad (1)$$

$$\mathrm{R} = \mathrm{G} \times \mathrm{P} + \mathrm{S} \quad (2)$$

$$\mathrm{NP} = \mathrm{R} - \mathrm{C} \quad (3)$$

$$\mathrm{PM} = (\mathrm{NP}/\mathrm{C}) \quad (4)$$

where G, DS, P, S and C are the grain production, straw weight, grain marketing price, governmental subsidy and total production cost, respectively.

$$SCP = 1/5 \sum_j (x_i - x_{max})/(x_{max} - x_{min}) \quad (5)$$

$$SHM = 1/6 \sum_j (x_i - x_{max})/(x_{max} - x_{min}) \quad (6)$$

where x_i, x_{max} and x_{min} are the raw data of each experiment plot and the maximum and minimum raw data of each replication, respectively; *i* is the experimental plot; *j* is the order of pH, SOM, N, P and K for SCP and Pb, Hg, Cr, Cd, Cu and Zn for SHM.

The indicators of benefit assessment, including profit, profit margin, revenue, yield, HI, SCP, SHM and weight of dry straw, were determined in line with the documents [32,33] and the expert and farmer consultations as described previously [31]. In total, fifteen experts from Fujian Agriculture and Forestry University, Fujian Academy of Agricultural Sciences, Jiangsu Academy of Agricultural Sciences and the departments of agriculture technology in Fujian and Jiangsu provinces and 25 farmers across the five towns of Fuqing city were consulted for the matters. The indexes (Table S1) of the benefits were weighted using the Analytic Hierarchy Process (AHP) [34] according to their relative importance on the basis of the experiment and the consultation results from the expert and farmer interviews.

To obtain normalization data for the benefits assessment, the raw values of the indicators were converted to dimensionless values x_i' by min-max normalization (Formula (7)) [35]. The benefits index (BI_i) of rice production within the different cropping systems was calculated using the Formula (8) [36]:

$$x_i' = (x_i - x_{max})/(x_{max} - x_{min}) \quad (7)$$

where x_i, x_{max} and x_{min} are the raw data of indicators from each experiment unit and the maximum and minimum raw data of the corresponding indicators of each replication, respectively; i is the random order of these experimental plots.

$$BI = 1/3\sum_j w_i\, x_i' \quad (8)$$

where w_i and x_i' are the weighted and dimensionless values of the ith indicator, respectively; j is the random order of the replications. Farmland rent (692 USD/ha) was not included in the economic benefit analysis of the FR practice.

The externality values were calculated as: externality value = profit × (social and ecological benefit index/profit weight × comprehensive benefit index).

2.4. Statistical Analysis

The contributions of cropping system, cultivar and their interactions with yield, harvest index, profits and soil properties including pH value and contents of organic matter, minerals and toxin chemicals were assessed using a multivariate analysis of variance (MANOVA), while the contributions of these independent variables to economic, social and ecological benefits as well as externalities were assessed by a one-way ANOVA. In the ANOVA and MANOVA, cultivar was treated as a fixed variable while cropping system was treated as a random variable. Duncan's Multiple Range Test was used to compare means of rice yield, harvest index, soil physical and chemical properties, profits, benefits and externality within dependent variables at the 0.05 probability level. All of the statistical analyses were performed using IBM SPSS 19.0 software (IBM Corp., Armonk, NY, USA).

3. Results

3.1. Cropping System Significantly Impacts the Socioeconomic Benefits of Rice Production

The ANOVA revealed a significant impact of cropping system on the yield, profit and profit margin of rice production ($p < 0.05$). No cropping system impact on harvest index ($p = 0.335$) was found. Similarly, cultivar and its interaction with cropping system did not have any biological and economic influences on rice production in the current study (Table 1).

Table 1. Analysis of variance evaluating the effect of cropping systems, cultivar and their interaction with yield, harvest index and profits of rice production.

Parameter	Yield			Harvest Index			Profit			Profit Margin		
	DF	F	P	DF	F	P	DF	F	P	DF	F	P
Cultivar	1	0.427	0.517	1	0.603	0.442	1	0.360	0.552	1	0.360	0.552
Cropping system	3	3.193	0.034	3	1.166	0.335	3	2.967	0.043	3	2.967	0.043
Cultivar × Cropping system	3	0.295	0.829	3	0.511	0.677	3	0.291	0.831	3	0.291	0.831
Error	40			40			40			40		

3.2. Difference in Production and Socioeconomic Benefits among Rice Cropping Systems

The yield, profit and profit margin of PR (potato rice rotation), FR (fallow followed by rice) and WR (watermelon rice rotation) were higher than those of RR (successive cropping of rice) (Table 2). Compared with the other three cropping systems, PR achieved the highest yield, profit and profit margin. It was followed by FR while RR performed worst. Rice yield from PR was significantly higher ($p < 0.05$) than that from RR and WR but was only marginally higher than that from FW (Table 2). Profits from PR were also significantly higher than those from all other cropping systems.

Table 2. Effect of cropping systems on yield and socioeconomic benefits of rice production.

Cropping System	Yield t/ha	Harvest Index %	Profit US Dollar/ha	Profit Margin %
RR	5.2 b	42.2 a	162 c	8.5 c
FR	6.1 ab	45.1 a	465 b	24.4 b
PR	7.1 a	42.8 a	826 a	43.4 a
WR	5.9 b	45.8 a	385 b	20.2 b

Note: The different letters following the values in a column indicate a significant difference ($p < 0.05$). The same letter means it is not significantly different. RR = successive cropping of rice; FR = fallow followed by rice; PR = potato rice rotation; WR = watermelon rice rotation.

3.3. Effects of Cropping Systems on the Chemical and Physical Properties of Soils

Chemical and physical properties including pH value, organic matter, mineral and heavy metal contents fluctuated greatly over the sampling times within the growing season (Figures 1 and 2, Table 3) in all cropping systems. Overall, the soils were acidified in the paddy fields and the most acidic soil was found in the RR experiment. Soil organic matter showed a downward trend, especially in WR. N, P and K contents were richest in the soil from PR, leading to the highest SCP index. With the exception of organic matter, WR also yielded better soil fertility (N, P and K) and SCP than those of RR and FR. Regarding the contents of harmful heavy metals, levels under RR were always the highest, although some of the differences were not significant from other cropping systems, leading to the highest SHM (Table 4). The temporal dynamics of the heavy metals in the soils from RR, PR and WR showed a similar trend of slightly increasing over the growing season. This pattern was more obvious in RR (Figure 2, Table 4). Except for Zn, the heavy metal contents in the soils from PR were higher than in those from FR and WR (Table 4).

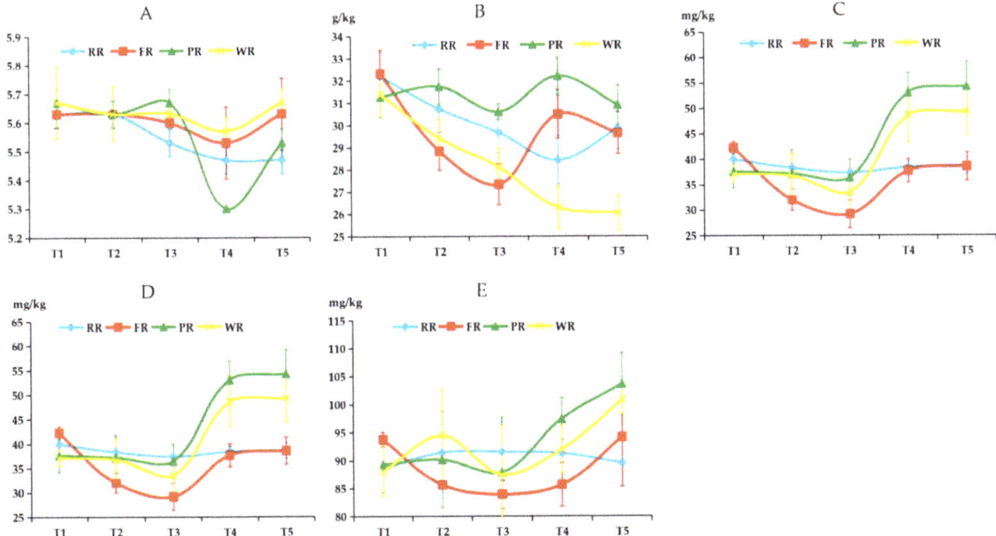

Figure 1. The temporal dynamics of soil chemical and physical properties. (**A**) pH value, (**B**) soil organic matter (SOM), (**C**) available nitrogen content (N), (**D**) available phosphorus (P) content and (**E**) available potassium (K) level. RR = successive rice cropping; FR = fallow followed by rice; PR = potato rice rotation; WR = watermelon rice rotation. Sampling date: T1 = September 2016; T2 = March 2017; T3 = September 2017; T4 = March 2018; T5 = September 2018.

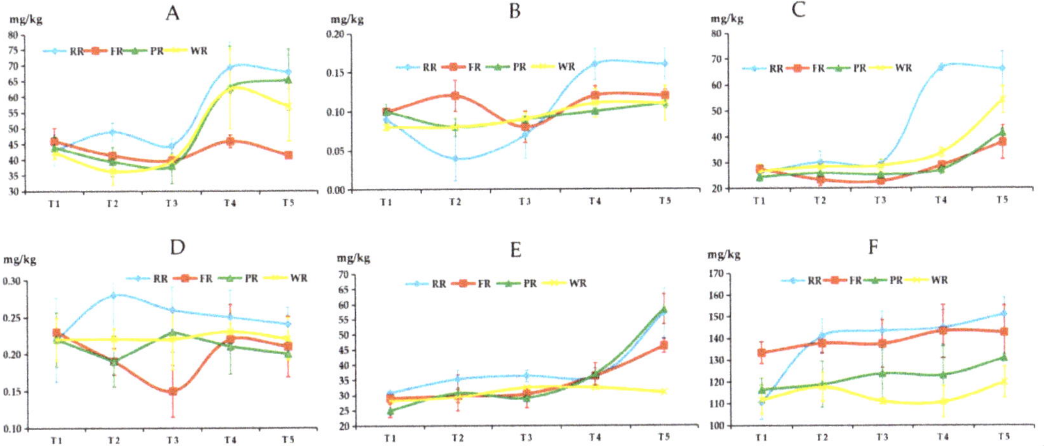

Figure 2. The temporal dynamics of soil heavy metal content: (**A**) Pb, (**B**) Hg, (**C**) Cr, (**D**) Cd, (**E**) Cu and (**F**) Zn. RR = successive rice cropping; FR = fallow followed by rice; PR = potato rice rotation; WR = watermelon rice rotation. Sampling date: T1 = September 2016; T2 = March 2017; T3 = September 2017; T4 = March 2018; T5 = September 2018.

Table 3. Effect of rice cropping systems on soil pH value, available nitrogen (N), available phosphorus (P), available potassium (K) and organic matter (SOM) level.

Cropping System	pH	SOM g/kg	N mg/kg	P mg/kg	K mg/kg	SCP
RR	5.50 b	29.75 a	127.00 b	38.03 ab	90.57 a	0.3850 b
FR	5.62 a	28.48 ab	131.83 b	33.87 b	89.05 a	0.3833 b
PR	5.60 a	30.75 a	156.55 a	45.33 a	95.80 a	0.6850 a
WR	5.65 a	27.07 b	132.62 b	41.37 ab	94.13 a	0.5050 ab

Note: The different letters following the values in a column indicate a significant difference ($p < 0.05$). The same letter means it is not significantly different. RR = successive cropping of rice; FR = fallow followed by rice; PR = potato rice rotation; WR = watermelon rice rotation. SCP is the weighted dimensionless values of pH, SOM, N, P and K. The values presented in the table were calculated from the average of the T1, T2, T3, T4 and T5 values presented in Figure 1.

Table 4. Effect of rice cropping systems on the heavy metal contents of soil.

Cropping System	Pb mg/kg	Hg mg/kg	Cr mg/kg	Cd mg/kg	Cu mg/kg	Zn mg/kg	SHM
RR	56.18 a	0.11 a	47.82 a	0.25 a	46.70 a	147.15 a	0.7650 a
FR	40.60 b	0.10 a	30.21 b	0.18 b	38.25 ab	139.97 ab	0.4000 b
PR	51.70 ab	0.10 a	33.40 b	0.22 ab	43.67 ab	127.22 bc	0.4850 b
WR	48.27 ab	0.10 a	38.07 b	0.22 ab	31.63 b	115.23 c	0.4333 b

Note: The different letters following the values in a column indicate a significant difference ($p < 0.05$). The same letter means it is not significantly different. RR = successive cropping of rice; FR = fallow followed by rice; PR = potato rice rotation; WR = watermelon rice rotation. SHM is the weighted dimensionless values of Pb, Hg, Cr, Cd, Cu and Zn. The values presented in the table were calculated from the average of the T1, T2, T3, T4 and T5 values presented in Figure 2.

3.4. Effects of Cropping Systems on Benefits and Externalities of Rice Production

The economic, social, ecological and comprehensive benefits and externalities generated by PR were always higher, significantly or marginally, than those generated by the other cropping systems, while RR always generated the least benefits (Table 5). FR also generated higher benefits in all aspects, except ecological, than those generated by WR and RR. Relative to RR, we estimated that PR, FR and WR generated 348, 157 and 133 USD/ha externality, respectively (Table 5).

Table 5. Effect of rice cropping systems on social, economic and ecological benefits.

Cropping System	Economic Benefit	Social Benefit	Ecological Benefit	Comprehensive Benefit	Externality Value US Dollar/ha
RR	0.1735 b	0.0537 b	0.0047 b	0.2319 b	0
FR	0.3028 ab	0.0872 ab	0.0083 ab	0.3984 ab	157
PR	0.4009 a	0.1105 a	0.0155 a	0.5269 a	348
WR	0.2286 b	0.0708 b	0.0100 ab	0.3094 b	133

Note: The different letters following the values in a column indicate a significant difference ($p < 0.05$). The same letter means it is not significantly different. RR = successive cropping of rice; FR = fallow followed by rice; PR = potato rice rotation; WR = watermelon rice rotation. The benefits were estimated according to the indicators and weights presented in Table S1 and the dimensionless values converted from the raw data using Formula (7). The externality of the RR practice was set to zero (CK) and the externalities of other practices were calculated relative to the RR externality.

4. Discussion

4.1. PR is Worth Implementing on the Basis of Rice Production Benefits

The cropping system significantly impacts the economic and ecological benefits of rice production (Table 1), and all of the paddy-upland rotations we studied generated better social returns including higher yield, higher profits, higher soil fertility and ameliorated soil contamination than those generated by successive rice cropping (Tables 2–5).

Among them, PR is the best cropping system, supported by the highest economic, social, ecological and comprehensive benefits that it generated (Table 5), consistent with previous reports [37–39]. The farm gate price of potato in the winter cropping areas of Southern China was >0.3 USD/kg over the past 10 years, with an average yield of ~33.5 tons/ha, while the cost of producing potatoes in the same period of time was ~6900 USD/ha, generating a much higher net income in the preceding seasons than growing rice, which was estimated to be 3400 USD/ha [40]. However, yield and economic benefits declined substantially when potatoes were consecutively grown for some years [41]. Taken together, these results indicate that the economic benefit of a PR cropping system outperforms that from an RR or a potato−potato system and could be adopted widely, particularly in Southern China where millions of hectares of arable lands are available in winters after rice crop is harvested [42], and the dry winter there is suboptimal for potato disease epidemics. Ecologically, rice rotation with legumes could be another option in this region, but this practice could not be widely accepted by local farmers due to the small contribution of legumes to the economics of the region. To further enhance the socioeconomic as well as ecological (see below) benefits of rice production, some green manure crops should be intermittently grown after a few cycles of PR practice [43,44].

It was reported that growing watermelon decreased soil fertility [45], but we did not find a general pattern in this regard. WR slightly increased N, P and K levels but marginally or significantly decreased organic matter level in the soils compared to FR and RR (Table 3). In this case, intermittently growing green manure plants after WR practice could, to some extent, compensate the organic matter loss [46,47] WR also generated a mixture of ecological benefits and costs relative to RR.

4.2. *The Amortized Cost of Fallow Should Be Considered in Production Analysis*

FR increased yield, direct farmer income and soil pollution (Tables 2, 4 and 5) but did not impact overall soil fertility (Table 3) compared to the RR system. This falsifies the theoretical expectation of soil fertility restoration associated with the practice. However, fallow can affect the entire soil community structure above and below ground, and its externality cannot be robustly evaluated without a comprehensive study covering a range of topics such as soil fertility, biodiversity, resource consumption, etc. In the current study, we only evaluated the impact of FR on soil nutrient and pollutants using two rice cultivars and further research involving more rice cultivars may be required for a more robust conclusion on the benefits of fallow. Furthermore, fallow practice abandons entire production for one or more seasons and significantly decreases the imminent economic benefit of farmers. This amortized cost should be factored into impact evaluation, resulting in a dilemma between economic and ecological benefits of fallow practice [48]. In spite of some economic and ecological benefits in the production season, the amortized cost of fallow should be considered. Therefore, a substantial government subsidy may be a prerequisite for the practice [49], which may not be sustainable for the countries with limited arable lands and floating cashes to compensate farmers while importing foods in the meantime.

4.3. *Accurate Management of Water and Fertilizer Could Constitute Supplementary Measures for Rice Production Following Crop Rotation*

Rice production heavily relies on high inputs of natural resources such as water and energy required to produce mineral fertilizers [50], greatly threatening the sustainable development of human society. Crop diversification through rotation can improve water as well as nutrient efficiency of rice production as a consequence of increased complementarity in the modes and forms of mineral elements consumed by different crops or crop genotypes [51]. Crop diversification through rotation may also alter soil chemical, physiological and/or biological properties, supporting large and sustainable production [52]. To materialize this advantage, nutritional requirement profiles and preferences of succeeding and preceding crops should be considered jointly. If nutrient residues from the preceding crops are high, the application of fertilizer and other forms of nutrient should be reduced in succeeding production, and vice versa [51,53]. Together with an appropriate water management strategy, this consideration can reduce ineffective tillers and straw biomass, leading to both improved harvest index and grain yields [54]. The organic matter and mineral element levels in the PR soil were significantly higher than in the other cropping systems (Table 3), suggesting that more nutrient residues are retained in the rotation fields with PR in particular. Therefore, accurate management of water and fertilizer use should constitute important elements of rice production following crop rotation. The highest level of heavy metals in RR (Table 4) also suggests that rotation could ameliorate metal contaminations in paddy soil generated by successive rice cropping and benefit the restoration of soil ecosystems. However, it is not clear whether the heavy metal reduction is due to the enhanced take-up by preceding crops and other biological factors, or using more fertilizers and pesticides in rice or contaminated water for rice irrigation. These issues are worthy of further processing.

4.4. *Externalities and Sunk Costs Are an Important Basis for Making Agricultural Policies*

Farmers usually do not clearly understand the complex quantitative interactions among primary production, input, profit, land use and sustainability [55,56]. The practices they adopt are mainly driven by purely economic factors, particularly the income measured by total production [57]. The risk of production, impacts on following crop and sunk costs associated with short- and long-term externalities such as soil resiliency and ecological sustainability of their lands and surroundings are largely ignored but should be included in decision making. Externalities, regardless of benefits or penalties, will eventually be directed back to producers and societies. As a regulator, governments should use an array of incentives or taxation policies to promote production systems with optimized comprehensive benefits by taking farmer incomes, soil fertility, environment pollution, ecological sustainability and socioeconomic development, etc., into account. In this study, we evaluated the synergistic impact of rice cropping systems on social economics and ecology and found that PR, FR and WR generated 348, 157 and 133 USD/ha externality, respectively. Although we found that rotations helped farmers to generate more profits, they also need to additionally invest in equipment required by different crops. Governments could use some of the externalities generated by rotation to top up the economic benefit of farmers for adopting these cropping systems. In the long term, a subsidy policy can ensure food safety and the protection of ecosystem services [58]. Externalities and sunk costs are an important basis for making agricultural policies; therefore, the inclusion of an externalities subsidy policy was also recommended for ecological production of crops [31]. However, economic policy-makers should evaluate the threshold of the subsidy according to the ecological and social benefits of the practices.

5. Conclusions

The overemphasis of farmers on direct output leads to a significant knowledge gap among farmers, governments and researchers [59] and unsustainable socioeconomic systems. This problem could be overcome by creating a dynamic economic policy for the adoption of more reasonable cropping systems by taking into account production externalities [60]. Adopting a cropping system with high positive externalities (ecological and social benefits) would increase natural resource use efficiency and social welfare [61]. Regarding rice production, we showed that yields, profits, benefits and externalities varied significantly among cropping strategies. Paddy-upland rotations, especially PR, showed a clear advantage over successive rice cropping and created substantially positive externalities. Some of the externalities could be directed back to farmers through a subsidy system to compensate their additional investments for equipment. Therefore, externalities and sunk costs should be considered in policy making. The internalization of externalities could be achieved by three ways: (1) cultivation intensification and/or technological advances, such as the precise management of water and fertilizer to increase per unit yield, (2) the appropriate dissemination of information regarding ecological practices and an improvement to the information symmetry of public and private stakeholders, including producers, consumers and material supply services, and (3) the provision of a sufficient subsidy to increase farmers' income to encourage farmers to adopt rational cropping systems.

Supplementary Materials: The following are available online at https://www.mdpi.com/2077-0472/11/2/91/s1, Table S1: The indicators and weights of the benefit assessment for different cropping systems.

Author Contributions: Conceptualization, Z.-B.C. and J.Z.; data curation, D.-C.H. and Z.-Z.L.; formal analysis, D.-C.H., Y.-L.M., Z.-Z.L. and J.Z.; funding acquisition, D.-C.H.; investigation, D.-C.H., C.-S.Z. and Y.-L.M.; methodology, D.-C.H. and Y.-L.M.; software, Y.-L.M. and C.-S.Z.; project administration, C.-S.Z.; writing—original draft preparation, D.-C.H.; writing—review and editing, D.-C.H., Z.-B.C., J.Z. and C.-S.Z.; supervision, Z.-B.C. and J.Z. All authors have read and agreed to the published version of the manuscript.

Funding: This research was funded by the National Natural Science Foundation of China, grant number 72073028, the Natural Science Foundation of Fujian province, China, grant number 2018J01707, and the Scientific Innovation Foundation of Fujian Agriculture and Forestry University, China, grant number CXZX2018100.

Institutional Review Board Statement: Not applicable.

Informed Consent Statement: Not applicable.

Data Availability Statement: The data presented in this study are available on request from the corresponding author.

Acknowledgments: The authors are grateful to the assistance of scientists in Fujian Agriculture and Forestry University, the Institute of Soil and Fertilizer of Fujian Academy of Agricultural Sciences, and the Soil and Fertilizer Station of Fuqing City in Fujian Province, China, in data collection and analysis. We thank Zi-Shuai Wang for his assistance with data collection and experiment implementation. We also thank the experts and farmers for returning the questionnaires and the anonymous reviewers for their valuable comments on the manuscript.

Conflicts of Interest: The authors declare no conflict of interest. The funders had no role in the design of the study; in the collection, analyses, or interpretation of data; in the writing of the manuscript, or in the decision to publish the results.

References

1. Burdon, J.J.; Barrett, L.G.; Yang, L.N.; He, D.C.; Zhan, J. Maximizing world food production through disease control. *BioScience* **2020**, *70*, 126–128. [CrossRef]
2. Ray, D.K.; West, P.C.; Clark, M.; Gerber, J.S.; Prishchepov, A.V.; Chatterjee, S. Climate change has likely already affected global food production. *PLoS ONE* **2019**, *14*, e0217148. [CrossRef]
3. Liu, M.; Pan, T.; Allakhverdiev, S.I.; Yu, M.; Shabala, S. Crop halophytism: An environmentally sustainable solution for global food security. *Trends Plant Sci.* **2020**, *25*, 630–634. [CrossRef]
4. Yildiz, T. An input-output energy analysis of wheat production in Carsamba district of Samsun province. *J. Agric. Fac. Gaziosmanpasa Univ.* **2016**, *33*, 10–20. [CrossRef]
5. Schepers, H.; Cooke, L. Potato Production in Northern and Western Europe. In *Fungicide Resistance in Plant Pathogens: Principles and a Guide to Practical Management*; Ishii, H., Hollomon, D.W., Eds.; Springer: Tokyo, Japan, 2015; pp. 237–278.
6. Gullino, M.L.; Garibaldi, A. Disease of roses: Evolution of problems and new approaches for their control. *Acta Hortic.* **1996**, *422*, 195–201. [CrossRef]
7. Köller, W.; Wilcox, W.F. Evidence for the predisposition of fungicide-resistant Isolates of Venturia inaequalis to a preferential selection for resistance to other fungicides. *Phytopathology* **2001**, *91*, 776–781. [CrossRef]
8. Chatzidimopoulos, M.; Lioliopoulou, F.; Sotiropoulos, T.; Vellios, E. Efficient Control of Apple Scab with Targeted Spray Applications. *Agronomy* **2020**, *10*, 217. [CrossRef]
9. Moradi, M.; Nematollahi, M.A.; Mousavi Khaneghah, A.; Pishgar-Komleh, S.H.; Rajabi, M.R. Comparison of energy consumption of wheat production in conservation and conventional agriculture using DEA. *Environ. Sci. Pollut. Res. Int.* **2018**, *25*, 35200–35209. [CrossRef]
10. Yang, L.N.; Pan, Z.C.; Zhu, W.; Wu, E.J.; He, D.C.; Yuan, X.; Qin, Y.Y.; Wang, Y.; Chen, R.S.; Thrall, P.H.; et al. Enhanced agricultural sustainability through within-species diversification. *Nat. Sustain.* **2019**, *2*, 46–52. [CrossRef]

11. Francaviglia, R.; Álvaro-Fuentes, J.; Di Bene, C.; Gai, L.; Regina, K.; Turtola, E. Diversified arable cropping systems and management schemes in selected European regions have positive effects on soil organic carbon content. *Agriculture* **2019**, *9*, 261. [CrossRef]
12. Döring, T.F.; Annicchiarico, P.; Clarke, S.; Haigh, Z.; Jones, H.E.; Pearce, H.; Snape, J.; Zhan, J.; Wolfe, M.S. Comparative analysis of performance and stability among composite cross populations, variety mixtures and pure lines of winter wheat in organic and conventional cropping systems. *Field Crops Res.* **2015**, *183*, 235–245. [CrossRef]
13. Hondrade, R.F.; Hondrade, E.; Zheng, L.; Elazegui, F.; Duque, J.L.J.E.; Mundt, C.C.; Cruz, C.M.V.; Garrett, K.A. Cropping system diversification for food production in Mindanao rubber plantations: A rice cultivar mixture and rice intercropped with mungbean. *PeerJ* **2017**, *5*, e2975. [CrossRef]
14. Sogoba, B.; Traoré, B.; Safia, A.; Samaké, O.B.; Dembélé, G.; Diallo, S.; Kaboré, R.; Benié, G.B.; Zougmoré, R.B.; Goïta, K. On-farm evaluation on yield and economic performance of cereal-cowpea intercropping to support the smallholder farming system in the Soudano-Sahelian zone of mali. *Agriculture* **2020**, *10*, 214. [CrossRef]
15. Tamburini, G.; Bommarco, R.; Wanger, T.C.; Kremen, C.; van der Heijden, M.G.A.; Liebman, M.; Hallin, S. Agricultural diversification promotes multiple ecosystem services without compromising yield. *Sci. Adv.* **2020**, *6*, eaba1715. [CrossRef]
16. Cardinale, B.J.; Duffy, J.E.; Gonzalez, A.; Hooper, D.U.; Perrings, C.; Venail, P.; Narwani, A.; Mace, G.M.; Tilman, D.; Wardle, D.A.; et al. Biodiversity loss and its impact on humanity. *Nature* **2012**, *486*, 59–67. [CrossRef]
17. Wagg, C.; Bender, S.F.; Widmer, F.; van der Heijden, M.G. Soil biodiversity and soil community composition determine ecosystem multifunctionality. *Proc. Natl. Acad. Sci. USA* **2014**, *111*, 5266–5270. [CrossRef]
18. Renard, D.; Tilman, D. National food production stabilized by crop diversity. *Nature* **2019**, *571*, 257–260. [CrossRef]
19. He, D.C.; Zhan, J.S.; Xie, L.H. Problems, challenges and future of plant disease management: From an ecological point of view. *J. Integr. Agric.* **2016**, *15*, 705–715. [CrossRef]
20. Muehe, E.M.; Wang, T.; Kerl, C.F.; Planer-Friedrich, B.; Fendorf, S. Rice production threatened by coupled stresses of climate and soil arsenic. *Nat. Commun.* **2019**, *10*, 4985. [CrossRef]
21. Tunc, T.; Sahin, U.; Evren, S.; Dasci, E.; Guney, E.; Aslantas, R.J.S.H. The deficit irrigation productivity and economy in strawberry in the different drip irrigation practices in a high plain with semi-arid climate. *Sci. Hortic.* **2019**, *245*, 47–56. [CrossRef]
22. Wehmeyer, H.; de Guia, A.H.; Connor, M. Reduction of fertilizer use in South China—impacts and implications on smallholder rice farmers. *Sustainability* **2020**, *12*, 2240. [CrossRef]
23. Azman Halimi, R.; Barkla, B.J.; Andrés-Hernandéz, L.; Mayes, S.; King, G.J. Bridging the food security gap: An information-led approach to connect dietary nutrition, food composition and crop production. *J. Sci. Food Agric.* **2020**, *100*, 1495–1504. [CrossRef] [PubMed]
24. Wang, D.; Huang, J.; Nie, L.; Wang, F.; Ling, X.; Cui, K.; Li, Y.; Peng, S. Integrated crop management practices for maximizing grain yield of double-season rice crop. *Sci. Rep.* **2017**, *7*, 38982. [CrossRef] [PubMed]
25. Schneider, M.; Pereira, É.R.; Castilho, I.N.B.; Carasek, E.; Welz, B.; Martens, I.B.G. A simple sample preparation procedure for the fast screening of selenium species in soil samples using alkaline extraction and hydride-generation graphite furnace atomic absorption spectrometry. *Microchem. J.* **2016**, *125*, 50–55. [CrossRef]
26. Gholizadeh, A.; Borůvka, L.; Saberioon, M.; Vašát, R. Visible, near-infrared, and mid-infrared spectroscopy applications for soil assessment with emphasis on soil organic matter content and quality: State-of-the-art and key issues. *Appl. Spectrosc.* **2013**, *67*, 1349–1362. [CrossRef]
27. Yang, J.S.; Li, P.; Ding, Y. Comparison with determing methods of organic matter for sludge compost in different treatments. *Appl. Mech. Mater.* **2012**, *1802*, 1070–1074. [CrossRef]
28. Iatrou, M.; Papadopoulos, A.; Papadopoulos, F.; Dichala, O.; Psoma, P.; Bountla, A. Determination of soil available phosphorus using the olsen and mehlich 3 methods for greek soils having variable amounts of calcium carbonate. *Commun. Soil Sci. Plant Anal.* **2014**, *45*, 2207–2214. [CrossRef]
29. Ogunbileje, J.O.; Sadagoparamanujam, V.M.; Anetor, J.I.; Farombi, E.O.; Akinosun, O.M.; Okorodudu, A.O. Lead, mercury, cadmium, chromium, nickel, copper, zinc, calcium, iron, manganese and chromium (VI) levels in Nigeria and United States of America cement dust. *Chemosphere* **2013**, *90*, 2743–2749. [CrossRef]
30. Stenberg, B.; Viscarra Rossel, R.A.; Mouazen, A.M.; Wetterlind, J. Visible and near infrared spectroscopy in soil science. *Adv. Agron.* **2010**, *107*, 163–215.
31. Zheng, R.; Zhan, J.; Liu, L.; Ma, Y.; Wang, Z.; Xie, L.; He, D. Factors and minimal subsidy associated with tea farmers' willingness to adopt ecological pest management. *Sustainability* **2019**, *11*, 6190. [CrossRef]
32. Suma, P.; Chinnici, G.; La Pergola, A.; Russo, A.; Bella, S.; Pecorino, B.; Pappalardo, G. Assessing the technical effectiveness and economic feasibility of pest management through structural heat treatment: An entomological and economic analysis in four mills in Sicily (Italy). *J. Econ. Entomol.* **2019**, *112*, 957–962. [CrossRef] [PubMed]
33. Janssens de Bisthoven, L.; Vanhove, M.P.M.; Rochette, A.J.; Hugé, J.; Verbesselt, S.; Machunda, R.; Munishi, L.; Wynants, M.; Steensels, A.; Malan-Meerkotter, M.; et al. Social-ecological assessment of lake manyara basin, tanzania: A mixed method approach. *J. Environ. Manag.* **2020**, *267*, 110594. [CrossRef] [PubMed]
34. De Marinis, P.; Sali, G. Participatory analytic hierarchy process for resource allocation in agricultural development projects. *Eval. Program Plan.* **2020**, *80*, 101793. [CrossRef] [PubMed]

35. Eliazar, I.; Metzler, R.; Reuveni, S. Gumbel central limit theorem for max-min and min-max. *Phys. Rev. E* **2019**, *100*, 020104. [CrossRef]
36. Sun, L.; Deng, J.; Fan, C.; Li, J.; Liu, Y. Combined effects of nitrogen fertilizer and biochar on greenhouse gas emissions and net ecosystem economic budget from a coastal saline rice field in southeastern China. *Environ. Sci. Pollut. Res. Int.* **2020**, *27*, 17013–17022. [CrossRef]
37. Costa, M.P.; Chadwick, D.; Saget, S.; Rees, R.M.; Styles, D. Representing crop rotations in life cycle assessment: A review of legume lca studies. *Int. J. Life Cycle Assess.* **2020**, *25*, 1–15. [CrossRef]
38. Laik, R.; Sharma, S.; Idris, M.; Singh, A.K.; Singh, S.S.; Bhatt, B.P.; Saharawat, Y.; Humphreys, E.; Ladha, J.K. Integration of conservation agriculture with best management practices for improving system performance of the rice–wheat rotation in the Eastern Indo-Gangetic Plains of India. *Agric. Ecosyst. Environ.* **2014**, *195*, 68–82. [CrossRef]
39. Meetei, T.T.; Kundu, M.C.; Devi, Y.B. Long-term effect of rice-based cropping systems on pools of soil organic carbon in farmer's field in hilly agroecosystem of Manipur, India. *Environ. Monit. Assess.* **2020**, *192*, 209. [CrossRef]
40. Liang, S.M.; Ren, C.; Wang, P.J.; Wang, X.T.; Li, Y.S.; Xu, F.H.; Wang, Y.; Dai, Y.Q.; Zhang, L.; Li, X.P.; et al. Improvements of emergence and tuber yield of potato in a seasonal spring arid region using plastic film mulching only on the ridge. *Field Crops Res.* **2018**, *223*, 57–65. [CrossRef]
41. Wheeler, R.M.; Fitzpatrick, A.H.; Tibbitts, T.W. Potatoes as a Crop for Space Life Support: Effect of CO (2), Irradiance, and Photoperiod on Leaf Photosynthesis and Stomatal Conductance. *Front. Plant Sci.* **2019**, *10*, 1632. [CrossRef]
42. Zhao, J.; Zhan, X.; Jiang, Y.; Xu, J. Variations in climatic suitability and planting regionalization for potato in northern China under climate change. *PLoS ONE* **2018**, *13*, e0203538. [CrossRef] [PubMed]
43. Kang, S.W.; Seo, D.C.; Kim, S.Y.; Cho, J.S. Utilization of liquid pig manure for resource cycling agriculture in rice-green manure crop rotation in South Korea. *Environ. Monit. Assess.* **2020**, *192*, 323. [CrossRef] [PubMed]
44. Gao, S.; Zhou, G.; Rees, R.M.; Cao, W. Green manuring inhibits nitrification in a typical paddy soil by changing the contributions of ammonia-oxidizing archaea and bacteria. *Appl. Soil. Ecol.* **2020**, *156*, 103698. [CrossRef]
45. Sawe, T.; Eldegard, K.; Totland, Ø.; Macrice, S.; Nielsen, A. Enhancing pollination is more effective than increased conventional agriculture inputs for improving watermelon yields. *Ecol. Evol.* **2020**, *10*, 5343–5353. [CrossRef] [PubMed]
46. Brown, L.K.; Kazas, C.; Stockan, J.; Hawes, C.; Stutter, M.; Ryan, C.M.; Squire, G.R.; George, T.S. Is Green Manure from Riparian Buffer Strip Species an Effective Nutrient Source for Crops? *J. Environ. Qual.* **2019**, *48*, 385–393. [CrossRef]
47. Ai, Y.J.; Li, F.P.; Gu, H.H.; Chi, X.J.; Yuan, X.T.; Han, D.Y. Combined effects of green manure returning and addition of sewage sludge compost on plant growth and microorganism communities in gold tailings. *Environ. Sci. Pollut. Res. Int.* **2020**, *27*, 31686–31698. [CrossRef]
48. Kumar, R.; Mishra, J.S.; Rao, K.K.; Mondal, S.; Hazra, K.K.; Choudhary, J.S.; Hans, H.; Bhatt, B.P. Crop rotation and tillage management options for sustainable intensification of rice-fallow agro-ecosystem in eastern India. *Sci. Rep.* **2020**, *10*, 11146. [CrossRef]
49. Ray, K.; Sen, P.; Goswami, R.; Sarkar, S.; Brahmachari, K.; Ghosh, A.; Nanda, M.K.; Mainuddin, M. Profitability, energetics and GHGs emission estimation from rice-based cropping systems in the coastal saline zone of West Bengal, India. *PLoS ONE* **2020**, *15*, e0233303. [CrossRef]
50. Khan, F.; Hussain, S.; Tanveer, M.; Khan, S.; Hussain, H.A.; Iqbal, B.; Geng, M. Coordinated effects of lead toxicity and nutrient deprivation on growth, oxidative status, and elemental composition of primed and non-primed rice seedlings. *Environ. Sci. Pollut. Res. Int.* **2018**, *25*, 21185–21194. [CrossRef]
51. Yang, J.; Zhou, Q.; Zhang, J. Moderate wetting and drying increases rice yield and reduces water use, grain arsenic level, and methane emission. *Crops J.* **2017**, *5*, 151–158. [CrossRef]
52. Francaviglia, R.; Álvaro-Fuentes, J.; Di Bene, C.; Gai, L.; Regina, K.; Turtola, E. Diversification and Management Practices in Selected European Regions. A Data Analysis of Arable Crops Production. *Agronomy* **2020**, *10*, 297. [CrossRef]
53. Hou, P.; Yu, Y.; Xue, L.; Petropoulos, E.; He, S.; Zhang, Y.; Pandey, A.; Xue, L.; Yang, L.; Chen, D. Effect of long term fertilization management strategies on methane emissions and rice yield. *Sci. Total Environ.* **2020**, *725*, 138261. [CrossRef] [PubMed]
54. Nguyen Van, H.; Sander, B.O.; Quilty, J.; Balingbing, C.; Castalone, A.G.; Romasanta, R.; Alberto, M.C.R.; Sandro, J.M.; Jamieson, C.; Gummert, M. An assessment of irrigated rice production energy efficiency and environmental footprint with in-field and off-field rice straw management practices. *Sci. Rep.* **2019**, *9*, 16887. [CrossRef] [PubMed]
55. Yanakittkul, P.; Aungvaravong, C. A model of farmers intentions towards organic farming: A case study on rice farming in Thailand. *Heliyon* **2020**, *6*, e03039. [CrossRef]
56. Zeng, Y.; Zhang, J.; He, K.; Cheng, L. Who cares what parents think or do? Observational learning and experience-based learning through communication in rice farmers' willingness to adopt sustainable agricultural technologies in Hubei Province, China. *Environ. Sci. Pollut. Res. Int.* **2019**, *26*, 12522–12536. [CrossRef]
57. Ito, S. Contemporary global rice economies: Structural changes of rice production, consumption and trade. *J. Nutr. Sci. Vitaminol.* **2019**, *65*, S23–S25. [CrossRef]
58. Chang, Y.; Zhang, Z.; Yoshino, K.; Zhou, S. Farmers' tea and nation's trees: A framework for eco-compensation assessment based on a subjective-objective combination analysis. *J. Environ. Manag.* **2020**, *269*, 110775. [CrossRef]
59. Senapati, A.K. Evaluation of risk preferences and coping strategies to manage with various agricultural risks: Evidence from India. *Heliyon* **2020**, *6*, 6. [CrossRef]

60. Cui, Z.; Zhang, H.; Chen, X.; Zhang, C.; Ma, W.; Huang, C.; Zhang, W.; Mi, G.; Miao, Y.; Li, X.; et al. Pursuing sustainable productivity with millions of smallholder farmers. *Nature* **2018**, *555*, 363–366. [CrossRef]
61. Caicedo Solano, N.E.; García Llinás, G.A.; Montoya-Torres, J.R. Towards the integration of lean principles and optimization for agricultural production systems: A conceptual review proposition. *J. Sci. Food Agric.* **2020**, *100*, 453–464. [CrossRef]

 *agriculture*

MDPI

Article

Crop Diversification in Viticulture with Aromatic Plants: Effects of Intercropping on Grapevine Productivity in a Steep-Slope Vineyard in the Mosel Area, Germany

Felix Dittrich [1], Thomas Iserloh [2], Cord-Henrich Treseler [3], Roman Hüppi [4], Sophie Ogan [5], Manuel Seeger [2] and Sören Thiele-Bruhn [1,*]

1 Department of Soil Science, Trier University, 54296 Trier, Germany; dittrich@uni-trier.de
2 Department of Physical Geography, Trier University, 54296 Trier, Germany; iserloh@uni-trier.de (T.I.); seeger@uni-trier.de (M.S.)
3 Winery Dr. Frey, 54441 Kanzem, Germany; flaschenpost@weingutdrfrey.de
4 Department of Environmental Systems Science, ETH Zürich, 8092 Zürich, Switzerland; roman.hueppi@usys.ethz.ch
5 Department of Biogeography, Trier University, 54296 Trier, Germany; ogan@uni-trier.de
* Correspondence: thiele@uni-trier.de

Abstract: The effects of intercropping grapevine with aromatic plants are investigated using a multi-disciplinary approach. Selected results are presented that address the extent to which crop diversification by intercropping impacts grapevine yield and must quality, as well as soil water and mineral nutrients (NO_3-N, NH_4-N, plant-available K and P). The experimental field was a commercial steep-slope vineyard with shallow soils characterized by a high presence of coarse rock fragments in the Mosel area of Germany. The field experiment was set up as randomized block design. Rows were either cultivated with Riesling (*Vitis vinifera* L.) as a monocrop or intercropped with *Origanum vulgare* or *Thymus vulgaris*. Regarding soil moisture and nutrient levels, the topsoil (0–0.1 m) was more affected by intercropping than the subsoil (0.1–0.3 m). Gravimetric moisture was consistently lower in the intercropped topsoil. While NO_3-N was almost unaffected by crop diversification, NH_4-N, K, and P were uniformly reduced in topsoil. Significant differences in grapevine yield and must quality were dominantly attributable to climate variables, rather than to the treatments. Yield stabilization due to intercropping with thyme and oregano seems possible with sufficient rainfall or by irrigation. The long-term effects of intercropping on grapevine growth need further monitoring.

Keywords: perennial cropping systems; grape production; medicinal and aromatic plants; grapevine yield; must quality; experimental design

Citation: Dittrich, F.; Iserloh, T.; Treseler, C.-H.; Hüppi, R.; Ogan, S.; Seeger, M.; Thiele-Bruhn, S. Crop Diversification in Viticulture with Aromatic Plants: Effects of Intercropping on Grapevine Productivity in a Steep-Slope Vineyard in the Mosel Area, Germany. *Agriculture* **2021**, *11*, 95. https://doi.org/10.3390/agriculture11020095

Academic Editors: Claudia Di Bene, Rosa Francaviglia, Roberta Farina, Jorge Álvaro-Fuentes and Raúl Zornoza

Received: 17 December 2020
Accepted: 20 January 2021
Published: 23 January 2021

1. Introduction

Grapevine (*Vitis vinifera* L.) cultivation covers 7.4 million hectares worldwide and has reached a high degree of agronomic specialization [1]. Producers that exclusively cultivate grapevine face increasing economic risks, as climate change may impact vineyard productivity [2]. In addition, many producers strive for a reduction of adverse environmental impacts such as soil degradation, biodiversity decline, and contamination of groundwater and surface water caused by intensive and eventually non-sustainable management practices (i.e., frequent tillage, and intensive fertilizer and pesticide use) [3–8]. Agricultural diversification has been proposed to combine both economic and environmental sustainability, and can be realized by an increase of the crop species diversity (e.g., by intercropping or crop rotation) or noncrop (e.g., by cover-cropping or hedgerows) [9–11]. In viticulture, an increase in plant species diversity, abundance, and soil cover is implemented by the use of cover crops, and this has been frequently reported to mitigate environmental impacts [5,12–15]. Cover crops provide several services for the vineyard ecosystem: protection from soil erosion, water purification, nutrient retention, and improved soil structure, and thus, enhanced water infiltration, increased soil quality, above- and below-ground biological diversity, and a significant contribution to weed, pest, and disease control [6,16–19].

Some authors have also considered regulative effects on grapevine growth as service [12,13], since competition between grapevines and cover crops for soil resources may have beneficial effects on grape yield and quality indices. For example, competition has been shown to limit vigor and vegetative growth, resulting in reduced canopy density, pest incidence, and berry size with increased must quality [12,18]. On the other hand, cover cropping can exert undesired disservices through severe competition or the provision of habitats for pests and pathogens, with significant reductions of grape yield and quality [14,20]. A proper choice and management of cover crops is therefore critical to facilitate services while preventing disservices. Beside the technical and pedoclimatic context, economic concerns (i.e., the risk of lower yields, missing short-term returns, and extra costs for managing cover crops) are most decisive, and limit a systematic adoption to variable spatiotemporal extents (i.e., from alternating row to complete respectively temporary to permanent cover) [15,21,22].

In vineyards, predominantly inter-rows are cover-cropped with purposely seeded or resident species, whereas the grapevine row (i.e., the space underneath and close to the grapevine plants) is still most commonly kept free of vegetation by mechanical or chemical means in order to prevent severe competition and diseases [8,23]. As a result, 20 to 25% of the total vineyard surface (assuming a 2 m row distance and a 0.4–0.5 m row width) remains uncovered, and constitute linear structures that are especially prone to soil erosion. On the other hand, this vacant space bears excellent options for the cultivation of other crops.

Aromatic plants have not yet been considered as viable intercropping option for vineyards, though characteristic traits (e.g., perennial, flat-growing, shade-tolerant and adopted to dry and warm pedoclimatic conditions) and increasing economic demands for products derived from aromatic plants make them suitable to combine short-term returns with environmental benefits [24]. Agronomic cultivation handbooks for aromatic plants describe a low to moderate need for soil resources, plant heights of 0.3 up to 1.0 m (during blossom), and profitable cultivation periods of five to 10 years with up to two harvest cuts per year under favorable climatic conditions [25,26]. Like grapevines, aromatic plants synthesize considerable amounts of secondary metabolites (in response to abiotic and biotic stress), and harvested plant materials, either raw or processed, provide various application possibilities in food, pharmaceutical, and cosmetics industries [24,27], and may act as novel agents for plant protection [28]. Aromatic plants are successfully grown as lower-strata species in multistrata agroforestry systems (e.g., orchards) [29], and can substantially contribute to ecosystem services such as biodiversity and habitat quality, pest and disease control, aesthetical land valorization, soil erosion control, and enhanced resource-use efficiency [27,30,31]. Some aromatic plant species are also capable of tolerating adverse environmental conditions, and have been suggested to be cultivated on marginal (i.e., contaminated, eroded, and moisture-deficient) soils [32–34]. These attributes are applicable to a wide range of vineyards, as they are frequently located on medium to steep slopes where intensive management has led to severe soil erosion and contamination (e.g., with Cu-based fungicides) [3,4,35].

This work has been initiated because, to the best of our knowledge, no specific study has investigated the effects of intercropping grapevine and aromatic plants under field conditions. We define the grapevine row as a valuable production area, where a permanent cultivation of additional, marketable crops (i.e., intercropping) and the concomitant omission of tillage may have profound effects on the overall vineyard productivity, the provision of ecosystem services, and above- and below-ground biodiversity. In this article, we present our experimental design and the development of grapevine yields and must quality, as well as soil water and plant-available nutrient levels over three years after implementing aromatic plants in a steep-slope vineyard. With respect to grapevine productivity, we aim to evaluate impacts of intercropping to assess its potential as appropriate diversification measure in vineyards. To this end, the effects of intercropping underneath grapevines using two aromatic plants (oregano and thyme) on the selected properties of grapevine yield and soil were investigated. Diversified cropping was compared to regular cultivation as a control that goes along with bare soil underneath grapevines.

2. Materials and Methods

2.1. Study Site

The experimental field is a commercial vineyard ('Wawerner Jesuitenberg') in the Mosel area of Germany (Figure 1) that is managed according to organic principles. Standard cultural practices encompass mulching, harrowing of grapevine rows, organic fertilization, and plant protection with Cu-based compounds. According to the Köppen classification, the climate is temperate oceanic, and the mean annual temperature, precipitation, and potential evapotranspiration are 9.1 °C, 722 mm, and 687 mm, respectively (www.am.rlp.de; meteorological station 'Kanzem'). Grapevines (*Vitis vinifera* L. cv. 'Riesling') were grafted on SO4 rootstocks and established in 2008 using wire-framed rows oriented along the slope. The spacing is 2 m between rows and 1 m within rows. The south-exposed vineyard plot has a size of 0.3 ha and a steep inclination (~45%), and has developed from Devonian argillaceous schist (Hunsrück Devonian strata [36]), as well as Pleistocene terrace sediments. Prior to grapevine planting, soil melioration by deep cultivation and amendments of organic and mineral origin modified the initial soil properties. The shallow (<0.5 m) and highly permeable (mean K_f-values in 2019: 2.5×10^{-5} ms^{-1}) soil profile is characterized by a high presence of coarse rock fragments (>50%), mainly ranging from 2 to 20 mm. The fine soil (<2 mm) shows a slightly acidic reaction (6.6 in $CaCl_2$, 1:2.5) and has a sandy loamy texture, and is composed of 60% sand, 25% silt, and 15% clay. A continuous supply of organic matter, via organic fertilization and mulching (pruning residues, cover crops), established a distinct topsoil horizon (0–0.1 m) that is enriched in soil organic carbon (SOC = 3.1%) and shows an effective cation exchange capacity (ECEC) of 12.1 cmol/kg^{-1}. SOC and ECEC in the subsoil horizon (0.1–0.3 m) are 2.2% and 9.0 cmol/kg^{-1}, respectively. According to the world reference base for soil resources, the soil is classified as Eutric Skeletic Regosol (Aric, Humic) [37].

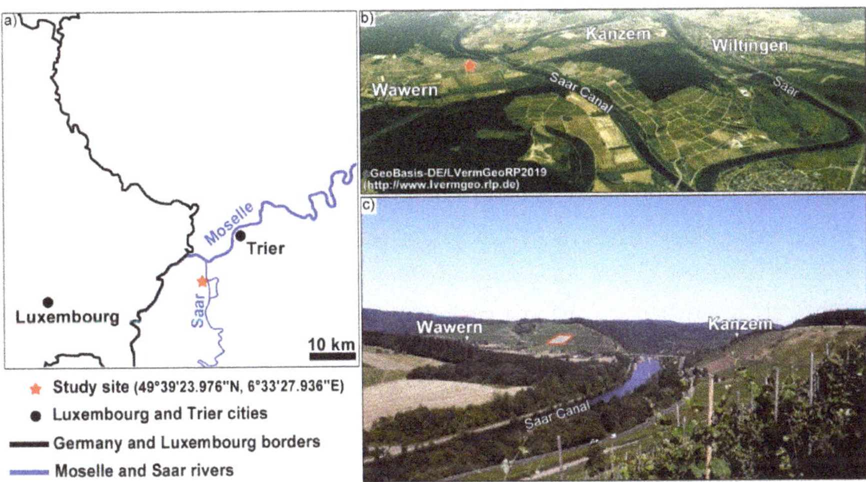

Figure 1. Map indicating the location of the study site (**a**), as well as an aerial view (**b**), and a photo of the study area close to the Saar Canal (**c**).

2.2. Experimental Design

The field experiment was set up as randomized block design (Figure 2) with three blocks, each consisting of two grapevine rows per treatment:

- Control (*Vitis vinifera* L. cv. 'Riesling' monocrop with regular mechanical tillage),
- Oregano (*Vitis vinifera* L. cv. 'Riesling' intercropped with *Origanum vulgare*),
- Thyme (*Vitis vinifera* L. cv. 'Riesling' intercropped with *Thymus vulgaris*).

In May 2018, aromatic plants were manually planted in one row per block as seedlings. The plant material was obtained from Pharma Saat GmbH (www.pharmasaat.de), and the soil was prepared using hand-held tools. A plant density of four (oregano) and five (thyme) seedlings between two grapevines was chosen to achieve proper soil cover for weed suppression and soil erosion control. In April 2019, a further implementation of aromatic plants in the second row per treatment was conducted. The intercropped rows were occasionally irrigated with ~2.6 L/m of grapevine row in 2018 and 2019, in order to prevent withering of seedlings. The total amount of supplied water was 2340 L for each intercropping treatment in 2018 (five applications starting from the planting date until the end of August) and 1400 L in 2019 (three applications in July/August). An evaluation of the performance of aromatic plants without associated grapevines was carried out in a nearby field that was well prepared prior to planting.

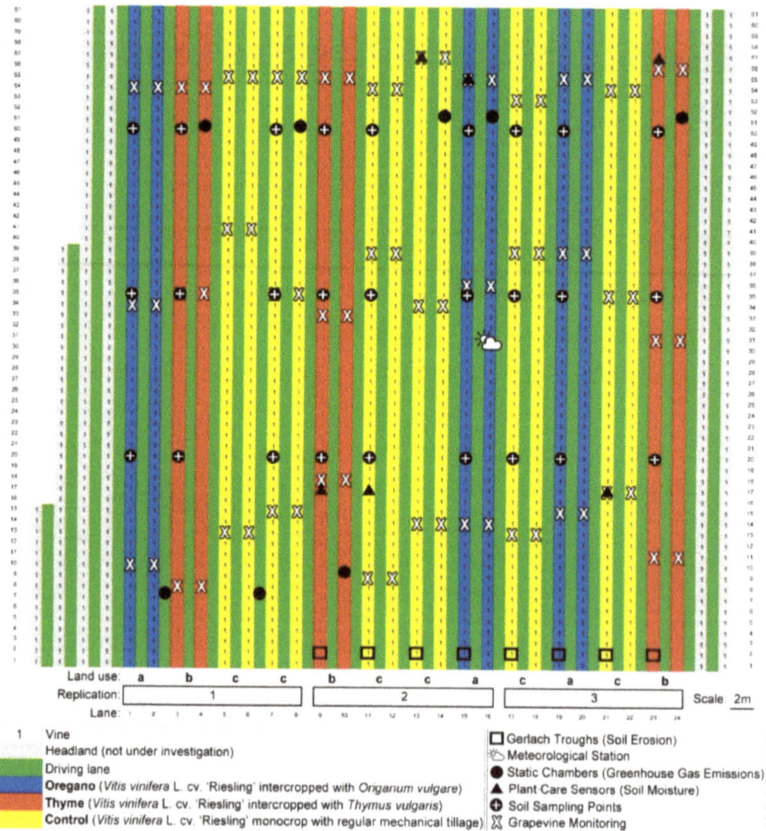

Figure 2. Experimental set-up showing grapevine rows colored according to treatments, and the positioning of monitoring equipment.

2.3. Crop Monitoring

Six grapevines per block and treatment were selected to monitor grapevine performance, yield, and must quality (Figure 2). For each grapevine, the total weight of grapes, the weight of selected berry clusters, and the number of produced clusters were quantified shortly prior to harvest. In addition, the total grape yields per row were determined and expressed per hectare. Must quality was evaluated using pH value (with a digital pH-meter, GPRT 1400 AN, Greisinger, Regenstauf, Germany) titratable acidity (in g/L with Neustädter titration cylinder; Wagner, Merkel, Sulfacor), and the concentration of total soluble solids, an indirect measure of sugar content (measured optically with a refractometer and expressed as °Brix). The incidence of fungal diseases on grape leaves and berries were visually assessed. The vegetative development of aromatic plants was assessed close to monitored grapevines by measuring plant height, length, and width. The aromatic plant root biomass was determined at the end of the experiment in 2020 on a total number of nine individuals per species that were planted in 2018 to estimate their below-ground impact.

2.4. Vineyard Soil Sampling and Analytical Protocols

A comprehensive monitoring of vineyard soil quality in both topsoil and subsoil (0–0.1 and 0.1–0.3 m, respectively) began in October 2018. Sampling was confined to rows diversified in May 2018. A minimum of two samples per block and treatment was continuously taken close (<0.1 m) to the grapevine row, at the beginning and the end of the crop cycle until October 2020 (see Figure 2). Coarse rock fragments were removed from the surface prior to sampling. Gravimetric soil moisture was determined by weighing before and after drying at 105 °C for 24 h. Plant-available forms of potassium (K) and phosphorus (P) were extracted using the calcium-acetate-lactate (CAL) method [38] and expressed as K and P. Briefly, 5 g of air-dried and sieved (<2 mm) soil was agitated for 2 h with 100 mL of CAL solution and subsequently filtered. Filtrates were quantified for K using a flame AAS (SpectrAA-10 Varian, USA) and P using a photospectrometric approach (λ = 710 nm; Shimadzu UV-1650 PC, Shimadzu, Japan) with ammonium molybdate as staining reagent. Ammonium (NH_4-N) and nitrate (NO_3-N) were simultaneously quantified using a continuous segmented flow analyzer (Bran+Luebbe GmbH, Norderstedt, Germany). Briefly, 5 g of frozen, field moist aliquots of aforementioned samples was agitated with 40 mL of 2 m KCl for 1 h and filtrated prior to analysis.

2.5. Soil Erosion Measurements

Continuous soil erosion measurements were conducted at the bottom of each treatment [39]. Gerlach troughs were built, installed, and utilized as sediment collectors [40]. These open soil-erosion plots give information about soil losses, but the contributing area is not defined and may be variable. Consequently, soil-erosion results are shown in kg m^{-1} of slope width. The sediment output of a definable field section was measured under real agricultural conditions. The collected material provided basic data on the transported grain sizes and nutrients of the particular field [41].

2.6. Statistical Analysis

Data analyses were conducted using the R statistical package version 3.3.2. [42]. Yield and must quality differences over time were assessed using one-way analysis of variance (ANOVA), applying a repeated measures design, followed by a Tukeys HSD post-hoc test. One-way ANOVA was used to differentiate between experimental treatments of individual years. In addition, the separation of grapevine yield and quality indices was tested using principal component analysis. Soil moisture and nutrient levels were evaluated using two-way ANOVA with time and depth as focal variables. For given sampling dates and depths, treatment effects were determined using one-way ANOVA. Values were considered significantly different at p-values < 0.05. Prior to analysis, normality and homoscedasticity were assessed using Shapiro-Wilk and Levene's test.

3. Results

3.1. Climatic Phenomena and Effects on Intercrop Growth

Generally, the experimental period was drier than long-term average values, and precipitation sums from calendar quarter (Q) I to III were lowest in 2019 (Table 1). However, considering QI to III, a distinct seasonal variability of precipitation was observed between years: the highest precipitation sums were recorded in QII of 2018, in QIII of 2019, and QI of 2020.

Table 1. Precipitation patterns and vegetation days (TØ > = 5 °C) over the experimental period (2018–2020) obtained from the Kanzem meteorological station (www.am.rlp.de). Long-term precipitation sums (1990–2020 from Trier-Petrisberg, and 2005–2020 from Kanzem) and the precipitation in 2017 were included as supplementary information.

		Precipitation Sum						Vegetation Days		
		1990–2020	**2005–2020**	**2017**	**2018**	**2019**	**2020**	**2018**	**2019**	**2020**
Quarter	Months			(mm)					(days)	
I	Jan-Mar	166	153	113	207	149	265	40	51	68
II	Apr-Jun	176	189	113	243	149	128	91	89	91
III	Jul-Sep	193	187	251	124	164	100	92	92	92
IV	Oct-Dec	205	193	250	185	267	234	71	69	66
$\sum$ I-III		535	529	477	574	462	493	223	232	251
$\sum$ I-IV		740	722	727	759	729	727	294	301	317

On 1 June, 2018, a heavy rain event substantially raised the recorded precipitation sums in QII and resulted in a translocation of soil and recently planted intercrop seedlings. A rain peak of 55.6 mm was recorded from 00:00 h–00:59 h, and had its highest intensity of 117.6 mm h^{-1} from 00:05 h–00:10 h. Total soil losses were highest in grapevine rows diversified with thyme (13.7 kg m^{-1}). Substantially lower amounts of soil were collected from Gerlach troughs at the bottom of rows diversified with oregano (0.4 kg m^{-1}) and control rows (0.1 kg m^{-1}). This event and the rather dry conditions in QIII in 2018 caused a poor intercrop establishment. Re-planting of translocated and withered seedlings and manual irrigation were necessary, and increased the management intensity for the diversification treatments. However, the aromatic plant stands recovered and grew even in periods of grapevine dormancy. A steady increase in aromatic plant width and height is shown in Figure 3. The harvest was conducted at blossom, occurring approximately four weeks earlier in case of thyme, and hence restricted their vegetative growth in summer periods when air temperatures and soil water limitation peaked. Overall, oregano plants grew wider and higher in 2019 and 2020, indicating a stronger impact on soil resources. This was confirmed by a notably higher below-ground root biomass of oregano (27 $\pm$ 8 g/plant) as compared to thyme (17 $\pm$ 11 g/plant) determined at the end of the experiment in 2020.

3.2. Soil Resource Availability and Development

Gravimetric soil moisture and nutrient levels as a function of time, soil depth, and intercrop and monocrop management at given sampling dates are presented in Table 2. Overall, the topsoil (0–0.1 m) was more affected by intercropping than the subsoil (0.1–0.3 m). Soil moisture was consistently lower in the topsoil due to intercropping, and was statistically significant at the end of each crop cycle. Subsoil samples showed inconsistent effects of intercropping on gravimetric water contents: from October 2018 until October 2019; slight increases were present for oregano and thyme, followed by slight decreases.

Table 2. Mean values (± standard deviation in brackets) of soil moisture and nutrient levels as affected by time and treatments. The row total considers all observations on the respective sampling date. Numbers followed by capital letters indicate significant differences between years and depths, whereas numbers followed by lowercase letters indicate significant differences within one year between the experimental treatments. Significance was given at *p*-level < 0.05.

		Soil Moisture (wt.%)		NO_3-N (mg/kg)		NH_4-N (mg/kg)		Available K (mg/kg)		Available P (mg/kg)	
	Depth [m]	0–0.1	0.1–0.3	0–0.1	0.1–0.3	0–0.1	0.1–0.3	0–0.1	0.1–0.3	0–0.1	0.1–0.3
October 2018	Total	9.0 (±1.4) F	9.2 (±1.1) F	6.9 (±4.0) A	na	4.5 (±1.9) B	na	689 (±120) A	458 (±82) B	195 (±25) A	165 (±17) BC
	Control	10.0 (±1.1) a	9.0 (±1.1) a	9.6 (±3.7) a	na	5.0 (±1.8) a	na	689 (±57) a	508 (±63) a	212 (±23) a	177 (±15) a
	Oregano	8.6 (±1.4) b	9.1 (±0.9) a	4.7 (±2.3) b	na	4.0 (±1.7) a	na	710 (±160) a	412 (±53) b	179 (±25) b	156 (±15) b
	Thyme	8.5 (±1.2) b	9.3 (±1.3) a	6.3 (±2.9) ab	na	4.6 (±2.2) a	na	669 (±129) a	453 (±100) ab	194 (±16) ab	162 (±15) ab
April 2019	Total	17.3 (±1.9) CD	15.4 (±1.4) D	nd0	2.8 (±0.9) BC	1.3 (±0.8) DE	6.0 (±1.8) A	236 (±55) CD	240 (±45) CD	184 (±38) AB	186 (±25) AB
	Control	18.8 (±1.3) a	14.9 (±1.7) a	nd0	2.8 (±1.0) a	2.0 (±0.9) a	5.9 (±2.7) a	275 (±19) a	247 (±62) a	200 (±13) a	181 (±15) a
	Oregano	16.4 (±2.1) a	15.6 (±1.2) a	nd0	3.1 (±1.1) a	0.8 (±0.3) b	5.8 (±1.3) a	222 (±66) a	239 (±29) a	180 (±41) a	194 (±33) a
	Thyme	16.8 (±1.5) a	15.7 (±1.2) a	nd0	2.6 (±0.6) a	1.0 (±0.4) b	6.2 (±1.6) a	212 (±54) a	233 (±46) a	173 (±50) a	182 (±27) a
October 2019	Total	20.3 (±1.5) B	17.9 (±1.3) C	1.7 (±1.0) BC	3.5 (±3.4) B	0.6 (±0.6) E	0.6 (±1.3) E	244 (±41) CD	283 (±90) C	142 (±9) C	158 (±11) BC
	Control	21.5 (±1.4) a	17.8 (±1.6) a	1.1 (±0.4) a	3.6 (±3.8) a	0.8 (±0.8) a	0.1 (±0.2) a	293 (±26) a	288 (±30) a	147 (±7) a	162 (±6) a
	Oregano	20.2 (±1.1) ab	18.0 (±1.3) a	2.3 (±1.3) a	2.5 (±3.1) a	0.3 (±0.3) a	0.1 (±0.2) a	225 (±21) b	289 (±115) a	135 (±7) a	155 (±7) a
	Thyme	19.2 (±1.0) b	17.9 (±1.1) a	1.7 (±0.7) a	4.3 (±3.6) a	0.7 (±0.5) a	1.4 (±2.0) a	215 (±19) b	272 (±114) a	142 (±10) a	157 (±17) a
April 2020	Total	11.9 (±1.6) E	11.7 (±1.2) E	0.5 (±0.6) BC	0.1 (±0.3) C	1.9 (±0.6) CD	2.3 (±0.8) C	200 (±28) D	211 (±38) D	151 (±15) C	157 (±42) C
	Control	12.8 (±1.8) a	12.3 (±1.4) a	0.4 (±0.5) a	0.3 (±0.3) a	2.0 (±0.4) a	2.6 (±0.9) a	225 (±28) a	232 (±47) a	156 (±16) a	168 (±62) a
	Oregano	11.4 (±1.5) a	11.3 (±1.0) a	0.7 (±0.8) a	0.3 (±0.3) a	1.9 (±0.9) a	2.1 (±0.7) a	186 (±9) b	197 (±16) a	149 (±17) a	144 (±25) a
	Thyme	11.4 (±1.1) a	11.6 (±0.9) a	0.3 (±0.2) a	0.4 (±0.2) a	1.8 (±0.3) a	2.4 (±0.6) a	189 (±23) b	204 (±36) a	148 (±13) a	159 (±31) a
October 2020	Total	26.1 (±3.1) A	20.2 (±2.5) B	6.9 (±3.6) A	7.9 (±5.2) A	2.3 (±0.8) CD	1.7 (±0.9) CD	278 (±35) C	276 (±47) C	143 (±19) C	181 (±42) AB
	Control	28.8 (±1.7) a	20.7 (±2.3) a	7.1 (±4.0) a	8.6 (±6.2) a	2.9 (±0.8) a	1.9 (±0.8) a	298 (±27) a	276 (±31) a	152 (±12) a	167 (±33) a
	Oregano	25.4 (±3.1) b	19.7 (±2.5) a	8.3 (±3.9) a	6.9 (±3.4) a	1.8 (±0.6) b	1.5 (±1.0) a	287 (±33) a	266 (±41) a	140 (±26) a	182 (±35) a
	Thyme	24.0 (±2.1) b	20.1 (±2.8) a	5.4 (±2.6) a	8.1 (±6.1) a	2.2 (±0.6) ab	1.7 (±0.9) a	249 (±26) b	286 (±66) a	137 (±16) a	194 (±53) a

Note: na = data not available; nd = soil content not detectable.

Soil nitrate (NO_3-N) was highest at extreme soil moisture status (i.e., extremely dry) in October 2018, and moist in October 2020. In topsoil samples taken in October 2018, the cultivation of aromatic plants caused a consistent reduction of NO_3-N. In contrast, a slight increase of NO_3-N across both intercropping treatments was detected in October 2019. Soil ammonium (NH_4-N) was uniformly reduced in topsoil due to intercropping, whereas statistical significance was given only in April 2019 and October 2020. Plant-available potassium (K) in both topsoil and subsoil was highest in October 2018, and dropped afterwards. Intercropping consistently reduced available K in topsoil samples from April 2019 onward, and was uniformly significant in October 2019 and April 2020. Plant-available phosphorus (P) was consistently lower in topsoil samples due to intercropping throughout the experiment. Despite a significant decrease in October 2018, K and P in the subsoil were largely unaffected by intercropping.

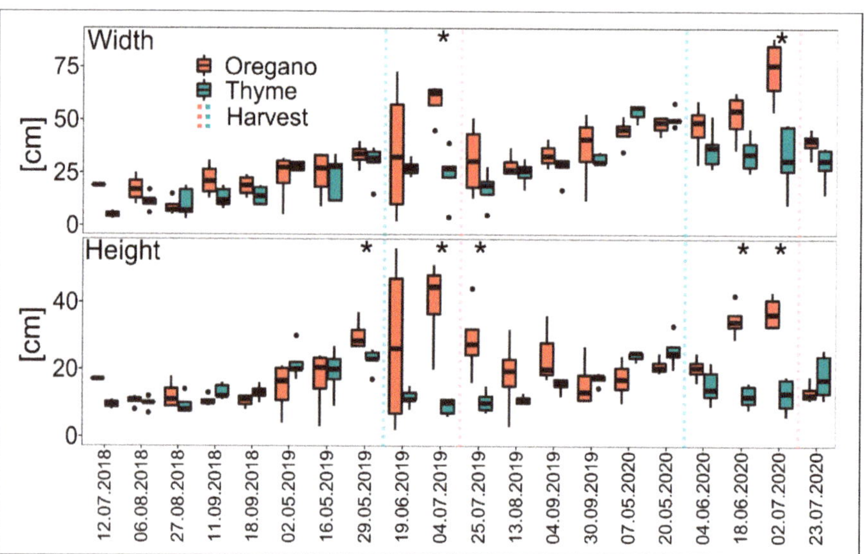

Figure 3. Vegetative development of oregano (red) and thyme (blue) intercrops over two years after planting as indicated by plant width (upper panel) and height (lower panel). Plant width was measured orthogonal to the grapevine row direction. The vertical dashed lines represent harvest dates of aromatic plants. Asterisks (*) indicate significant differences between aromatic plants at p-levels < 0.05.

3.3. *Grapevine Performance and Harvest Properties*

Grapevine yield and must quality indices as affected by time and treatment are shown in Table 3. Generally, significant differences were dominantly attributable to the different years, rather than to the treatments. Overall crop yields per plant and hectare did not significantly differ between 2018 and 2020. However, in 2019, a clear tendency toward reduced productivity was observed, indicating significantly lower (~20%) yields per hectare. Although more cluster numbers were produced, their lower weight negatively affected yields as compared to the other years. The principal component analysis revealed a clear separation according the three experimental years (Figure 4), whereas no clear separation could be detected by grouping according to the experimental treatments. Furthermore, yields per hectare were closely associated with cluster weights, and opposed the quality indices juice pH and concentrations of total soluble solids (TSS), which were highest in 2019. Concomitantly, must pressed in 2019 showed the lowest amounts of titratable acidity (TA) (Table 3).

Table 3. Mean values (± standard deviation in brackets) of the grapevine yield and must quality indices as affected by time and treatment. The row total considers all observations from the respective year. Numbers followed by capital letters indicate significant differences between years, whereas numbers followed by lowercase letters indicate significant differences within one year between the experimental treatments. Significance was given at p-level < 0.05.

Indices	Treatment	2018	2019	2020
Crop yield (kg/plant)	Total	1.6 (±0.5) A	1.3 (±0.5) A	1.6 (±0.8) A
	Control	1.6 (±0.3) a	1.2 (±0.3) a	1.8 (±0.9) a
	Oregano	1.8 (±0.4) a	1.3 (±0.4) a	1.3 (±0.6) a
	Thyme	1.4 (±0.6) a	1.4 (±0.7) a	1.6 (±1.0) a
Crop yield (kg/ha)	Total	6749 (±536) A	5393 (±698) B	6901 (±1118) A
	Control	6632 (±327) a	5059 (±1108) a	7249 (±1105) a
	Oregano	7113 (±802) a	5329 (±244) a	5952 (±1236) a
	Thyme	6501 (±297) a	5791 (±498) a	7501 (±426) a

Table 3. *Cont.*

Indices	Treatment	2018	2019	2020
Produced clusters (number/plant)	Total	20.9 (±4.8) AB	24.5 (±5.8) A	18.2 (±7.8) B
	Control	22.2 (±5.3) a	23.7 (±5.2) a	20.0 (±5.2) a
	Oregano	22.4 (±3.7) a	25.1 (±7.0) a	13.5 (±6.7) a
	Thyme	18.1 (±4.2) a	24.7 (±5.9) a	19.6 (±9.9) a
Cluster weight (g)	Total	94B (±20) B	75 (±21) C	111 (±32) A
	Control	89 (±22) a	77 (±22) a	107 (±36) a
	Oregano	96 (±23) a	72 (±21) a	114 (±18) a
	Thyme	96 (±15) a	75 (±21) a	112 (±41) a
Juice pH	Total	2.9 (±0.05) C	3.3 (±0.12) A	3.2 (±0.10) B
	Control	2.8 (±0.06) a	3.3 (±0.20) a	3.1 (±0.10) a
	Oregano	2.9 (±0.05) a	3.2 (±0.05) a	3.1 (±0.09) a
	Thyme	2.9 (±0.04) a	3.2 (±0.04) a	3.2 (±0.11) a
Titratable acidity (g/L)	Total	9.0 (±0.7) A	7.8 (±0.5) B	9.2 (±1.4) A
	Control	9.3 (±0.7) a	7.3 (±0.2) b	9.5 (±1.9) a
	Oregano	9.0 (±0.6) a	8.0 (±0.4) a	8.7 (±0.7) a
	Thyme	8.6 (±0.5) a	8.0 (±0.4) a	9.4 (±1.3) a
Total soluble solids (°Brix)	Total	21.4 (±1.3) A	22.1 (±1.5) A	19.3 (±1.6) B
	Control	20.9 (±1.7) a	22.2 (±1.2) a	18.7 (±1.2) a
	Oregano	21.7 (±1.1) a	22.3 (±2.2) a	20.1 (±2.2) a
	Thyme	21.5 (±1.2) a	21.9 (±0.9) a	19.2 (±1.0) a

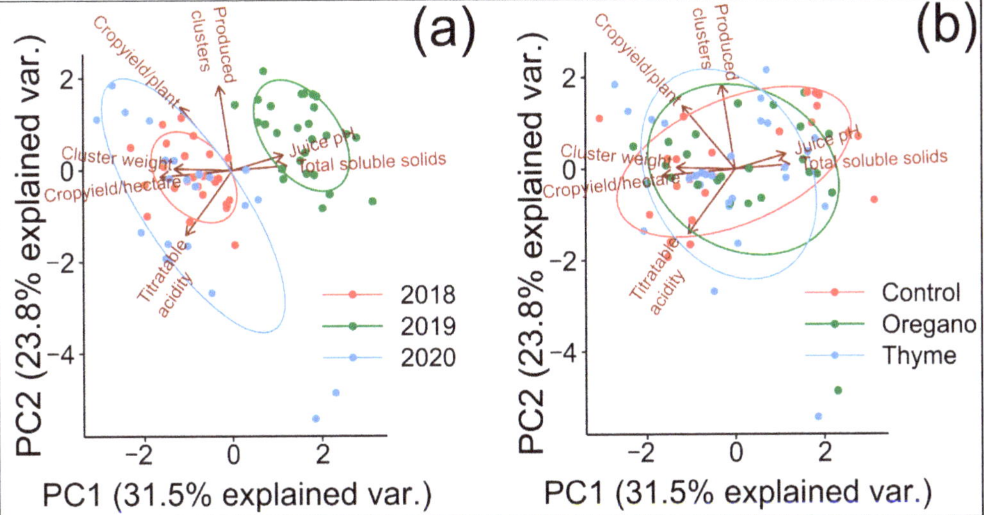

Figure 4. The principal component analysis of the grapevine yield and must quality indices grouped according to the three experimental years (**a**) and the experimental treatments (**b**) considering monocropped and intercropped grapevines.

In 2018, grapevines intercropped with thyme tended to produce a lower number of clusters, corresponding to lower overall yields, whereas the opposite was true for grapevines intercropped with oregano. The trend to depressed yields of grapevines intercropped with thyme was more pronounced in terms of yield per plant (−17%) than for yields per hectare (−5%). Apparently, the average cluster weight did not respond accordingly, and intercropped grapevines tendentially produced heavier clusters when compared to control rows. This consistent response of intercropped grapevines is in line with a trend to higher juice pH values, lower amounts of TA, and increased concentrations of TSS.

In 2019, the productivity of intercropped grapevines showed a uniform trend toward more produced clusters and slightly increased (~10%) overall crop yields per plant and hectare. With regard to this, grapevines intercropped with thyme tended to be slightly more productive as compared to oregano. In contrast, control grapevines tendentially produced heavier clusters. In accordance with the previous year, intercropping mixed cultivation similarly affected must quality in 2019: slightly lower juice pH values corresponded to significantly higher (8.0 vs. 7.3 g/L) amounts of TA.

In 2020, grapevines intercropped with oregano showed a clear trend to reduced (−20%) yields per plant and hectare, following the trends of least numbers, but heaviest weights of clusters. Suitably, juice quality of grapevines intercropped with oregano tended to show lower amounts of TA and higher concentrations of TSS.

4. Discussion

4.1. Effects of Weather Conditions and Crop Plant Diversification

The presented results approve the vital importance of water as a key driver for both biotic and abiotic processes relevant to vineyard productivity. Generally, grapevine yield is composed of several components (e.g., the number of clusters and their weight), and is shaped by the temporal expression of interacting climatic (amount and distribution of precipitation, evaporation), pedological (ability to store and supply water and nutrients), and viticultural (choice of cultivar and rootstock, training system, pruning level, and irrigation) factors [43–45].

Overall, grapevine yields in our study were comparable to those reported from other vineyards [3]. The total amount and distribution of precipitation varied markedly and shaped the general conditions for plant growth. As yield formation was determined over two crop cycles [46,47], the yields seemed to largely depend on the water availability in QIII and QIV of the previous year, when grapevines usually replenish resources, supporting early-phase development in the following year [48]. Additionally, the distinct differences in precipitation recorded during QI restored soil water resources, which were subsequently supplied to grapevines in QII and enabled vigorous growth and a lush canopy development. Consequently, high precipitation sums in QIII and QIV in 2017 and 2019 pre-determined high yield levels in 2018 and 2020, which were realized by comparably high precipitation sums during QI in 2018 and 2020. In addition, significantly more days with an average temperature above 5 °C were recorded in QI of 2020. This threshold temperature is considered as the lower baseline at which grapevine vegetative growth is induced [49,50]. Hence, considerably more vegetation days in QI of 2020 enabled early vegetative grapevine development, and were finally contributing to the highest yields in 2020 measured during the experimental period. Expression of yield components was inversely related in 2019 and 2020, i.e., most numbers, but lightest weights of clusters in 2019, and vice versa in 2020. This response confirmed the yield component compensation principle [51], which states that grapevines compensate modifications of one yield component by changing levels of another yield component, and may offset the loss of yield potential. Must quality responded accordingly and revealed a measurable vintage effect [44], with highest pH values and lowest amounts of titratable acidity found in must from berries harvested in 2019. Apparently, the intense precipitation in QIII of 2019 raised soil moisture levels and favored high nutrient-uptake rates, resulting in a prominent depletion of all measured nutrients in the topsoil. Particularly, the K uptake, due to its neutralizing effect on organic acids [52,53], directly affected must quality. In addition, lower cluster weights and presumably lighter and smaller berries containing relatively higher concentrations of total soluble solids also suggested an indirect impact on must quality [3].

Intercropping and associated cultural practices showed both beneficial and detrimental effects (mostly insignificant) on grapevine yields. Generally, a reducing effect on grapevine growth and yield can be expected due to resource competition, particularly in case of complete and permanent vineyard soil cover, and was also reported from cover-cropping studies conducted in vineyards all over the world [12,54,55]. Clear tendencies toward decreased yields were observed for intercropping with thyme in 2018 and oregano in 2020. In both cases, lower yields were associated with a clear trend toward the production of smallest numbers of clusters. However, clusters tended to be even heavier compared to those from monocropped grapevines. These contrasting effects on yield parameters might be attributable to the compensation principle stated above, but yield losses were not compensated in the case of intercropping with thyme (2018) and oregano (2020), respectively. It must be noted that soil moisture and nutrient levels, determined shortly after harvest, were slightly lower in rows diversified with oregano and thyme compared to control rows. However, they were largely similar between both intercropping treatments so that additional, presumably dynamic factors contributed to the diverse response of yield indices. In this context, water deficits and other stresses during early stages of grapevine development (i.e., around blossom) may induce embryo abortion, poor fruit set, and reduced cluster numbers [48,51]. It must be assumed that stress events of intercropped grapevine rows occurred during QII and/or QIII in 2018 and 2020.

The extreme erosion event in June 2018 resulted in soil losses that were manifoldly higher in rows intercropped with thyme. With respect to the remarkable difference in soil loss between rows intercropped with thyme and oregano, it is likely that minor pedological and topographical variabilities between the recently prepared and planted rows induced a concentrated flow of surface water during the heavy rain event. It is assumed that this event, just at the time of grapevine blossom, induced short-term physiological stress, because a typical consequence of soil erosion is root exposure [56]. Thus, it appears reasonable that the soil-erosion event exposed near-surface secondary site roots of the grapevines and affected the soil-root-shoot-fruit pathway, with negative implications for cluster numbers and yields, as well as for the amount of titratable acidity. In a comprehensive study on the effects of soil erosion on vineyard production across Europe [3], reduced productivity (in terms of overall yields, cluster numbers, and weights) and higher levels of maturity (as indicated by lower amounts of titratable acidity and excessive concentrations of sugar) have been reported for grapevines grown on degraded vineyard plots. However, there is good reason to assume that once diversification crop plants are established, they will substantially contribute to a reduction in soil erosion [57].

Furthermore, the fact that grapevines intercropped with thyme in 2018 were finally capable of producing cluster weights and concentrations of total soluble solids comparable to those of grapevines intercropped with oregano suggests that stress in the early stage of development was of minor importance during berry ripening (veraison). Sugar accumulation and berry growth (due to water import) rapidly increased with the beginning of veraison, and water supply during this developmental phase is even more critical for cluster weights and must quality [51]. In this context, intercropped grapevines apparently profited from manual irrigation applied from the beginning of veraison to prevent withering of intercrop seedlings. On one hand, this measure increased intercrop management intensity and inputs, but in turn, was effective in partially redeeming the developmental drawbacks of grapevines intercropped with thyme, and even increased yields of grapevines intercropped with oregano in comparison to monocropped grapevines managed without additional irrigation.

4.2. Competition between Grapevine and Diversification Crops

Increased yields were also observed for both intercropping treatments in 2019. Because consistent reductions of mineral nutrients (i.e., NH_4-N, K (significantly in October), and P) were observed in topsoil samples of both intercropping treatments throughout 2019, availability of water, rather than nutrients, seemed to be the driving factor for overall vineyard productivity. This finding is in line with several studies that considered soil water availability as the most influential soil component in vineyards (rather than nutrient availability or composition) [43,45]. Another indicator that nutrient competition between grapevines and intercrops was of minor importance was the differentiated response of NO_3-N and NH_4-N to intercropping. NO_3-N is considered the primary nitrogen source for grapevines [51,53,58]. Our results rejected the concern of competition for NO_3-N among grapevines and aromatic plants, which is explained by the strong overall regulation of NO_3-N availability by soil moisture levels. In contrast, NH_4-N with lower mobility in soil was consistently reduced by both species of aromatic plants, indicating their higher affinity and demand toward NH_4-N. Additionally, differences in nutrient status between control (monocropping) and the diversified cropping systems (oregano, thyme) further declined with soil depth (soil layer from 0.1–0.3 m; Table 2). Generally, the rooting depth of grapevine is substantially deeper than that of aromatic plants [59–61]. Hence, we concluded that differences in the nutrient status of the top layer (0–0.1 m) were mostly related to the impact of the diversification plants, while nutrient uptake in the deeper soil was dominated by grapevine and largely unaffected by monocropping and diversified cropping.

Following the necessity of irrigation measures to prevent intercrop withering in 2018 and 2019, the intercropped rows did not receive additional water in 2020. Hence, the substantially lower yields per hectare of grapevines intercropped with oregano suggest severe water limitation by competition. Considering that the oregano was harvested about six weeks later than the thyme in 2020, increased/longer water consumption during periods of rare precipitation appears to be the critical driver for lower yields of grapevines intercropped with oregano. In a comparative study on oregano and thyme performance under open-field and shade-enclosure conditions [62], oregano showed a higher leaf area and increased transpiration, and produced significantly more below- and above-ground biomass. Although the aforementioned study did not report effects on soil resources, we assumed that the higher primary production was associated with a higher consumption of soil resources. However, on both sampling dates in 2020, no significant or distinct differences regarding nutrient concentrations were found among the intercropping treatments. Consequently, oregano seems to be more competitive than thyme, due to an assumed higher consumption of soil water. This assumption was furthermore underlined by a higher root biomass of oregano found at the end of the experiment in 2020.

Interestingly, the amounts of titratable acidity of must obtained from both intercropping treatments were significantly higher in 2019 than for must obtained from control grapevines. Again, the slightly lower cluster weights of intercropped grapevines may have had indirect effects on must quality. However, as the cultivation of aromatic plants also significantly lowered K levels in topsoil samples determined soon after harvest, a direct effect on must quality was most likely attributable to competition for K between intercrops and grapevines. In a review on cover-cropping impacts on grapevine growth and must quality [12], mostly decreased amounts of titratable acidity were found in must from cover-cropped vineyards. However, given the desired wine style in the area, aiming for a well-balanced ratio of sugar and acidity, the higher acidity level maintained in musts from intercropped grapevines is considered positive, when comparing the low level of acidity with musts from the other experimental years.

5. Conclusions

The results of the experimental field study showed that crop-plant diversification using aromatic plants in vineyards can be successfully established. This comports with impacts of intercropping grapevines with aromatic plants on grapevine yield and must quality, as well as soil water and nutrient levels. Our study revealed the potential, but also the possible vulnerabilities, of crop diversification in vineyards. We conclude that climatic variability between the years was the most important factor determining yields, and extreme weather events can induce a significant reduction in productivity. Additionally, we also observed some insignificant yield losses due to intercropping, particularly induced by water competition. With respect to this, thyme appears to be less competitive due to an earlier harvest date and a lower respectively shorter consumption of soil water during the crop cycle. Generally, water competition will be less pronounced in soils with a higher water-storage capacity. Management measures such as irrigation are an option to alleviate competition between grapevines and aromatic plants to ensure long-term vineyard productivity. As irrigation is already widely applied in many viticultural areas around the world, and its further implementation in vineyards, especially in the Mosel region, may become a necessary management tool in the near future due to global climate change, intercrop marketing can be a viable cross-financing option for irrigation investment. However, we found that competition is not necessarily detrimental, and beneficial effects on must quality due to intercropping were found. Especially under high moisture regimes during veraison, additional competition and nutrient uptake by intercrops may enhance final must and wine quality. On sites that are prone to soil erosion, the timing of intercrop establishment needs to be carefully considered (preferably in periods of moist soil, for better infiltration and rapid juvenile development of seedlings). Soil preparation prior to diversified crop establishment may increase soil vulnerability for erosion compared to non-tilled rows, thus counteracting the expected erosion-reducing effect of diversified cropping in the long term. Furthermore, the long-term effects of intercropping on grapevine growth need to be monitored. An overarching evaluation of crop diversification by intercropping in steep-slope vineyards requires the ongoing assessments of viticultural inputs, economic revenues, soil erosion, infiltration capacity, chemical and biological soil quality, greenhouse gas emissions, and pollinator occurrence.

Author Contributions: Conceptualization, all authors; methodology, all authors; validation, all authors; formal analysis, F.D., T.I.; investigation, F.D., T.I., C.-H.T., R.H., S.O.; resources, C.-H.T., M.S., S.T.-B.; data curation, F.D., T.I., C.-H.T., R.H., S.O.; writing—original draft preparation, F.D.; writing—review and editing, all authors; visualization, F.D., T.I.; supervision, C.-H.T., M.S., S.T.-B.; project administration, C.-H.T., M.S., S.T.-B.; funding acquisition, M.S., S.T.-B. All authors have read and agreed to the published version of the manuscript.

Funding: This research was funded by the European Commission Horizon 2020 project Diverfarming (grant agreement 728003). The funders had no role in the design of the study; in the collection, analyses, or interpretation of data; in the writing of the manuscript; or in the decision to publish the results.

Institutional Review Board Statement: Not applicable.

Informed Consent Statement: Not applicable.

Data Availability Statement: All data will be made available via Zenodo.

Conflicts of Interest: The authors declare no conflict of interest.

References

1. International Organization of Vine and Wine. 2019: Statistical Report on World Vitiviniculture. Available online: http://oiv.int/public/medias/6782/oiv-2019-statistical-report-on-world-vitiviniculture.pdf (accessed on 5 August 2020).
2. Mozell, M.R.; Thach, L. The Impact of Climate Change on the Global Wine Industry: Challenges & Solutions. *Wine Econ. Policy* **2014**, *3*, 81–89. [CrossRef]
3. Costantini, E.A.C.; Castaldini, M.; Diago, M.P.; Giffard, B.; Lagomarsino, A.; Schroers, H.-J.; Priori, S.; Valboa, G.; Agnelli, A.E.; Akça, E.; et al. Effects of Soil Erosion on Agro-Ecosystem Services and Soil Functions: A Multidisciplinary Study in Nineteen Organically Farmed European and Turkish Vineyards. *J. Environ. Manag.* **2018**, *223*, 614–624. [CrossRef]
4. Komárek, M.; Čadková, E.; Chrastný, V.; Bordas, F.; Bollinger, J.-C. Contamination of Vineyard Soils with Fungicides: A Review of Environmental and Toxicological Aspects. *Environ. Int.* **2010**, *36*, 138–151. [CrossRef] [PubMed]
5. Hall, R.M.; Penke, N.; Kriechbaum, M.; Kratschmer, S.; Jung, V.; Chollet, S.; Guernion, M.; Nicolai, A.; Burel, F.; Fertil, A.; et al. Vegetation Management Intensity and Landscape Diversity Alter Plant Species Richness, Functional Traits and Community Composition across European Vineyards. *Agric. Syst.* **2020**, *177*, 102706. [CrossRef]
6. Paiola, A.; Assandri, G.; Brambilla, M.; Zottini, M.; Pedrini, P.; Nascimbene, J. Exploring the Potential of Vineyards for Biodiversity Conservation and Delivery of Biodiversity-Mediated Ecosystem Services: A Global-Scale Systematic Review. *Sci. Total Environ.* **2020**, *706*, 135839. [CrossRef] [PubMed]
7. Schaller, K. Intensive Viticulture and Its Environmental Impacts: Nitrogen as a Case Study. *Acta Hortic.* **2000**, *502*. [CrossRef]
8. Steenwerth, K.L.; Belina, K.M. Vineyard Weed Management Practices Influence Nitrate Leaching and Nitrous Oxide Emissions. *Agric. Ecosyst. Environ.* **2010**, *138*, 127–131. [CrossRef]

9. Wezel, A.; Casagrande, M.; Celette, F.; Vian, J.-F.; Ferrer, A.; Peigné, J. Agroecological Practices for Sustainable Agriculture. A Review. *Agron. Sustain. Dev.* **2014**, *34*, 1–20. [CrossRef]
10. Brooker, R.W.; Bennett, A.E.; Cong, W.-F.; Daniell, T.J.; George, T.S.; Hallett, P.D.; Hawes, C.; Iannetta, P.P.M.; Jones, H.G.; Karley, A.J.; et al. Improving Intercropping: A Synthesis of Research in Agronomy, Plant Physiology and Ecology. *New Phytol.* **2015**, *206*, 107–117. [CrossRef]
11. Tamburini, G.; Bommarco, R.; Wanger, T.C.; Kremen, C.; van der Heijden, M.G.A.; Liebman, M.; Hallin, S. Agricultural Diversification Promotes Multiple Ecosystem Services without Compromising Yield. *Sci. Adv.* **2020**, *6*, eaba1715. [CrossRef]
12. Guerra, B.; Steenwerth, K. Influence of Floor Management Technique on Grapevine Growth, Disease Pressure, and Juice and Wine Composition: A Review. *Am. J. Enol. Vitic.* **2012**, *63*, 149–164. [CrossRef]
13. Mcgourty, G.; Reganold, J. Managing Vineyard Soil Organic Matter with Cover Crops. In Proceedings of the Soil Environment and Vine Mineral Nutrition Symposium, San Diego, CA, USA, 29–30 June 2004; American Society for Enology and Viticulture: Davis, CA, USA, 2005; pp. 145–151.
14. Smith, R.; Bettiga, L.; Cahn, M.; Baumgartner, K.; Jackson, L.E.; Bensen, T. Vineyard Floor Management Affects Soil, Plant Nutrition, and Grape Yield and Quality. *Calif. Agric.* **2008**, *62*, 184–190. [CrossRef]
15. Garcia, L.; Celette, F.; Gary, C.; Ripoche, A.; Valdés-Gómez, H.; Metay, A. Management of Service Crops for the Provision of Ecosystem Services in Vineyards: A Review. *Agric. Ecosyst. Environ.* **2018**, *251*, 158–170. [CrossRef]
16. Steenwerth, K.; Belina, K.M. Cover Crops Enhance Soil Organic Matter, Carbon Dynamics and Microbiological Function in a Vineyard Agroecosystem. *Appl. Soil Ecol.* **2008**, *40*, 359–369. [CrossRef]
17. Peregrina, F.; Pérez-Álvarez, E.P.; García-Escudero, E. Soil Microbiological Properties and Its Stratification Ratios for Soil Quality Assessment under Different Cover Crop Management Systems in a Semiarid Vineyard. *J. Plant Nutr. Soil Sci.* **2014**, *177*, 548–559. [CrossRef]
18. Mercenaro, L.; Nieddu, G.; Pulina, P.; Porqueddu, C. Sustainable Management of an Intercropped Mediterranean Vineyard. *Agric. Ecosyst. Environ.* **2014**, *192*, 95–104. [CrossRef]
19. Fiera, C.; Ulrich, W.; Popescu, D.; Bunea, C.-I.; Manu, M.; Nae, I.; Stan, M.; Markó, B.; Urák, I.; Giurginca, A.; et al. Effects of Vineyard Inter-Row Management on the Diversity and Abundance of Plants and Surface-Dwelling Invertebrates in Central Romania. *J. Insect Conserv.* **2020**, *24*, 175–185. [CrossRef] [PubMed]
20. Tesic, D.; Keller, M.; Hutton, R.J. Influence of Vineyard Floor Management Practices on Grapevine Vegetative Growth, Yield, and Fruit Composition. *Am. J. Enol. Vitic.* **2007**, *58*, 1–11.
21. Leibar, U.; Pascual, I.; Aizpurua, A.; Morales, F.; Unamunzaga, O. Grapevine Nutritional Status and K Concentration of Must under Future Expected Climatic Conditions Texturally Different Soils. *J. Soil Sci. Plant Nutr.* **2017**, *17*, 385–397. [CrossRef]
22. Bell, S.-J.; Henschke, P.A. Implications of Nitrogen Nutrition for Grapes, Fermentation and Wine. *Aust. J. Grape Wine Res.* **2005**, *11*, 242–295. [CrossRef]
23. Ingels, C.A.; Scow, K.M.; Whisson, D.A.; Drenovsky, R.E. Effects of Cover Crops on Grapevines, Yield, Juice Composition, Soil Microbial Ecology, and Gopher Activity. *Am. J. Enol. Vitic.* **2005**, *56*, 19–29.
24. EIP-AGRI Focus Group. Plant-Based Medicinal and Cosmetic Products—EIP-AGRI—European Commission. Available online: Ec.europa.eu/eip/agriculture/en/publications/eip-agri-focus-group-plant-based-medicinal-and (accessed on 11 August 2020).
25. Hoppe, B. *Handbuch des Arznei- und Gewürzpflanzenanbaus, Band 5: Arznei- und Gewürzpflanzen L-Z;2013*; Verein für Arznei- und Gewürzpflanzen Saluplanta e.V., Bernburg: Bernburg, Germany, 2013; ISBN 978-3-935971-64-5.
26. Handbook for Cultivation of Medicinal and Aromatic Plants | Project Cult: Strengthening Capacities of Farmers for Cultivation of Medicinal and Aromatic Plants through Transfer of Knowledge and Good Practices. Available online: http://projectcult.com/wp-content/uploads/2016/11/PRIRACNIK_EN_.pdf (accessed on 21 January 2021).
27. Rao, E.V.S.P. Economic and Ecological Aspects of Aromatic-Plant-Based Cropping Systems. *CAB Rev.* **2015**, *10*, 1–8. [CrossRef]
28. Campos, E.V.R.; Proença, P.L.F.; Oliveira, J.L.; Bakshi, M.; Abhilash, P.C.; Fraceto, L.F. Use of Botanical Insecticides for Sustainable Agriculture: Future Perspectives. *Ecol. Indic.* **2019**, *105*, 483–495. [CrossRef]
29. Rao, M.R.; Palada, M.C.; Becker, B.N. Medicinal and Aromatic Plants in Agroforestry Systems. *Agrofor. Syst.* **2004**, *61*, 107–122. [CrossRef]
30. Song, B.; Han, Z. Assessment of the Biocontrol Effects of Three Aromatic Plants on Multiple Trophic Levels of the Arthropod Community in an Agroforestry Ecosystem. *Ecol. Entomol.* **2020**, *45*, 831–839. [CrossRef]
31. Chen, X.; Song, B.; Yao, Y.; Wu, H.; Hu, J.; Zhao, L. Aromatic Plants Play an Important Role in Promoting Soil Biological Activity Related to Nitrogen Cycling in an Orchard Ecosystem. *Sci. Total Environ.* **2014**, *472*, 939–946. [CrossRef]
32. Rao, E.V.S.P. *Aromatic Plant Species in Agricultural Production Systems Based on Marginal Soils*; CAB Int.: Wallingford, UK, 2012.
33. Zheljazkov, V.D.; Craker, L.E.; Xing, B.; Nielsen, N.E.; Wilcox, A. Aromatic Plant Production on Metal Contaminated Soils. *Sci. Total Environ.* **2008**, *395*, 51–62. [CrossRef]
34. Pandey, J.; Verma, R.K.; Singh, S. Suitability of Aromatic Plants for Phytoremediation of Heavy Metal Contaminated Areas: A Review. *Int. J. Phytoremediat.* **2019**, *21*, 405–418. [CrossRef]
35. Ballabio, C.; Panagos, P.; Lugato, E.; Huang, J.-H.; Orgiazzi, A.; Jones, A.; Fernández-Ugalde, O.; Borrelli, P.; Montanarella, L. Copper Distribution in European Topsoils: An Assessment Based on LUCAS Soil Survey. *Sci. Total Environ.* **2018**, *636*, 282–298. [CrossRef]

36. Cordier, S.; Harmand, D.; Lauer, T.; Voinchet, P.; Bahain, J.-J.; Frechen, M. Geochronological Reconstruction of the Pleistocene Evolution of the Sarre Valley (France and Germany) Using OSL and ESR Dating Techniques. *Geomorphology* **2012**, *165–166*, 91–106. [CrossRef]
37. FAO. *World Reference Base for Soil Resources 2014: International Soil Classification System for Naming Soils and Creating Legends for Soil Maps*; FAO: Rome, Italy, 2014; ISBN 978-92-5-108369-7.
38. Schüller, H. Die CAL-Methode, Eine Neue Methode Zur Bestimmung des Pflanzenverfügbaren Phosphates in Böden. *Z. Für Pflanzenernähr. Bodenkd.* **1969**, *123*, 48–63. [CrossRef]
39. Iserloh, T.; Seeger, M. Sediment load and concentration. In *Handbook of Plant and Soil Analysis for Agricultural Systems*; Álvaro-Fuentes, J., Loczy, D., Thiele-Bruhn, S., Zornoza, R., Eds.; CRAI: Boston, MA, USA, 2019; pp. 214–216. ISBN 978-84-16325-86-3.
40. Gerlach, T. Hillslope Troughs for Measuring Sediment Movement. *Rev. Geomorphol. Dyn.* **1967**, *17*, 173–174.
41. Schmidt, R.-G. Probleme Der Erfassung Und Quantifizierung von Ausmass Und Prozessen Der Aktuellen Bodenerosion (Abspülung) Auf Ackerflächen. In *Physiogeographica, Baseler Beiträge zur Physiogeographie*; Geographisch-Ethnologischen Gesellschaft: Basel, Switzerland, 1979; Volume 1.
42. R Core. *Team R: A Language and Environment for Statistical Computing*; R Core: Vienna, Austria, 2016.
43. Cortell, J.M.; Halbleib, M.; Gallagher, A.V.; Righetti, T.L.; Kennedy, J.A. Influence of Vine Vigor on Grape (*Vitis vinifera*, L. Cv. Pinot Noir) and Wine Proanthocyanidins. *J. Agric. Food Chem.* **2005**, *53*, 5798–5808. [CrossRef] [PubMed]
44. Van Leeuwen, C. 9—Terroir: The effect of the physical environment on vine growth, grape ripening and wine sensory attributes. In *Managing Wine Quality*; Reynolds, A.G., Ed.; Woodhead Publishing Series in Food Science, Technology and Nutrition; Woodhead Publishing: Cambridge, UK, 2010; pp. 273–315, ISBN 978-1-84569-484-5.
45. Lebon, E.; Dumas, V.; Pieri, P.; Schultz, H.R. Modelling the Seasonal Dynamics of the Soil Water Balance of Vineyards. *Funct. Plant Biol.* **2003**, *30*, 699–710. [CrossRef] [PubMed]
46. Guilpart, N.; Metay, A.; Gary, C. Grapevine Bud Fertility and Number of Berries per Bunch are Determined by Water and Nitrogen Stress around Flowering in the Previous Year. *Eur. J. Agron.* **2014**, *54*, 9–20. [CrossRef]
47. Fourie, J.C. Soil Management in the Breede River Valley Wine Grape Region, South Africa. 3. Grapevine Performance. *S. Afr. J. Enol. Vitic.* **2011**, *32*, 60–70. [CrossRef]
48. Lebon, G.; Wojnarowiez, G.; Holzapfel, B.; Fontaine, F.; Vaillant-Gaveau, N.; Clément, C. Sugars and Flowering in the Grapevine (*Vitis vinifera* L.). *J. Exp. Bot.* **2008**, *59*, 2565–2578. [CrossRef]
49. Molitor, D.; Junk, J.; Evers, D.; Hoffmann, L.; Beyer, M. A High-Resolution Cumulative Degree Day-Based Model to Simulate Phenological Development of Grapevine. *Am. J. Enol. Vitic.* **2014**, *65*, 72–80. [CrossRef]
50. García de Cortázar-Atauri, I.; Brisson, N.; Gaudillere, J.P. Performance of Several Models for Predicting Budburst Date of Grapevine (*Vitis vinifera* L.). *Int. J. Biometeorol.* **2009**, *53*, 317–326. [CrossRef]
51. Keller, M. *The Science of Grapevines*; Elsevier: Amsterdam, The Netherlands, 2020; ISBN 978-0-12-816365-8.
52. Spayd, S.E.; Wample, R.L.; Evans, R.G.; Stevens, R.G.; Seymour, B.J.; Nagel, C.W. Nitrogen Fertilization of White Riesling Grapes in Washington. Must and Wine Composition. *Am. J. Enol. Vitic.* **1994**, *45*, 34–42.
53. Keller, M.; Pool, R.M.; Henick-Kling, T. Excessive Nitrogen Supply and Shoot Trimming Can Impair Colour Development in Pinot Noir Grapes and Wine. *Aust. J. Grape Wine Res.* **1999**, *5*, 45–55. [CrossRef]
54. Giese, G.; Velasco-Cruz, C.; Roberts, L.; Heitman, J.; Wolf, T.K. Complete Vineyard Floor Cover Crops Favorably Limit Grapevine Vegetative Growth. *Sci. Hortic.* **2014**, *170*, 256–266. [CrossRef]
55. Sulas, L.; Mercenaro, L.; Campesi, G.; Nieddu, G. Different Cover Crops Affect Nitrogen Fluxes in Mediterranean Vineyard. *Agron. J.* **2017**, *109*, 2579–2585. [CrossRef]
56. Gärtner, H. Tree Roots—Methodological Review and New Development in Dating and Quantifying Erosive Processes. *Geomorphology* **2007**, *86*, 243–251. [CrossRef]
57. Durán Zuazo, V.H.; Rodríguez Pleguezuelo, C.R. Soil-Erosion and Runoff Prevention by Plant Covers. A Review. *Agron. Sustain. Dev.* **2008**, *28*, 65–86. [CrossRef]
58. Keller, M. Deficit Irrigation and Vine Mineral Nutrition. *Am. J. Enol. Vitic.* **2005**, *56*, 267–283.
59. Lehnart, R.; Michel, H.; Löhnertz, O.; Linsenmeier, A. Root Dynamics and Pattern of "Riesling" on 5C Rootstock Using Minirhizotrons. *VITIS J. Grapevine Res.* **2008**, *47*, 197. [CrossRef]
60. De Baets, S.; Poesen, J.; Knapen, A.; Barberá, G.G.; Navarro, J.A. Root Characteristics of Representative Mediterranean Plant Species and Their Erosion-Reducing Potential during Concentrated Runoff. *Plant Soil* **2007**, *294*, 169–183. [CrossRef]
61. Sotiropoulou, D.E.; Karamanos, A.J. Field Studies of Nitrogen Application on Growth and Yield of Greek Oregano (Origanum Vulgare Ssp. Hirtum (Link) Ietswaart). *Ind. Crops Prod.* **2010**, *32*, 450–457. [CrossRef]
62. Murillo-Amador, B.; Nieto-Garibay, A.; López-Aguilar, R.; Troyo-Diéguez, E.; Rueda-Puente, E.O.; Flores-Hernández, A.; Ruiz-Espinoza, F.H. Physiological, Morphometric Characteristics and Yield of *Origanum Vulgare* L. and *Thymus Vulgaris* L. Exposed to Open-Field and Shade-Enclosure. *Ind. Crops Prod.* **2013**, *49*, 659–667. [CrossRef]

MDPI

Article

Competition Effects and Productivity in Oat–Forage Legume Relay Intercropping Systems under Organic Farming Conditions

Viktorija Gecaitė [1,*], Aušra Arlauskienė [1] and Jurgita Cesevičienė [2]

[1] Joniškėlis Experimental Station, Lithuanian Research Centre for Agriculture and Forestry, Pasvalys Distr., 39301 Joniškėlis, Lithuania; ausra.arlauskiene@lammc.lt

[2] Institute of Agriculture, Lithuanian Research Centre for Agriculture and Forestry, Instituto 1, Kėdainiai Distr., 58344 Akademija, Lithuania; jurgita.ceseviciene@lammc.lt

* Correspondence: viktorija.gecaite@lammc.lt; Tel.: +370-62851350

Abstract: Cereal-legume intercropping is important in many low-input agricultural systems. Interactions between combinations of different plant species vary widely. Field experiments were conducted to determine yield formation regularities and plant competition effects of oat (*Avena sativa* L.)–black medick (*Medicago lupulina* L.), oat–white clover (*Trifolium repens* L.), and oat–Egyptian clover (*T. alexandrinum* L.) under organic farming conditions. Oats and forage legumes were grown in mono- and intercrops. Aboveground dry matter (DM) measured at flowering, development of fruit and ripened grain, productivity indicators, oat grain yield and nutrient content were established. The results showed that oats dominated in the intercropping systems. Oat competitive performance (CP_O), which is characterized by forage legumes aboveground mass reduction compared to monocrops, were 91.4–98.9. As the oats ripened, its competitiveness tendency to declined. In oat–forage legume intercropping systems, the mass of weeds was significantly lower compared to the legume monocrops. Oats and forage legumes competed for P, but N and K accumulation in biomass was not significantly affected. We concluded that, in relay intercrop, under favourable conditions, the forage legumes easily adapted to the growth rhythm and intensity of oats and does not adverse effect on their grain yield.

Keywords: aboveground mass; black medick; Egyptian clover; grain yield; nutrients; white clover

Citation: Gecaitė, V.; Arlauskienė, A.; Cesevičienė, J. Competition Effects and Productivity in Oat–Forage Legume Relay Intercropping Systems under Organic Farming Conditions. *Agriculture* **2021**, *11*, 99. https://doi.org/10.3390/agriculture11020099

Academic Editors: Claudia Di Bene, Rosa Francaviglia, Roberta Farina, Jorge Álvaro-Fuentes and Raúl Zornoza

Received: 11 December 2020
Accepted: 20 January 2021
Published: 25 January 2021

1. Introduction

Enhancing crop diversity and growing legumes are increasingly recognised as a crucial lever for sustainable agroecological development [1]. This is the basis of organic arable farms. The choice of plant species in a stockless farm is small and the use of forage legumes is limited. Intercropping is important in many subsistence or low-input/resource-limited agricultural systems [2]. Intercropping, the simultaneous growth of more than one crop species or genotype in the same field [3], is the practical application of basic ecological principles [2]. Intercropping effects consist of competition (niche differentiation, resource sharing and weed control), diversity (insect and disease control), facilitation (physical support, nitrogen fixation and excretion of allelochemicals and modification of the rhizosphere) and associated diversity (habitats for natural predators, litter diversity and enhanced soil microbial diversity) [4]. Strip, mixed and relay intercropping can be used to increase crop yields through resource partitioning and facilitation. Relay intercropping involves the staggered planting of two or more crops together in a way whereby only parts of their life cycles overlap [5]. Farmers often intercrop forage legumes into winter or spring cereals as a way to increase crop diversity and increase labile nitrogen pools [6]. This intercropping system works particularly well because of the different phenologies of the two crops, which minimises light competition, as well as differences in nutrient acquisition [7]. Intercropping can allow better use of subsoil resources, and thereby decrease the need for resource input and help avoid nutrient losses [8,9]. Moreover, one crop can provide resources for the other one with a positive interspecific interaction, which is at the basis of facilitation processes [10]. Intercropping systems improve soil temperature and moisture regulation; erosion, nutrient run-off and leaching are reduced; weeds are controlled insect and disease development cycles are interrupted and soil organic matter content is improved with recycled nutrients being made available to subsequent crops [11,12].

There is competition between the plants in the intercropping system. It is one of many ecological processes shaping the composition, dynamics and productivity of the plant community [1]. Mainly, plants compete for soil resources and light. These interactions affect plant density and plant development rhythm as well as productivity and fertility [13]. Plant competition in spring cereal–forage legume intercropping systems can be regulated by proper selection of plant species [8], optimal plant seed rate and sowing time and methods [14]. Oat is grown in non-fertile soil regions and also in crop production farms globally. Oats are widely used cereal grains grown for its seed and are increasingly used every year. The growing population of health-conscious people is forcing oats manufacturers to increase growing demand. Therefore, the rising demand for oats has led to improve and increase their research. Relay intercropping systems are an important cropping strategy for sustainable agriculture in many countries as they create benefits in terms of better utilisation of soil resources, weed control and yield diversification. In Lithuania, combinations of legumes and non-legumes are a widely spread practice and several studies have been published on the subject, e.g., pea–spring cereal intercropping systems [15,16].

Red (*T. pratense* L.) and white clover are most often under sown with cereals. In order to increase the services provided by plants and their applicability in greening technologies (cover and catch crops, mixer, strip, relay intercropping systems, etc.), other types of forage legumes have been studied and adapted. Egyptian is a high-yielding, nutritious, cool-season forage crop that can grow on a wide range of soils, though it prefers fertile [17,18]. Egyptian clover can withstand some drought and short periods of waterlogging [19]. This type of clover has a short growing season, therefore, there is a wider range for its application compared to perennial clover.

Black medick is a self-seeding legume that has potential for pasture, green manure, cover cropping, intercropping, and phytoremediation throughout temperate and subtropical regions of the world. It is grown both for agronomic and environmental benefits [20]. More recently, black medick has been recognised for its heavy metal tolerance [21]. Its root leachates provide selective allelopathic suppression of weed growth [22]. The use of a self-seeding legume may be a solution to avoid the cost of seeding cover crops annually [23]. However, black medick can also spread like a weed [24]. Nitrogen fertiliser suppresses medick growth, so black medick cover would be beneficial only in low N or organic farming systems [25]. The aim of this study was to determine yield formation regularities and the yield and plant competition effects in different relay intercropping systems, namely, oat–black medic, oat–white clover and oat–Egyptian clover in clay loam Cambisol under organic farming conditions.

2. Material and Methods

2.1. Experimental Site

Field experiments were conducted at the Joniškėlis Experimental Station of the Lithuanian Research Centre for Agriculture and Forestry in the northern part of Central Lithuania's lowland. The soil of the experimental site is *Endocalcari Endohypogleyic Cambisol*, whose texture is clay loam on silty clay with deeper lying sandy loam. The topsoil (0–25 cm) is close to neutral (pH 6.1), medium in phosphorus (P_2O_5 146 mg kg^{-1}), high in potassium (K_2O 276 mg kg^{-1}) and moderate in humus (2.54%).

During plant development and growth in 2018, April was the wettest, however a similar amount of rainfall fell in May, July and August (Figure 1a). April and May were quite abundant in humidity, heat and sunlight which led to good plant development in the first stages of oat growth. In contrast, precipitation was considerably lesser compared to the standard climate normal (SCN) average data resulted in slower plant development. The year 2019 was slightly wetter and the monthly distribution of precipitation was significantly more even than 2018. In 2019, April was distinguished by a very low rainfall (Figure 1b). The drought, which began in the first 10-day period of April and extended to the end of May. A more abundant amount of precipitation fell only in the third 10-day period of May after which more intensive growth of the aboveground mass of plants began. June was unusually hot, and July was exceptionally wet.

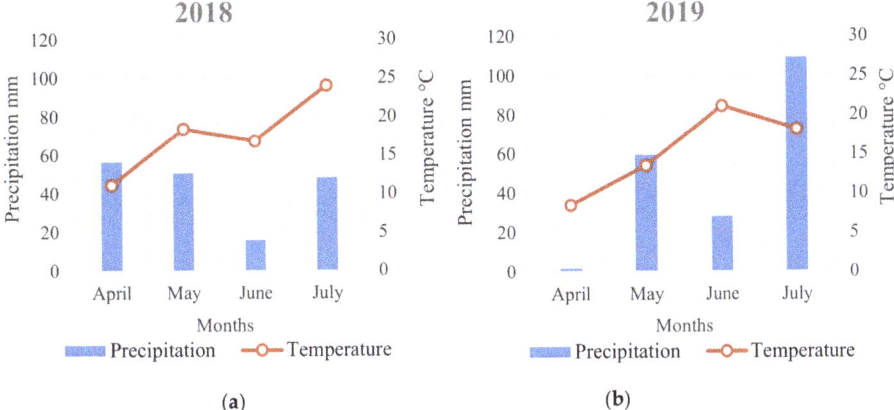

Figure 1. Distribution of monthly (April–July) precipitation and average temperature over 2018 (**a**) and 2019 (**b**) at Joniškėlis Experimental Station of the Lithuanian Research Centre for Agriculture and Forestry.

2.2. Experimental Design, Plant Sampling and Analysis

Two analogous field experiments were established and carried out in 2018 and 2019 in separate areas of the same field. The following experimental design was used for monocrops and relay intercropping systems: (1) Oat (O; cultivar 'Migla DS'); (2) black medick (BM; cultivar 'Arka 133 DS')); (3) white clover (WC; cultivar 'Nemuniai'); (4) Egyptian clover (EC; cultivar 'Cleopatra'); (5) oat–black medick (O+BM); (6) oat–white clover (O+WC); (7) oat–Egyptian clover (O+EC). The oat was sown on 23 April 2018 and 16 April 2019 at a seeding rate of 450 seeds m^{-2} using a drill at a 3 cm depth. The forage legume species were intercropped in oats on 25 April 2018 and 16 April 2019 at a seed rate of 50 seeds m^{-2}. The forage legume seeds were sown at a 2 cm depth using a drill. The experiment was laid out in a one-factor randomised complete block design in four replications and individual plot size was 6 × 20 m. Crops were cultivated according to organic management practices.

At full germination, the oats and forage legumes plants were accounted in 0.25 m^2 plots, in four places per plot. In order to evaluate the growth patterns, of plants, the aboveground mass of oat, forage legumes and weeds were determined when oats reached the flowering (BBCH 61–63), grain development (BBCH 71–73), and ripened grain (BBCH 87–89) growth stages. Sampling of aboveground biomass occurred at four randomly chosen squares of 0.25 m^2 in each plot, which were cut to ground level and weighed on each date. The aboveground dry matter (DM) mass was determined (dried to a constant weight at 105 °C in a forced-air oven), of oat, forage legumes, weeds and subsample values from each plot were averaged.

The oat crop was harvested, at complete maturity stage (plot size 2.3 × 18.0 m) with a small combine harvester on 2 August 2018, and 4 August 2019. Before combining, 25 oat plants per plot were collected to determine the number of panicles per unit area (panicle m^{-2}), number of grains per panicle (grain per panicle) and grain weight per panicle (g) of oats. Grain samples (1 kg) were taken from each plot for the determination of 1000-grain weight (TGW) and grain DM content. Oat grain and straw yield were measured by weighing. The grain yield was converted to standard moisture (14%) and straw to DM. Competitive performance (CP) was expressed as the percent reduction in aboveground dry mass as follows: $CP_o = [(P_{fl/s} - P_{fl/i}) / P_{fl/s}] \times 100$; where CPo is the relative competitive ability of the oat; $P_{fl/s}$ is the dry mass of the forage legume grown alone (control) and $P_{fl/i}$ is the mass of the forage legume grown in intercrop.

Oat grain and forage legume aboveground mass samples collected at the oat ripened grain stage were dried, milled and analysed for nitrogen (N), phosphorus (P) and potassium (K) content. The concentration was evaluated in the sulphuric acid digestates. Plant samples for N determination were analysed using the Kjeldahl method with a Kjeltec system 1002 (Foss Tecator, Hoganas, Sweden). The concentration of P was quantified spectrophotometrically by a coloured reaction with ammonium molybdate-vanadate at a wavelength of 430 nm on a Cary 50 UV-Vis spectrophotometer (Varian Inc., Palo Alto, CA, USA). Respective K concentration was evaluated by atomic absorption spectrometry with an Analyst 200 (Perkin Elmer, Waltham, MA, USA) in accordance with the manufacturer's instructions.

2.3. Statistical Analysis

The data were statistically processed using three-factor (year, assessment time and intercrop) for aboveground mass of oat, perennial legume and weed and two-factor (year and intercrop) for grain yield and its component of oat, nutrition concentration and content analysis of variance as well as correlation and regression methods. The data were analysed when the factual Fisher criterion ($F_{fact.}$) was higher than the theoretical one ($F_{theor.}$). The significance of differences among the treatment means was estimated at the 0.05 probability levels. Interrelationships among aboveground mass of weeds, forage legumes and oats in monocrops and intercropping systems and among P in oats grain and legume aboveground mass were estimated. Simple linear regression (SLR) was applied to the data. Statistical analysis of the experimental data was performed using the ANOVA version 3.1 software and STAT_ENG version 1.5 from the programme package SELEKCIJA [26].

3. Results

3.1. Oat and Forage Legume Mass

Statistical analysis showed that the oat aboveground dry mass yield was significantly ($p < 0.01$) influenced by interaction of year and assessment time (Table 1). The relay intercropping systems did not have any significant effect on the yield of the aboveground oat mass.

The first assessment of the aboveground mass of plants was performed at the beginning of oat flowering and did not differ significantly between the years. The assessment of the aboveground mass of oats during oat grain development revealed that the intensity of the aboveground mass increase of oats was as follows: in 2018 to 18.1% and in 2019 to 84.2%, compared to the first assessment. In general, a significantly higher aboveground mass of oats was found during grain development in 2019, compared to 2018. The increase in the aboveground mass of oats in relay intercrop has been less pronounced (grain development stage) (Table 2). During the fully ripe stage of oats grain, the changes in their aboveground mass were inconsistent (compared to the grain development stage). In 2019, the DM yield of oat aboveground mass was less compared to the second assessment; however, it was significantly greater on average compared to the corresponding data for 2018. In both years, the yields of oats and oats intercropped with legumes did not differ.

Table 1. Probability (p) level of factors for aboveground mass of oat, legumes and weeds.

Factor/Treatment	Aboveground Mass		
	Oat	Legumes	Weeds
Year (Y)	<0.01 **	<0.01 **	<0.01 **
Assessment time (Ta)	<0.01 **	<0.01 **	<0.01 **
Intercrop (Ic)	n.s.	<0.01 **	<0.01 **
Interaction Y × Ta	<0.01 **	<0.01 **	<0.01 **
Interaction Y × Ic	n.s.	<0.01 **	<0.01 **
Interaction Ta × Ic	n.s.	<0.01 **	<0.01 **
Interaction Y × Ta × Ic	n.s.	<0.01 **	<0.01 **

**—differences significant at 99% probability levels, n.s.—no significant.

Table 2. The variation in the aboveground mass (kg DM ha^{-1}) of oat during oat reproductive periods in 2018 and 2019.

Treatment	2018			2019		
	Oat Reproductive Growth Stage (BBCH)					
	Flowering (61–63)	Grain Development (71–73)	Ripened Grain (87–89)	Flowering (61–63)	Grain Development (71–73)	Ripened Grain (87–89)
O	4733	5154	5714	6047	10,138	8283
O+BM	4681	5538	5739	5612	10,514	8219
O+WC	4595	5842	5469	5423	9930	8023
O+EC	5096	6023	5729	5076	10,237	8048
Interaction Y × Ic	4776 a	5639 ab	5663 b	5540 ab	10,205 c	8143 c

O—oat monocrop; intercropping systems: O+BM—oat–black medick, O+WC—oat–white clover, O+EC—oat–Egyptian clover; means followed by the same letters are not significantly different at $p \leq 0.05$.

The legume aboveground mass yields were influenced by the interaction of all three factors (year, assessment time and treatment, Table 1). During oat flowering, the legume in oats–forage legume intercropping systems did not have any negative effect on the yield of oat aboveground mass in either year (Figure 2). According to the data recorded for 2018, the aboveground mass of legumes was 2.2 times higher on average and that of weeds was 70.6% lower on average compared to 2019. In both years, the aboveground mass was significantly higher for EC when grown as a monocrop compared to other legumes.

During oat grain development stage, the aboveground mass of the legume monocrop increased most in 2018 (4.1 times) compared to 2019 (2.7 times). In both years, the aboveground mass of legumes grown with oats increased similarly (2.7–2.8 times) compared to the first assessment. The mass of different legume species grown in the sole crop and together with oats varied significantly. In terms of the aboveground mass yield, the legumes grown as monocrops ranked as follows: EC > WC > BM. The variations between the yields of these species of legumes were significant. The aboveground mass yield of legumes grown as monocrops was greater both years compared to those grown together with oats.

As plants matured, the assimilated materials accumulated in the aboveground mass were transported from leaves to seeds. The oat aboveground mass dried up, exposing the lower crop level. In 2018, during the reproductive period, the mass of perennial legumes increased both in the monocrop and legumes intercropped with oats, with the exception of EC. In 2019, the aboveground mass of legumes increased by 4.7 times on average, this was due to better weather conditions. While assessing different legume species, it was found that the lowest yield of the aboveground mass was that of BM, and the highest was of WC and EC. There was no significant difference between the aboveground mass yields of the latter species, either as monocrops or intercrop.

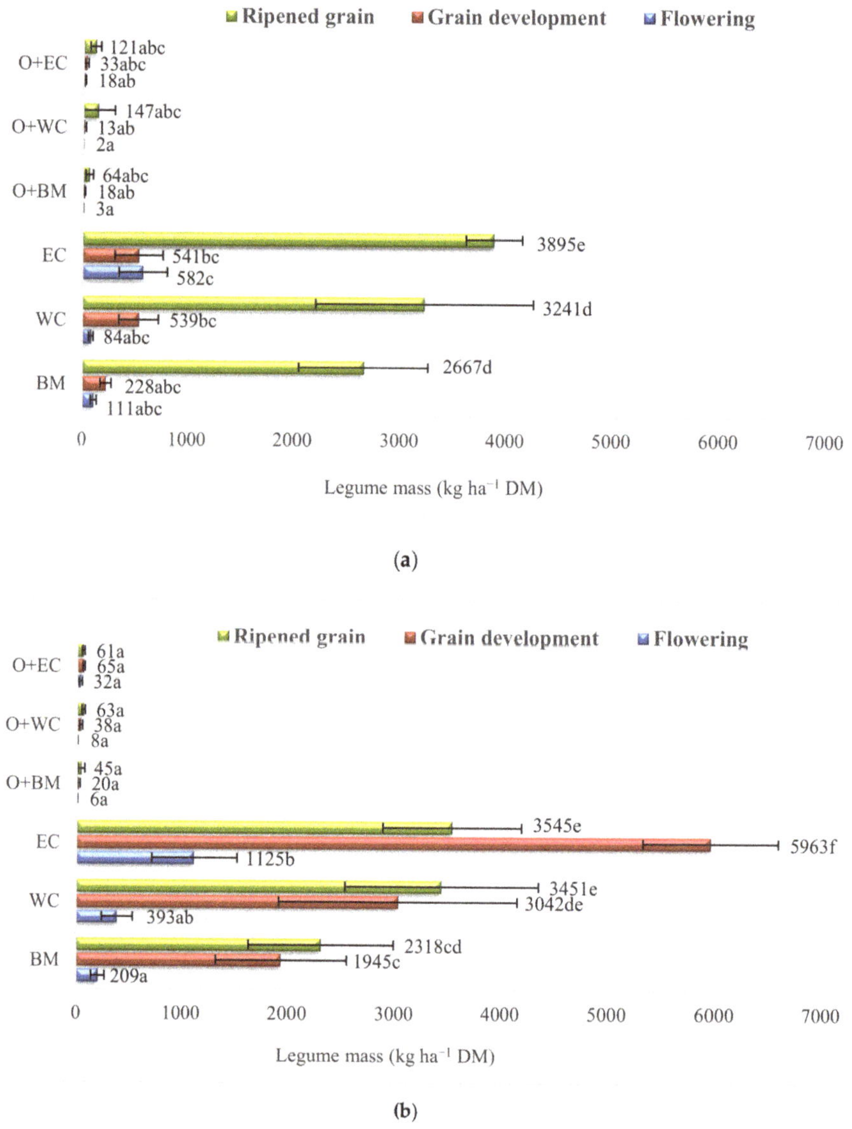

Figure 2. Forage legume aboveground mass during oat reproductive periods in 2018 (**a**) and 2019 (**b**). BM—black medick, WC—white clover, EC—Egyptian clover; intercropping systems: O+BM—oat–black medick, O+WC—oat–white clover, O+EC—oat–Egyptian clover; means followed by the same letters are not significantly different at $p \leq 0.05$.

3.2. Oat and Forage Legume Competition

Competitive oat (CP_o) results in the oat–forage legume relay intercrop were expressed as a percentage decrease in the aboveground mass of forage legume (Table 3). The competitive performance was significantly ($p < 0.01$) influenced by year. The relay intercropping systems assessment time did not have any significant effect on this indicator. In all assessment time, a strong decrease (91.4–98.9%) in the aboveground mass of forage legumes was found. On average, in 2019, the CP value was significantly 2.5 percentage points lower than in 2018. Plant competition in relay intercrop depended on the influence of meteorological conditions on the parallel germination and growth of oats and forage legumes.

Table 3. Competitive performance (CP_O %) of oats in oat–forage legume relay intercropping (mean ±).

Intercrop	2018			2019		
	Oat Reproductive Stage			Oat Reproductive Stage		
	Flowering	Grain Development	Ripened Grain	Flowering	Grain Development	Ripened Grain
O+BM	97.1 ± 1.31	98.9 ± 0.47	98.0 ± 0.81	96.3 ± 2.89	91.4 ± 3.17	97.7 ± 0.53
O+WC	97.7 ± 0.67	98.5 ± 057	98.1 ± 0.28	96.2 ± 2.98	97.5 ± 2.08	95.7 ± 2.11
O+EC	97.0 ± 1.07	98.9 ± 0.17	98.2 ± 0.34	95.9 ± 2.33	92.5 ± 3.95	97.0 ± 0.65
Mean of year			98.1 a			95.6 b

Intercropping systems: O+BM—oat–black medick, O+WC—oat–white clover, O+EC—oat–Egyptian clover; means followed by the same letters are not significantly different at $p \leq 0.01$.

The growth intensity of the aboveground mass of forage legumes varied from year to year (Figure 3). Forage legumes grew most intensively in the following oat growth stages: in 2018 (BBCH 61–73), and in 2019 (BBCH 71–89). This did not affect the legume yield. The growth of legumes was influenced not only by oat productivity (competition), but also by favourable environmental conditions. Under favourable conditions, the legumes easily adapted to the growth rhythm and intensity of oats. In extensive intercultural systems, the yield of legumes is low and the main growth takes place after the oats are harvested.

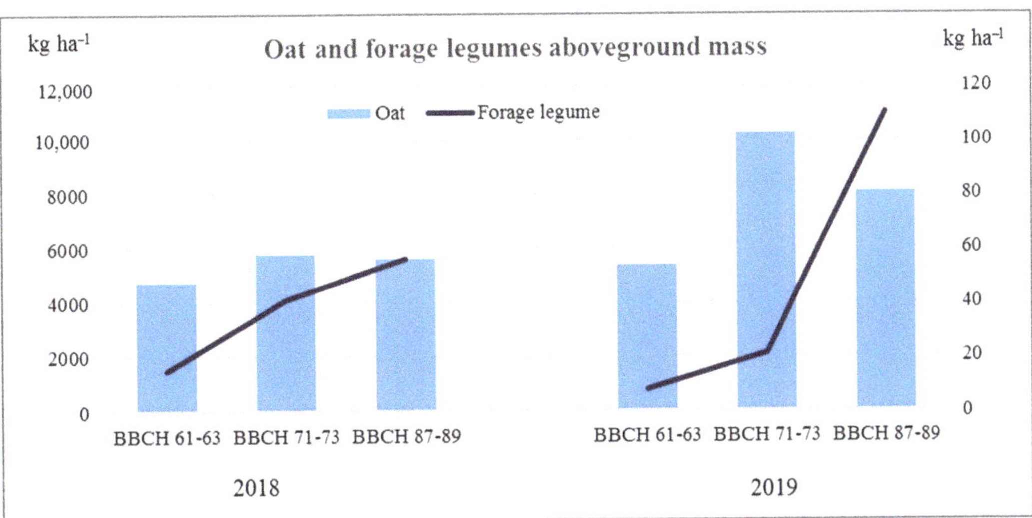

Figure 3. Dynamics of oat and forage legume aboveground mass change during the vegetation period in intercropping systems (average data). Oats growth stages: (BBCH 61–63) flowering, (BBCH 71–73) grain development, (BBCH 87–89) ripened grain.

3.3. Weeds Mass

Statistical analysis results showed that the legume aboveground mass yields were influenced by the interaction of all three factors (year, assessment time and treatment, Table 1). During oat flowering the greatest weed mass was found in all types of forage legume sole crops (Figure 4). During grain development, the weed mass increased 4.5 and 1.3 times (in 2018 and 2019, respectively) compared to the flowering stage. In both years, a significantly greater weed mass was found in legume monocrops, being inversely proportional to the legume mass. The greatest weed mass was found in the BM monocrop. In oat–forage legume intercropping systems the mass of weeds was significantly lower compared to the legume monocrops.

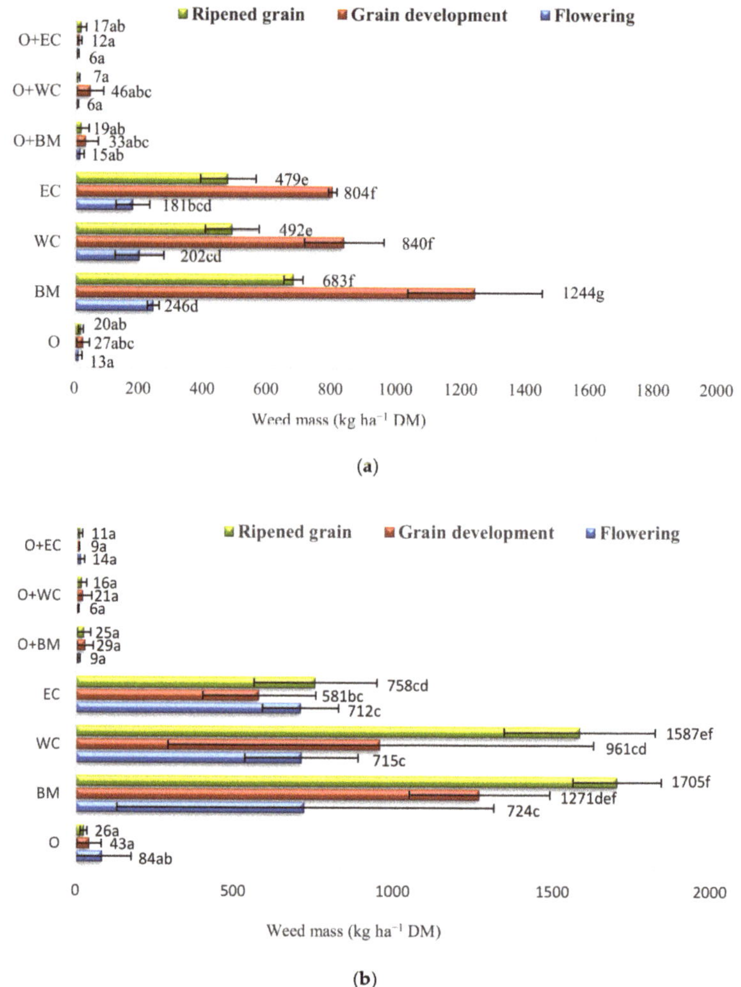

Figure 4. Weed aboveground mass during oat reproductive periods in 2018 (**a**) and 2019 (**b**).BM—black medick, WC—white clover, EC—Egyptian clover; intercropping systems: O+BM—oat–black medick, O+WC—oat–white clover, O+EC—oat–Egyptian clover; means followed by the same letters are not significantly different at $p \leq 0.05$.

In 2018, weed mass decreased compared to the second record (ripened grain). Significantly higher weed mass was found in legume monocrops and depended on the yield of aboveground mass of legumes. Meanwhile, the weed mass in 2019 varied less consistently. The increase in weed mass was greatest in legume monocrops. In both years, the weed mass decreased most in the oat monocrop and in O+BM and O+WC relay intercropping systems (compared to the second record). Hereupon, the weed mass tended to increase in growing oats intercropped with annual EC, compared to the second record. This species of annual clover matures earliest and exposes the soil surface, thus creating favourable conditions for weeds to grow.

According to the data from both study years, significant competitive relationships between the bottom level plants (legumes and weeds) and oat in intercropping systems were established at the beginning of oat reproduction. A moderate inverse linear relationship was obtained between the aboveground mass of oats and the mass of lower-level plants (forage legumes and weeds) (Figure 5). There were no consistent relationships during oat maturation. In intercropping systems, the correlations of weed aboveground mass with forage legume mass were weak and nonsignificant at all measurement dates.

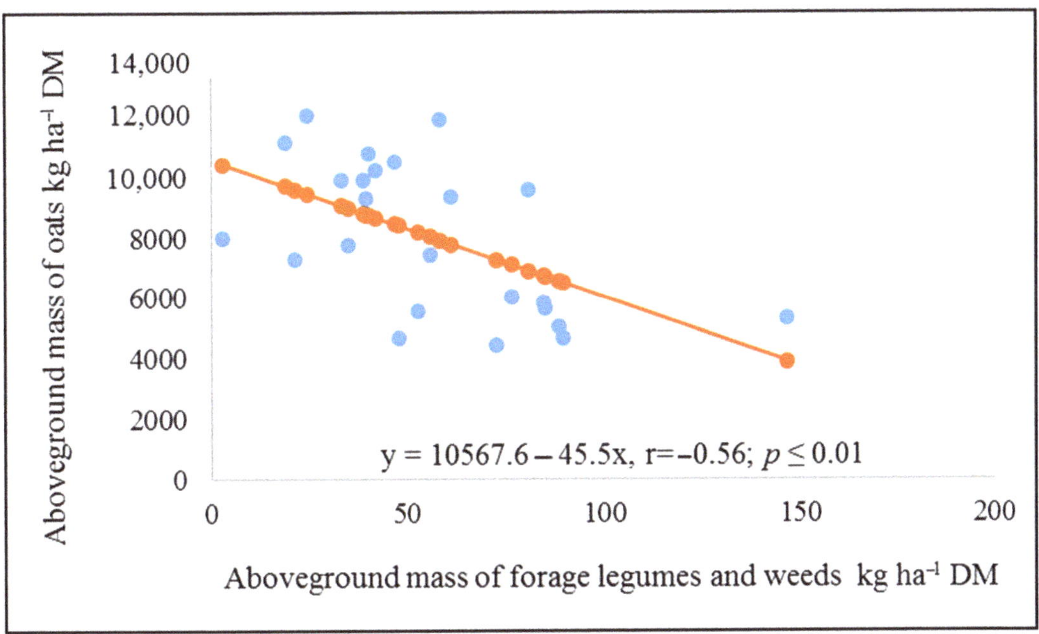

Figure 5. Dependence of aboveground mass yield of oats (grain development stage) on the total aboveground mass of forage legumes and weeds in oats–forage legume intercropping systems.

3.4. Oat Grain Yield and Its Components

Our results showed that the yield and its components were most affected by the meteorological conditions of the years. The influence of the forage legumes was not significant. Having compared the data for different growth periods it was found that the number of plants and panicles in 2019 was lower, and the number and weight of grains per panicle and TGW were higher compared to the data for 2018 (Table 4).

Table 4. The variation in grain yield components of oat growing with or without forage legumes.

Mono- and Intercrops	Crop Number (Plant m^{-2})		Number of Panicles (Panicle m^{-2})	Grain Number per Panicle	Grain Weight per Panicle g	TGW g	Grain Yield Kg ha^{-1}
	Forage Legumes	Oats					
			2018				
O		411	418	44.3	1.46	32.3	2889 ab
O+BM	33	408	445	34.3	1.11	32.9	2900 ab
O+WC	32	413	425	37	1.16	32.9	2932 ab
O+EC	29	396	451	36	1.15	32.7	3058 b
Mean	31 a	407 b	434 b	37.9 a	1.22 a	32.7 a	2945
			2019				
O		339	371	55	1.99	38.1	4087 b
O+BM	38	334	380	57	2.01	38.5	4075 ab
O+WC	42	328	382	55	2.02	39.2	3974 ab
O+EC	49	329	348	57	2.02	38.9	3892 ab
Mean	43b	333 a	370 a	55.8 b	2.01 b	38.7 b	4007

O—oat, BM—black medick, WC—white clover, EC—Egyptian clover; intercrop systems: O+BM—oat–black medick, O+WC—oat–white clover, O+EC—oat–Egyptian clover; means followed by the same letters are not significantly different at $p \leq 0.05$.

This was due to meteorological conditions in 2019 (Figure 1a). As far as legume and oat intercropping was concerned, in 2018, the number and weight of grains per oat panicle were significantly reduced compared to oat monocropping. In 2019, less favourable plant germination conditions led to a thinner oat crop and less consistent productivity indicators. Productivity rates of oat panicles were significantly higher in the thinner crop compared to 2018. There was no significant difference between treatments. The change in yield was due to the different distribution of productivity indicators.

3.5. Nutrient Content in Oat Grain and Legume Biomass

Oat grains were accumulated 15.0–15.80 and 20.79–21.54 g N kg^{-1} DM, 3.29–3.64 and 3.12–3.33 g P kg^{-1} DM in 6.62–7.09 and 3.04–3.30 g K kg^{-1} DM in 2018 and 2019, respectively. Concentration of nutrients (NK) in grains depended on the year ($p < 0.01$), legumes had no significant effect. The amount of nutrients accumulated in the oats grain was influenced by the yield. Intercropping of forage legumes with oats and annual conditions influenced the content of nutrients accumulated in grain. In 2019, the N concentration in grain was 38.7% higher on average, and nitrogen content accumulated in the yield was nearly two times higher compared to 2018. Due to the influence of the year, phosphorus concentration in oat grain varied slightly. In 2018, significantly more potassium was used to grow 1 kg of oat grain compared to 2019. These study data indicated a tendency for annual O+EC intercropping systems to increase competition with oats for nutrients, in contrast to perennial forage legumes.

More pronounced differences in NPK concentrations were found in the aboveground mass of forage legumes (Table 5). In 2018, significantly lesser nitrogen concentrations were found in the EC mass when intercropped with oats compared to other forage legume species, regardless of their cultivation method (in mono- and intercropping). In 2019, a significantly higher concentration in the WC aboveground mass was found in the O+WC relay intercrop compared to the monocrop. The greatest aboveground mass P concentration was measured in EC (2018) and BM (2019), regardless of the cultivation method. Correlation analysis showed that with increasing P concentration in grain of oat, its value in aboveground mass of forage legume decreased as well (Figure 6). The most adverse effects were found for EC. In 2019, favourable for plant growth, this relationship was nonsignificant (r = −0.49).

Table 5. Amount of nutrients accumulated in aboveground mass of forage legumes.

Treatment	Concentration of Nutrients, g kg^{-1} DM			Accumulated Nutrients, kg ha^{-1} DM		
	N	P	K	N	P	K
			2018			
BM	28.10 bc	2.84 ab	22.77 ab	65.13	6.58	52.32
WC	28.87 bcde	2.95 ab	32.40 cd	117.01	10.22	111.48
EC	28.50 bc	3.29 d	30.37 c	100.72	11.65	106.32
O+BM	27.80 b	2.99 abcd	40.17 def	1.33	0.14	1.60
O+WC	27.83 b	2.98 ab	41.27 f	1.75	0.19	2.61
O+EC	22.20 a	3.29 cd	33.87 cdef	1.36	0.20	2.06
			2019			
BM	30.9 bcde	2.97 ab	27.93 bc	82.51	7.95	74.51
WC	28.87 bcde	2.75 a	34.13 cdef	96.36	8.84	110.63
EC	29.63 bcde	2.88 ab	26.90 abc	115.74	11.25	105.12
O+BM	31.75 cde	3.10 bcd	20.93 a	2.09	0.20	1.33
O+WC	32.68 e	2.88 ab	31.68 c	4.86	0.43	4.56
O+EC	29.63 bcde	2.88 ab	26.90 abc	3.63	0.35	3.31
			Mean of I c			
BM				73.82 d	7.26 b	63.41 b
WC				106.69 c	9.53 c	105.72 c
EC				108.23 c	11.45 d	111.05 c
O+BM				1.71 a	0.17 a	1.47 a
O+WC				3.3 a	0.31 a	3.59 a
O+EC				2.50 a	0.28 a	2.69 a

BM—black medick, WC—white clover, EC—Egyptian clover; intercrop systems: O+BM—oat-black medick, O+WC—oat-white clover, O+EC—oat-Egyptian clover; N—nitrogen, P—phosphorus, K—potassium; means followed by the same letters are not significantly different at $p \leq 0.05$.

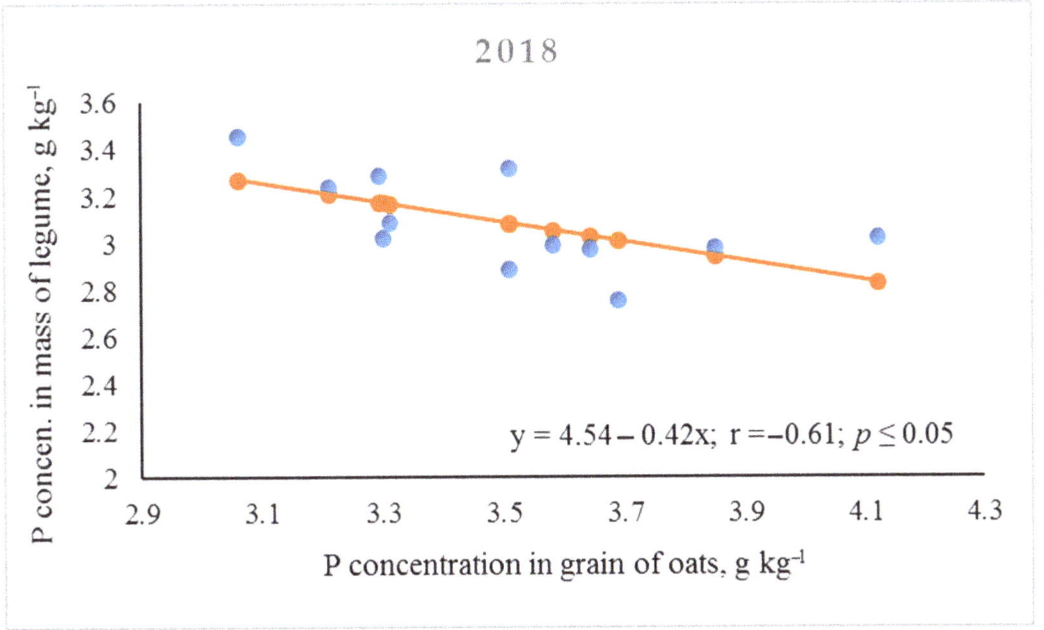

Figure 6. Dependence of P in grain yield of oat (ripened grain stage) on the P of aboveground mass of forage legume in intercropping systems.

The highest potassium concentration was found in the WC biomass, regardless of the year and cultivation method. Black medick grown in monocrop (2018 and intercrop (2019) also EC (regardless of their cultivation method in 2019) had significantly less concentration of K in aboveground mass.

The significant influence of forage legume cultivation methods (in mono- and intercrop) on nutrient accumulation in aboveground mass has been identified ($p < 0.01$). Having compared different legume monocultures, BM was found to have accumulated the lowest nitrogen amount in the aboveground mass. Forage legumes intercropped with oats demonstrated low nitrogen accumulation (1.33–1.75 kg ha^{-1} DM in 2018 on average, and 2.09–4.86 kg ha^{-1} in 2019), there was no significant difference between legume species. A very low phosphorus amount was accumulated in the legume mass in relay intercropping systems. WC and EC, grown as monocrops, demonstrated higher phosphorus content in biomass. Of the forage legumes grown in monoculture, BM accumulated the lowest levels of potassium (as well as nitrogen and phosphorus), and WC accumulated the highest ones. Similarly, the potassium content varied in the aboveground mass of other forage legumes intercropped with oats.

4. Discussion

Oats held the dominant position in oat–forage legume relay intercropping systems. The regularities of oat yield formation were determined by meteorological conditions and self-regulatory functions of crop productivity. In 2019, underdeveloped reproductive stems were compensated by a higher number of grains and their weight compared to a denser 2018 crop. Our studies suggest that the forage legume in relay intercropping systems may have a negative effect on later-emerging crop components (grain number and weight); nevertheless, it did not reduce yield. The data are consistent with those of other researchers who argue that legumes in relay intercropping systems have no significant effect on cereal yield [8]. According to Gaudin et al. [11] red clover can compete with cereals and even reduce their yield.

Studies of the dynamics of oat aboveground mass accumulation showed that the accumulation of DM took place most intensively in 2019 and continued even after the flowering of cereals. This year was characterised by cool weather and sufficient rainfall. The period from flowering to the beginning of maturity, when there is enough moisture, nutrients and maximum amount of solar energy, is of great importance for plant productivity [27]. During oat maturation (mid-July to early August), the redistribution of accumulated assimilated materials in plants resulted in a decrease (2019) or a marginal variation (2018) in the aboveground mass of oats. A number of researchers have observed a decline in cereal competition at the end of the plant growth period [7]. During this period, favourable conditions are created for the plants of the lower crop level to grow. It is argued that the dominance of cereals over forage legumes is necessary to avoid cereal yield losses [28].

In relay intercrops, the aboveground mass of forage legumes began to form more intensively after heading of oats. Annual EC developed most intensively, demonstrating the highest mass. A previous study [7] indicated that the highest competition between spring wheat and annual Persian clover (*Trifolium resupinatum* L.) occurs when cereals are in the stem elongation stage (BBCH 31–32); the mass of annual clover decreases twice during the entire growth period. Contrary to Barilli et al. [10] and Sharpe et al. [24] the aboveground mass of BM was the lowest in intercrop. This may have been determined by the genetic diversity of the species and varietal characteristics. According to other researchers, BM and red clover competed best with winter wheat, in contrast to BM and WC [8]. Our studies concur with Bybee-Finley and Ryan [5] that the accumulation of forage legume mass in a cereal crop is minimal and most of the biomass is formed after cereal harvest. It is believed that forage legume mass increases six times during the post-harvest period [29]. Environmental variables such as soil type, precipitation quantity and distribution during the growing season and day length also have an effect on biomass accumulation [11]. Fertile clay loam soils have a higher supply of resources and lower competition between plants compared to less fertile ones, where oats are usually grown. Clover is considered to be an under sown crop that poorly competes with cereals [30]. Our study substantiated this argument. In relay intercrops, WC, BM and EC were dominated by oats for all growing periods.

Under sown forage legumes establish in the lower crop level and compete with weeds. This is especially important in the second half of summer when an under sown clover with drying crop leaves covers the soil surface [11]. Plants and weeds compete for light, water and nutrient resources [31,32]. Den Hollander et al. [33] state that the relative growth rate is determined by the characteristics of clover, such as light extinction coefficient, light use efficiency and specific leaf area. It was determined that the fastest soil surface cover was demonstrated by Persian clover. [11] reported that forage legume species that produce a high yield of aboveground mass are considered to be effective competitors for local resources. The competitive properties of forage legumes also depend on the sowing rate and sowing time. Forage legumes under sown in cereals can suppress weeds, however, legumes can compete with cereals too [33]. Verret et al. [34] indicated that the use of legume companion plants generally seemed to enhance weed control without reducing crop yield. Our research shows that oat was the most weed suppressive, and the forage legumes in relay intercropping systems only enhanced their effect as oats matured. Egyptian clover competed best with weeds, as confirmed by other researchers [35]. The positive effect of low-mass forage legumes on weed reduction cannot be assessed unequivocally. Many weed species are adapted to spread in cereal crops. Therefore, as cereals mature and forage legumes grow intensively, some of the weeds have already matured their seeds and dried out.

In our studies, forage legumes accumulated low N, which is confirmed by other researchers [30]. A decrease in light and water resources under the cereal canopy may also directly reduce nodule formation and N fixation in clover species [36]. On the other hand, forage legumes that are intercropped with cereal also fix a greater proportion of nitrogen than legumes grown in monoculture [13]. Nitrogen uptake in under sown crops is affected by competition with main crops [37]. It is proposed that the competition for N can be detrimental when cereals and legume catch crops are sown simultaneously in spring [8]. Additionally, catch crops generally seem more suitable as post crops for P [38]. Plant species with different growth cycles can ensure a more efficient use of environmental resources. We can say that, in our studies, the drought after sowing postponed the competition for resources between the intercropped plants and oats. Forage legumes, that produced a small aboveground mass, accumulated low N levels, which is confirmed by other researchers. Our research has shown that EC with a growth period similar to oats can compete more intensively with oat for resources. Meanwhile, the intensity of WC and BM nutrient uptake increased with maturation of oats and the decrease in nutrient utilisation by the oat. The high yield of intercropping is connected to better exploitation of soil resources, and deep rooting of some species is a determinant factor for complementarity in competition for soil resources [9]. They may also have the ability to absorb different quantities of nutrients and produce distinct root exudates (organic acids) resulting in benefits both for the soil and organisms [10]. Finally, one crop can provide resources for the other one with a positive interspecific interaction, which is at the basis of facilitation processes [10].

5. Conclusions

Oats dominated in oat–forage legume relay intercropping systems. Meteorological conditions of the year and crop self-regulation functions had a significant impact on the yield components of oats and the regularities of aboveground mass formation. This determined the intensity of aboveground mass formation of forage legumes in relay intercrop. Forage legumes grew most intensively in the following oat growth stages: In 2018 (BBCH 61–73), and in 2019 (BBCH 71–89). Annual EC demonstrated the earliest aboveground mass formation. Oat competitive performance (CP_o), which is characterised by forage legumes aboveground mass reduction compared to monocrops, were 91.4–98.9%. As the

MDPI

Article

Foodshed, Agricultural Diversification and Self-Sufficiency Assessment: Beyond the Isotropic Circle Foodshed—A Case Study from Avignon (France)

José Luis Vicente-Vicente [1,*], Esther Sanz-Sanz [1,2], Claude Napoléone [2], Michel Moulery [2] and Annette Piorr [1]

[1] Leibniz Centre for Agricultural Landscape Research (ZALF), 15374 Müncheberg, Germany; Maria-Esther.Sanz_Sanz@zalf.de (E.S.-S.); apiorr@zalf.de (A.P.)

[2] UR Ecodeveloppement, French National Institute for Reasearch on Agriculture, Food and Environment (INRAE), 84914 Avignon, France; claude.napoleone@inrae.fr (C.N.); michel.moulery@inrae.fr (M.M.)

* Correspondence: vicente@zalf.de

Abstract: The regionalization of food systems in order to shorten supply chains and develop local agriculture to feed city regions presents particular challenges for food planning and policy. The existing foodshed approaches enable one to assess the theoretical capacity of the food self-sufficiency of a specific region, but they struggle to consider the diversity of existing crops in a way that could be usable to inform decisions and support urban food strategies. Most studies are based on the definition of the area required to meet local consumption, obtaining a map represented as an isotropic circle around the city, without considering the site-specific pedoclimatic, geographical, and socioeconomic conditions which are essential for the development of local food supply chains. In this study, we propose a first stage to fill this gap by combining the Metropolitan Foodshed and Self-sufficiency Scenario model, which already considers regional yields and specific land use covers, with spatially-explicit data on the cropping patterns, soil and topography. We use the available Europe-wide data and apply the methodology in the city region of Avignon (France), initially considering a foodshed with a radius of 30 km. Our results show that even though a theoretically-high potential self-sufficiency could be achieved for all of the food commodities consumed (>80%), when the specific pedological conditions of the area are considered, this could be suitable only for domestic plant-based products, whereas an expansion of the initial foodshed to a radius of 100 km was required for animal products to provide >70% self-sufficiency. We conclude that it is necessary to shift the analysis from the size assessment to the commodity-group–specific spatial configuration of the foodshed based on biophysical and socioeconomic features, and discuss avenues for further research to enable the development of a foodshed assessment as a complex of complementary pieces, i.e., the 'foodshed archipelago'.

Keywords: foodshed; archipelago; city region; food modelling; food self-sufficiency; self-reliance; food security; agricultural diversification; food planning; regional food system

Citation: Vicente-Vicente, J.L.; Sanz-Sanz, E.; Napoléone, C.; Moulery, M.; Piorr, A. Foodshed, Agricultural Diversification and Self-Sufficiency Assessment: Beyond the Isotropic Circle Foodshed—A Case Study from Avignon (France). *Agriculture* **2021**, *11*, 143. https://doi.org/10.3390/agriculture11020143

Academic Editor: Cornelia Flora
Received: 21 January 2021
Accepted: 5 February 2021
Published: 10 February 2021

1. Introduction

A lack of confidence in conventional market-based agriculture has arisen since the 1990s [1], together with a fear of long-distance food supply disruptions, emphasized by crises such as the covid-19 pandemic [2,3]. Feeding the city on sustainable and healthy agriculture became a local policy concern [4–6], and proximity is an effective way to enhance the confidence. Nevertheless, regional self-sufficiency has not been a focus of policy decision-making until recently [7,8]. In other words, social awareness about sustainable regional food security requests an increase in regional—or domestic—food self-sufficiency levels [6,9–14], where dietary patterns, consumer behaviors, and diversified farming play an important role [5,15]. In addition to implementing farming-related concepts, such as ecological intensification, the challenge is to enhance the efficiency of food chains, building upon proximity in all of the diversity of emerging concepts, and linking local agricultural supply to the urban final demand [10]. There is no consensus regarding a definition of 'local' and 'regional food systems' in terms of the distance between production and consumption, and the concept remains an elision implicitly contrasted to the 'global' [2]. Furthermore, the region is a social construct shaped by networks and connectivity, in which the formal territorial jurisdictional functions and capacities intersect with contingent interests [16]. Inspired by the relational approach of Clancy and Ruhf [17], the concept of a 'regional food system' is considered in this article as the system in which as much food as possible is produced, processed, distributed and purchased to meet the population's demands within a particular meaningful geographical area. Some existing methods analyze the main characteristics and drivers of regional food systems in a specific context. On the one hand, qualitative methods, such as socio-empirical surveys, are able to finely illustrate the stakeholders' behaviors [18]. On the other hand, quantitative food assessments can give an overview of the status of the food supply and demand [19–25], whereas other methodologies are focused on the current spatial distribution of crops and land use change dynamics (e.g., in urban areas) [26,27].

Specifically, quantitative foodshed approaches can assess the capacity or flows, or both approaches at the same time [28]. In the capacity assessments, to which the majority of the studies belong, the theoretical food land footprint and the potential self-sufficiency are evaluated by considering the population, current dietary patterns, farmland available, land use cover, and regional yields (e.g., the Metropolitan Foodshed and Self-sufficiency Scenario: MFSS [29]). Such an approach is very valuable to raise the urban residents' awareness of the spatial impact of their current food diet by highlighting theoretical changes in the extension of the land footprint depending on different scenarios (e.g., the change in the land footprint if one shifts to a more plant-based diet, or from a conventional to organic food diet) [30,31] or to assess the role of public procurement in food self-sufficiency [32]. As foodshed models use data on food consumption and production, and take into account the land cover, the result is the achievement of a theoretical self-sufficiency level for all food commodities, or at least for some of them. While the models addressing all food commodities do not consider the real land allocation to specific crops, but rather only the type of land cover and yield level [29], others focus on specific crops but are able to allocate them [33,34]. The second type of foodshed approaches, the ones assessing the flows [22,23,35], are especially valuable to the study of the distribution networks, as they place consumers and producers. Finally, the hybrid approaches combine the assessments of the capacity and the flows (e.g., [21,36,37]) and, thus, are aimed at comparing the potential food self-sufficiency with the current levels; therefore, they assess the dependencies on foreign food sources, the vulnerabilities of the food system, and the agricultural environmental impacts of the food system's relocalization [28]. The vast majority of the foodshed assessments are developed at a regional level, although some global-scale studies and models have recently appeared (e.g., [38,39]).

However, in order to enforce a local food policy responding to the willingness to establish regional food proximity, empirical evidence on the food self-sufficiency capacity is required, which takes into account the local agronomic heterogeneity of soils as well as various farming systems and marketing modes. In that way, public action can be located where it is most likely to be effective. Therefore, a foodshed is not a standard concept that could be applied to different cases in the same way; rather, different biophysical and socioeconomic conditions should be considered. Soil fertility features, for instance, are usually a key determinant defining the kind and intensity of the agricultural production at a specific location. They are very often not evenly distributed around the urban area in a gradient, as the theoretical concept by von Thünen would suggest, where the type of agriculture is determined by the distance to the city center [29]. By contrast, the spatial distribution of agricultural production responds to the biophysical constraints and the particular history of each place in terms of its urbanization, development of the agricultural sector, organization of activities (including agricultural sectors), and environmental protection [40,41]. Furthermore, the land use is influenced by farm structures and plot sizes, and thus different land covers coexist, especially in the surroundings of urban areas, while other land uses (e.g., extensive livestock farming) only take place in specific areas under suitable biophysical conditions. However, so far, the majority of regional foodshed assessments have been developed in an isotropic way, by considering administrative boundaries and biophysical constraints in a second step (notably, for the availability of monitoring data in high-density and identical quality, e.g., population data). Indeed, foodsheds are usually defined by a radius around the city (i.e., centroid), and are therefore represented as circles [28–30,38]. Accordingly, the foodshed concept represented by just one circle around the city must be reconsidered in order to consider the landscape heterogeneity, and furthermore, to include societal demands. Therefore, we modified the traditional foodshed concept in this study to address these limitations, and applied it to a specific Mediterranean city region, the area of Avignon (France). This specific area is surrounded by high-fertility soils dedicated mainly to commercial agriculture (vegetable, fruit trees and vineyards), and it has a high heterogeneous geomorphology as the distance from the city increases, where soils dedicated to extensive livestock farming appear.

The overall goal of this study is to develop a hybrid foodshed assessment aimed at evaluating the potential and current self-sufficiency of a proposed foodshed. The specific objectives of the study to achieve this end are threefold: (i) to propose and assess a foodshed with a radius of 30 km for the city region of Avignon, which could potentially provide a high degree of self-sufficiency; (ii) to assess the role of agricultural and livestock diversification in increasing the current self-sufficiency within the initial foodshed with a radius of 30 km; and (iii) to propose and discuss the expansion of the initial foodshed considering the landscape heterogeneity and anisotropy, in order to develop a more realistic scenario in terms of achieving a high degree of food self-sufficiency.

2. Materials and Methods

2.1. Study Aea

We first selected a foodshed with a radius of 30 km around Avignon. Thirty kilometers is a nonnormative distance set by the French Senate to define the maximum spatial distance between the site of production and the point of sale for fresh fruit and vegetable short circuits. The initial foodshed selected, formed by a total of 171 communes (i.e., municipalities), comprises two different administrative regions and three different departments (similar to counties) in South-East France: Bouches-du-Rhône and Vaucluse in the region of Provence-Alpes-Côte d'Azur, and Gard in the region of Occitanie (Figure 1A; Tables S1–S3 in the Supplementary Material). Furthermore, the foodshed is close to the administrative region of Auvergne-Rhône-Alpes, particularly the two southern departments, Ardèche and Drôme. The municipality of Avignon is located within the Vaucluse department (Figure 1B).

The area is relatively flat, typically between 0 and 400 m, and is crossed by the Rhône River from north to south. However, the altitudes become higher towards the west (Gard) and east (Vaucluse), and remain low towards the south (Bouches-du-Rhône), where the river flows into the Mediterranean Sea. The soils in the low altitudes are usually deeper, whereas the depth decreases significantly with higher altitudes and slopes. As such, almost all of the foodshed area in Bouches-du-Rhône is formed by deep or very deep soils, whereas this description applies to about half of the area in Vaucluse, and around a quarter of the Gard area (Figure 2).

A)

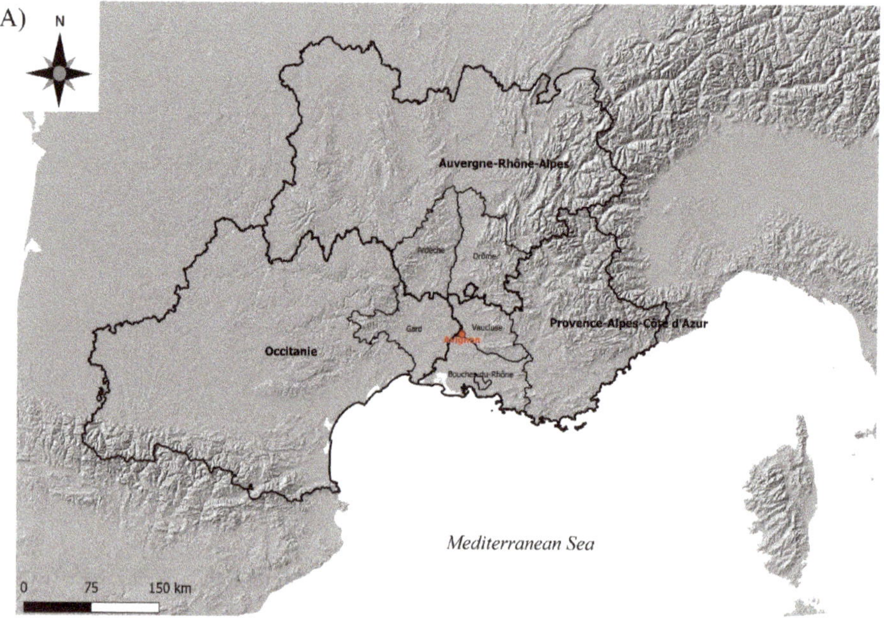

B)

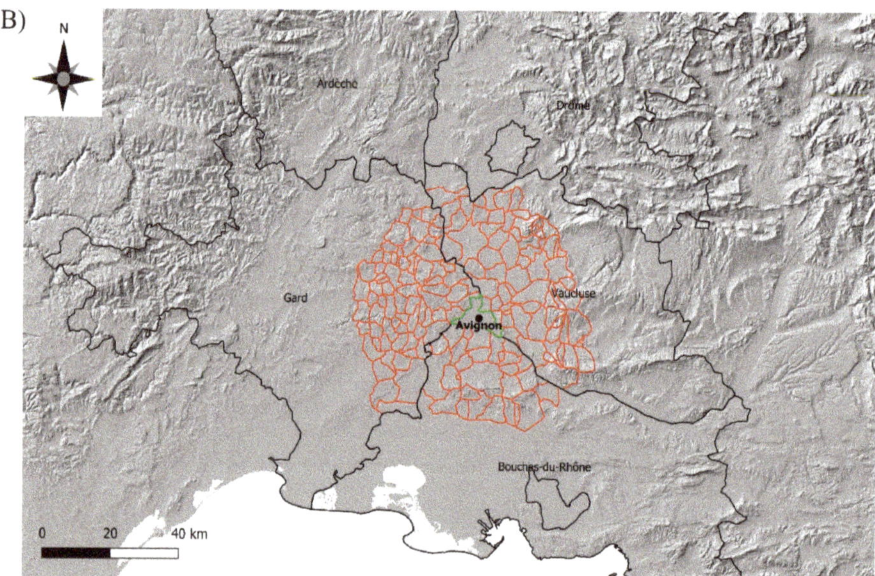

Figure 1. Location of Avignon (Vaucluse department) in South-East France, and the surrounding departments and regions (in bold) (**A**), and the location of the municipalities/communes (in red) forming the selected initial foodshed in a radius of 30 km around the municipality of Avignon (in green) (**B**). Note that the proposed foodshed belongs to two other departments (Gard and Bouches-du-Rhône), and is near the departments of Ardèche and Drôme to the North. Details on the population and surface area for each commune are given in the Supplementary Material (Tables S1–S3).

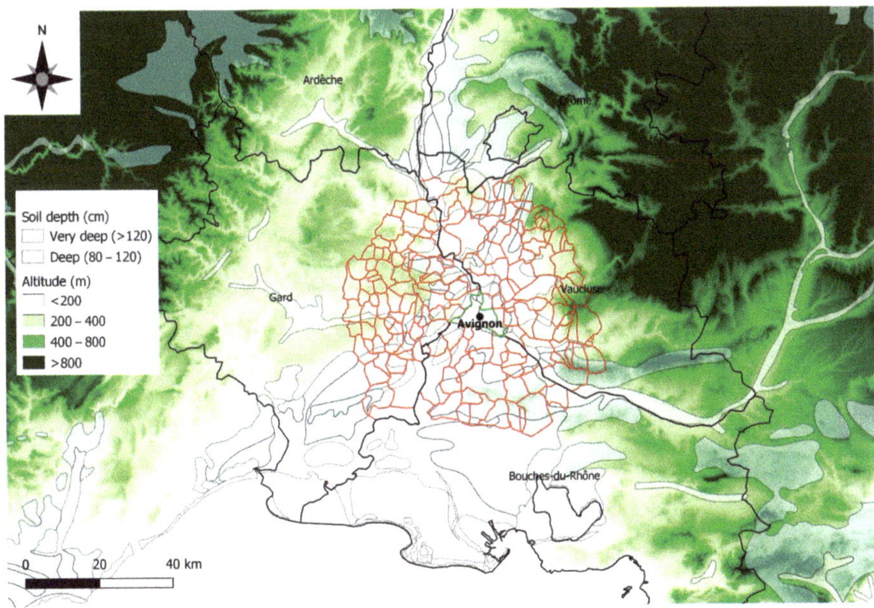

Figure 2. Location of the deep (80–120 cm) and very deep (>120 cm) soils in the foodshed area (in red), and in the surrounding areas of the different departments. The rest of the area is covered by soils with a shallow (0–40 cm) or moderate (40–80 cm) depth. The map also shows the altitudes in four categories (<200, 200–400, 400–800 and >800 m above sea level). Note that deep soils are usually located in low-altitude areas.

2.2. Application of the MFSS Model for the Avignon Foodshed Assessment: Food Land Footprint and the Potential Self-Sufficiency of the Foodshed

The MFSS model [29] incorporates the two dimensions driving the food self-sufficiency analysis: estimated demand and potential supply. The model also distinguishes between domestic and exotic products, and between organic and conventional production systems. However, only one scenario, the business as usual (i.e., conventional, and a mixture of regional- and import-based diets), has been used for the first stage of the Avignon foodshed assessment, which aimed to test whether the initial strategically defined foodshed is suitable for achieving a high degree of potential self-sufficiency.

Very briefly, the model considers the utilizable agricultural area (UAA), which represents the potential area available for agriculture. The Corine Land Cover map (2018) was used to estimate the UAA, and eight land uses were included: non-irrigated arable land, permanently-irrigated land, rice fields, fruit trees and berry plantations, olive groves, pastures, annual crops associated with permanent crops, and complex cultivation patterns. Vineyards were excluded from the UAA assessment, because we assume that their agronomic use will not change in the future due to the high profitability of the wine industry in the study area. In addition, areas formed totally or partially by natural vegetation (e.g., forests or crops with significant patches of natural vegetation) were excluded from the assessment.

The model estimates the area demand for the population within the foodshed—i.e., the area required to meet the food consumption—for each food commodity (i.e., food land footprint) by considering the yields and population. The data on food consumption for 2017 were taken from the Food and Agriculture Organization of the United Nations (FAO) statistics (http://www.fao.org/faostat/en/#data/FBS, accessed on 22 July 2020), whereas the data on the yields for domestic plant-based products are regional, and were taken from national and regional reports for 2017–2018 [42–44] (see Table S4 in the Supplementary Material). The data on the yields for animal products (beef, eggs, poultry, pork, milk and dairy products, mutton and goat) and non-domestic food commodities were taken from Zasada et al., who estimated the land footprint of the FAO's animal products categories by applying conversion factors from European studies [29].

When applying the MFSS model, the aggregated area demand per department is spatially represented by a circle—defined by a radius—with a centroid of the administrative boundary polygon, in this case the municipality of Avignon. The process can be summarized as the combination of the consideration of the UAA inside and outside the boundaries. The UAA for the whole region is represented as the overall agricultural area share of the region [29]. Therefore, the potential food self-sufficiency of the foodshed is estimated as the ratio between the area demand—or the food land footprint—and the current UAA to meet the regional food demand. Thus, food self-sufficiency values higher than 100% mean that the complete area demand for food production can be met within the boundaries of the foodshed. On the contrary, values lower than 100% would require food imports.

2.3. Materials Used for the Current Crop Production and Self-Sufficiency Level Assessment of Plant-Based Products

It is necessary to assess the current crop production in order to evaluate the role of agricultural diversification in increasing food self-sufficiency. This was carried out by using the 2014 Land Parcel Identification System (LPIS)—graphically represented in the French *Registre Parcellaire Graphique* (RPG)—which geo-locates and informs about areas under different European Union aid schemes of the Common Agricultural Policy (CAP). The current area dedicated to the different crops was then compared with the food land footprint, which was estimated previously for the foodshed by applying the MFSS model. Thus, the current level of self-sufficiency is determined from the current dedicated area:area demand ratio, and is expressed as a percentage (Figure 3).

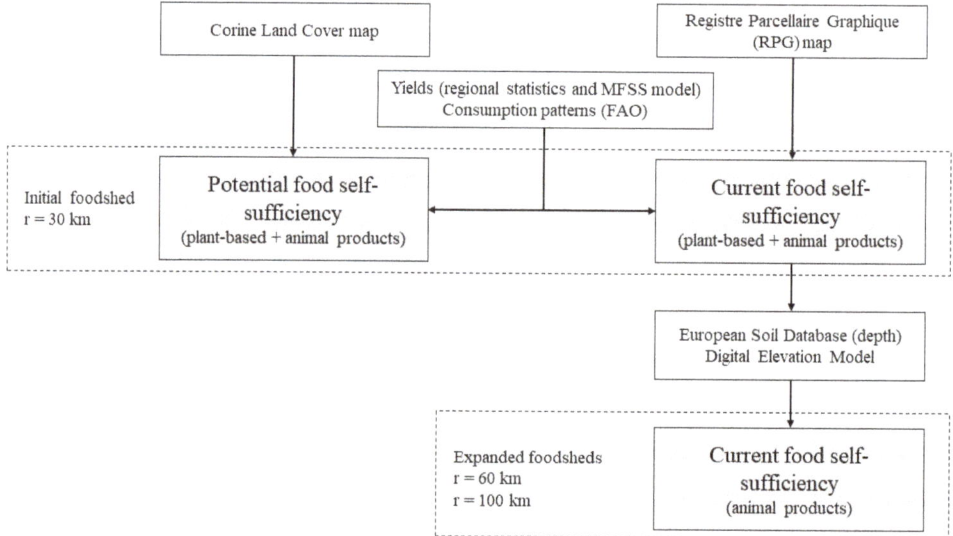

Figure 3. Scheme of the methodology followed in the study.

The assessment was carried out by grouping the plant-based products most commonly consumed following the LPIS (RPG)-MFSS model categories, excluding less relevant food products. Five food products were assessed: cereals, vegetables, pulses, fruits from temperate areas, and wine and grapes. Oilseeds and nuts were also assessed, but the results are shown only in the Table S5 (see Section 3.2).

2.4. Materials Used for the Food Land Footprint and Foodshed Assessment of Animal Products

The current area dedicated to the production of animal products was estimated by assessing the information provided by the RPG map. Three categories were selected for the assessment: fodder, temporary grasslands, and permanent grasslands. Summer pastures were excluded from the analysis, because they are available only during a short period of time in the study area. The area demand for the consumption of animal products was estimated by applying the organic scenario of the MFSS model. The selection of the organic product system instead of the conventional one is because organic livestock farming is more often linked to extensive farming systems (i.e., the use of grasslands or pastures as animal feed) in the areas near where the livestock farm is located. The current self-sufficiency for animal products within the foodshed radius of 30 km was estimated in the same way as that developed for the plant-based products (Figure 3).

Since the study area is dedicated mainly to growing commercial crops (predominantly vegetables and fruit; see Table S5 in the Supplementary Material), an expansion of the foodshed was assessed solely for animal products, considering only unsuitable soils for commercial crops (Figure 3). One explanatory variable, the soil depth, was selected in order to address this issue, as pastures and fodder for extensive agriculture are usually located in shallow-depth and weakly-developed soils (AC soil profile), whereas commercial crops are usually placed in deep and highly-developed soils (ABC soil profile). The soils closer to the river in the study area are classified as Luvisols or Cambisols, whereas Leptosols are the most common types in mountainous areas, followed by Cambisols [45]. The European Soil Database was used for this analysis. This database identifies soils according to different properties, and the category 'soil depth to rock' is one of them. Thus, four categories of soils are distinguished: (i) shallow (<40 cm), (ii) moderate (40–80 cm), (iii) deep (80–120 cm), and (iv) very deep (>120 cm). We considered that commercial crops are more likely to be grown in deep and very deep soils (>80 cm), whereas fodder and pastures are mostly located in shallow and moderate soils (<80 cm).

After selecting the areas currently dedicated to feeding livestock, and excluding those located in deep and very deep soils, two radii for the expanded foodshed were considered: (1) 60 km and (2) 100 km (Figure 3), and two other departments located in the Auvergne-Rhône-Alpes region to the north—but very close to the borders of the foodshed (Ardèche and Drôme)—were included in the assessment. The first expanded radius, 60 km, was selected in order to include only those mountainous areas that are very close to the initial foodshed of 30 km, whereas the purpose of the second expanded radius, 100 km, was to include the mountainous areas of the five departments surrounding the initial foodshed (Figure 2).

2.5. Methodology Used for the Assessment

A summary of the methodology followed for the development of the analysis is shown in Figure 3. The area demand for the different products was extracted from the MFSS model [29]. The yields for plant-based products were taken from regional statistics [42–44]. The potential self-sufficiency analysis is based on the Corine Land Cover Map [46] and FAO data on food consumption without considering food waste (http://www.fao.org/faostat/en/#data/FBS, accessed on 22 July 2020), whereas the assessment of the current self-sufficiency for the plant-based and animal products for the current and expanded foodsheds (60 and 100 km radius) were based on the LPIS database that is graphically represented in the RPG map [47], considering only people living within the initial foodshed of 30 km [48]. The assessment of the soil depth was carried out by using the European

Soil Database [49], whereas the elevation was taken from the Digital Elevation Model over Europe [50]. The land cover and crop area assessments, as well as the soil and expanded foodshed assessments, were developed using QGIS 3.12.1 [51].

3. Results

3.1. Foodshed Assessment and Potential Self-Sufficiency for the Proposed Foodshed

Table 1 and Figure 4 summarize the results of the area demand and potential food self-sufficiency for plant-based and animal products. The communes within the Bouches-du-Rhône department had the highest potential self-sufficiency, 189%, due to the high amount of UAA per capita (3861 m^2) compared to the area demand per capita (2047 m^2). However, the self-sufficiency values for the communes—or municipalities—belonging to the other two departments, Gard and Vaucluse, were lower than 100% (65 and 62%, respectively), due to the relatively low UAA per capita. However, while the main restriction for achieving a high degree of self-sufficiency in Gard was the low total UAA (around 26,000 ha), the UAA in Vaucluse was relatively high (around 53,000 ha), but the population density was much higher (278 inhabitants per km^2) compared to the other two departments (around 150 inhabitants per km^2), due mainly to the fact that Avignon, the main city in the study area, is located in Vaucluse department. The potential self-sufficiency estimated for the whole study area is around 83%, and the estimated radius to meet the theoretical 100% food self-sufficiency is 37 km, which is slightly higher than the initial radius of 30 km of the foodshed selected.

Table 1. Results of the area demand, radius and self-sufficiency for the three departments and the whole foodshed.

Department	Total Area	UAA	Population Density	Total Area Demand	UAA per Capita	Area Demand per Capita	Radius	Self-Sufficiency
Bouches-du-Rhône	77,556	44,792	150	23,752	3861	-	9	189
Gard	123,599	26,010	158	40,103	1328	-	17	65
Vaucluse	149,457	52,606	278	84,973	1268	-	25	62
Foodshed	350,613	123,408	207	148,827	1698	2047	37	83

Total area (ha), utilizable agricultural area (UAA) (ha), population density (inhabitants per km^2), total area demand (ha), UAA per capita (m^2 per capita), area demand per capita (m^2 per capita), radius (km), and food self-sufficiency (%) values for the municipalities belonging to the three departments, and for the whole foodshed (radius: 30 km).

However, whereas the area demand is a relatively accurate value, because it is based on the current consumption per capita, this is not the case for the UAA. The UAA represents the potential area that could be used for agriculture and livestock. Therefore, the food self-sufficiency values estimated do not show the current situation, but rather show a theoretical one, which we compared to the current situation of the agricultural cropping pattern determined by the specific regional pedoclimatic and socioeconomic characteristics (see the following subchapters).

The estimation of the potential food self-sufficiency in the business-as-usual scenario does not consider any change in food consumption patterns. This limitation must be pointed out regarding the fact that there are products currently consumed that cannot be produced regionally (e.g., bananas), and hence, importantly, the resulting food land footprint of the foodshed (2047 m^2 capita^{-1}) does not take place 100% regionally. Nevertheless, these products, all plant-based or drinks based on plants, only represent 156 m^2 capita^{-1} of the total 563 m^2 capita^{-1} of the plant-based products' land footprint, since the rest of the products could theoretically be produced in the region (Table S5 in the Supplementary Material). Therefore, adapting diets has not been considered as a key driver in achieving a high level of food self-sufficiency in the region, and the focus was on the role of the regional spatial crop diversification and its drivers.

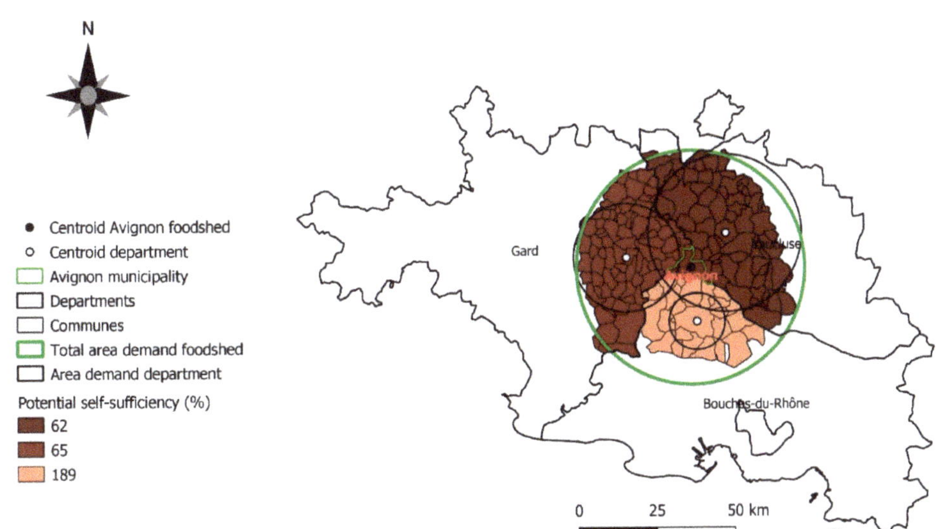

Figure 4. Mapping of the area demand (circles, in km) and food self-sufficiency (colors, expressed in %) for three departments (black) and for the whole foodshed proposed (green) based on the 30 km radius recommendation by the French Senate. See Table 1 for the specific values of the radius and area demand.

3.2. *Current Crop Production and Self-Sufficiency of Plant-Based Products*

The challenge is on the supply side, as consumption is not a key driver for increasing the food self-sufficiency level. According to the RPG map, the total agricultural area currently used within the foodshed is estimated to be around 110,000 ha. This area, which includes vineyards, is lower than the UAA estimated previously (Table 1). This is due to the different way of estimating the cultivated area. While the UAA comes from the Corine Land Cover map, an estimation from remote sensing, the LPIS database—and the RPG map—is constructed from cadastral data related to the CAP payments, and therefore some plots might not be included, leading to an underestimation of the real cultivated area. However, since the accuracy in terms of crop identification is greater in the LPIS database than in the Corine Land Cover map, the former—modified and adapted to the conditions of the study—was selected for this part of the assessment (Figure 5).

Considering the current consumption and production values in the foodshed area, we estimated the current level of food self-sufficiency for cereals, vegetables, pulses, fruit from temperate regions, and wine and grapes. The results show that only cereals achieve a value lower than 100% (Figure 6 and Table S5 in the Supplementary Material), whereas 100% food self-sufficiency is clearly achieved for the rest of the products. Fruit accounted for the highest value (761%), followed by wine and grapes (498%), pulses (455%), and vegetables (220%). Even if the food sufficiency capacity for plant-based products is very high, they account for only 38% of the food products forming the average diet, whereas the other 62% belong to the consumption of animal products. In the following section, we analyze the food self-sufficiency capacity for the animal products.

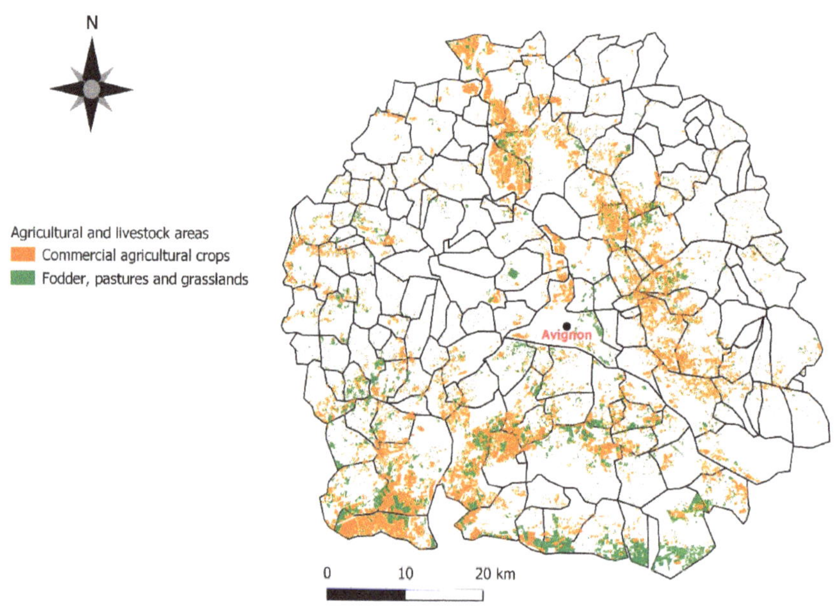

Figure 5. Distribution of commercial agricultural crops, excluding vineyards, and areas dedicated to feeding livestock (fodder, pastures and grasslands) in the foodshed. Our own elaboration based on the RPG map [47].

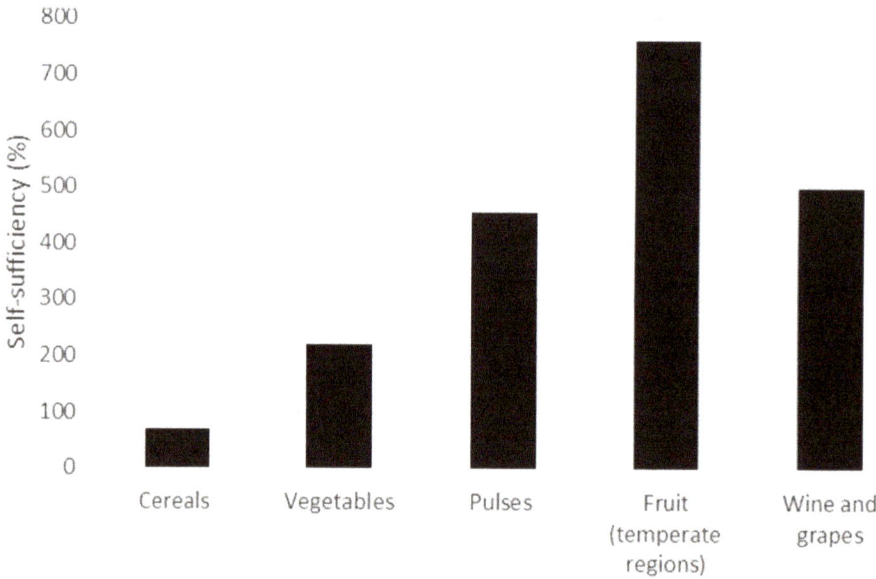

Figure 6. Current food self-sufficiency achieved by cereals, vegetables, pulses, fruit from temperate regions, and wine and grapes in the foodshed area. More information on the food groups and calculations is given in Table S5 in the Supplementary Material.

3.3. Current Livestock Production and Potential Self-Sufficiency of Animal Products

We estimated the total land footprint for organic animal products to be around 133,000 ha (Table 2). This value is 1.4 times higher than the one estimated for conventional farming (92,700 ha) due to the higher area demand estimated by the model for organic livestock farming. This area represents around two-thirds of the food land footprint of the whole diet. Of this area, around 39% is a consequence of the consumption of milk and dairy products, followed by beef consumption (26%), pork (18%), poultry (10%), eggs (5%), and mutton and goat (2%).

Table 2. Results of the area demand, current area used, and self-sufficiency of animal products for the three departments and the foodshed.

Product	Area Demand of Each Product	Total area Demand of Each Department	Current Area Used for Pastures and Fodder in the Foodshed	Current Self-Sufficiency
Beef	34,787	-	-	-
Eggs	6034	-	-	
Poultry	13,483	-	-	-
Pork	24,023	-	-	-
Milk and dairy	51,484	-	-	-
Mutton and goat	3138	-	-	-
Total	132,950	-	-	-
Department				
Bouches-du-Rhône	-	21,218	6826	32
Gard	-	35,824	4300	12
Vaucluse	-	75,908	2994	4
Total 30 km Foodshed	-	132,950	14,120	11

Area demand per capita of organic animal products (beef, eggs, poultry, pork, milk and dairy products, and mutton and goat) (ha) of the population living within the foodshed of 30 km, an estimation of the current area used for livestock farming within the foodshed (ha), and the current food self-sufficiency (%) for the whole foodshed and the municipalities located in the three departments.

However, the current surface dedicated to pastures and fodder in the proposed foodshed of 30 km is 14,120 ha, thus enabling the 11% self-sufficiency of animal food products. There is also an unbalance between the different departmental areas within the foodshed. While the highest area demand takes place in the Vaucluse area, Bouches-du-Rhône accounts for the highest current dedicated area for livestock. Consequently, the highest self-sufficiency for animal products is achieved in Bouches-du-Rhône (32%), followed by Gard (12%), and Vaucluse (4%) (Table 2).

An expansion of the foodshed proposed is considered in the following section, as the area available for the production of animal products within the foodshed of 30 km only covers around 11% of the total area demand, and extensive agriculture takes place in soils with medium-to-low fertility (e.g., high slope, shallow depth, high stoniness, low pH).

3.4. Assessment of the Expanded Foodshed for Animal Products

An expansion of the foodshed for animal products was simulated, considering the pedological conditions and geomorphology. Two buffers around the municipality of Avignon were considered for the expansion of the foodshed: 60 and 100 km. The immediate consequence is that the foodshed area must include other departments beyond the three considered so far. Geomorphologically, the foodshed is well connected to the two adjacent departments to the north, Ardèche and Drôme, in the Auvergne-Rhône-Alpes region, which also account for a high surface of area dedicated to extensive agriculture (Figure 7) and, at the same time, avoid competition with the neighboring city of Marseille (South-East). Plots under deep or very deep soils (i.e., >80 cm depth) were excluded from the assessment in order to avoid land-use conflicts, and to avoid including areas with a high aptitude for commercial agriculture in the study (Figure 8); therefore, only plots with a non-commercial agricultural suitability were included in the assessment.

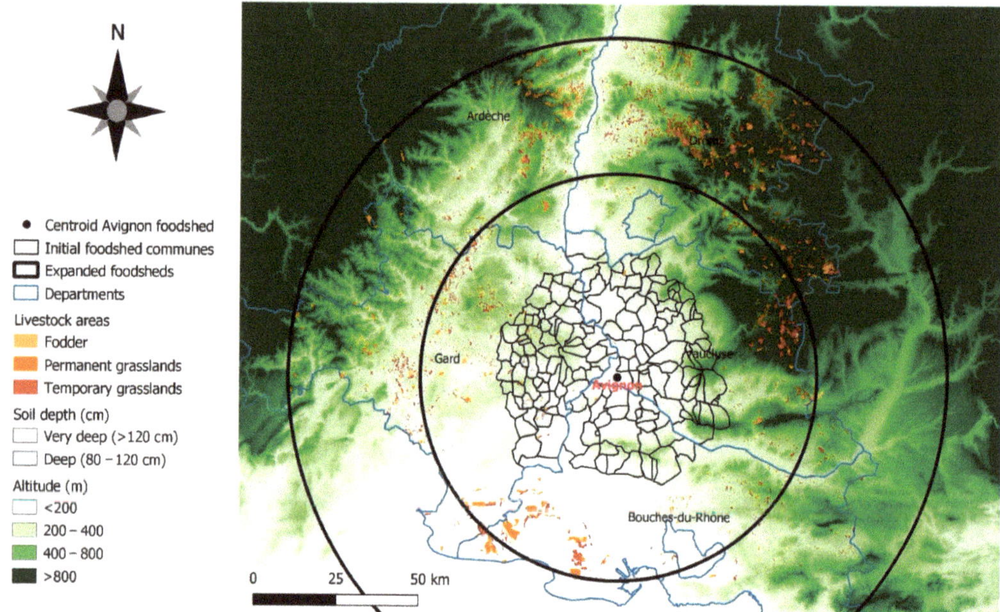

Figure 7. Current area used for extensive livestock farming in the five departments considered in the study (**blue**) in two radii around the city of Avignon: 60 and 100 km (**black bold**). Information on soil depth and altitude is also shown.

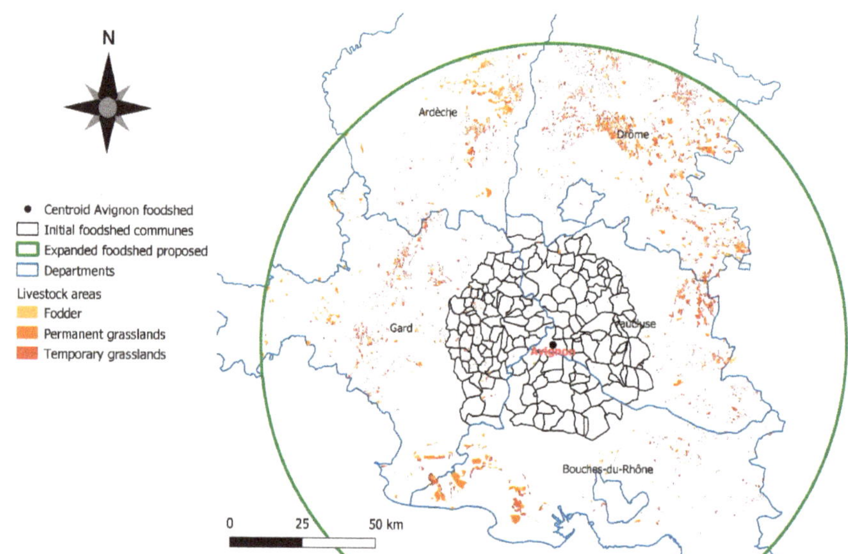

Figure 8. Current area used for livestock in the five departments considered in the study (blue) in the expanded foodshed proposed (radius of 100 km) (green bold). The areas under soils with >80 cm depth were excluded from the assessment.

Thus, the area available for extensive livestock farming resulting from the first buffer, a radius of 60 km, is only 38,000 ha, and is around 97,000 ha in the case considering a radius of 100 km, suggesting that it is especially after 60 km that the plots used for extensive agriculture appear, whereas the soils closer to the foodshed that was initially proposed are mostly used for commercial agriculture (Figure 7). As a result of the selection of the foodshed for animal products with a radius of 100 km, and the exclusion of deep and well-developed soils (Figure 8), the food self-sufficiency for these products would be around 73%, without considering the population of these mountainous areas, accounting typically for a much lower population density than the areas closer to the river.

4. Discussion

Our study shows that assessing the size of the foodshed both quantitatively—by applying the MFSS model and assessing the supply-demand balance—and qualitatively—by combining these outcomes with the different biophysical maps—is an appropriate and realistic way of evaluating the theoretical potential and current food self-sufficiency degree at a regional level (i.e., hybrid foodshed assessment [28]). The outcomes from the MFSS model can, thus, be combined with different types of maps to obtain a more accurate overview of the regional food system, enabling food self-sufficiency issues and foodshed assessment to be addressed realistically. This more realistic approach was partially addressed recently by some authors, for instance, by including the economic dimension [52] or food traceability [53], but lacking a more accurate assessment of the amount and balance of the regional domestic supply–demand, and especially a more realistic assessment of the regional food supply, which is, indeed, the final outcome of our assessment.

4.1. The Spatial Configuration of the Foodshed, Taking into Account Crop Diversification Questions

Higher food self-sufficiency is linked closely to the composition of crops (for instance, homogeneous vs. diversified) to provide sufficient diversity in marketed food products. However, crop diversification concerns not only agricultural cash crops providing food for human nutrition but also the pasture and fodder areas required for livestock farming. In our study, we found that increasing the pasture areas within the initial proposed foodshed of 30 km radius was not realistic (Table 1) due to the suitable soil conditions for commercial agricultural crops and, therefore, the lack of available area for extensive livestock farming (Table 2). This leaves two alternatives: (1) expanding the initial foodshed to incorporate the closest pastoral areas (Figures 7 and 8) and (2) considering the foodshed as a complex of complementary parts extending beyond the isotropic circle, thus shifting the discussion from the size to the spatial configuration of the foodshed.

Expanding the initial foodshed to a larger circle would make it possible to include the closest area suited to the targeted production. In our case study, while the foodshed of 30 km radius initially proposed is self-sufficient for many of the plant-based crops, the expanded foodshed of 100 km—including the surrounding mountainous areas (Figure 8)—would increase animal-product self-sufficiency values to >70%. The new radius is an interesting illustration of the theoretical extent of the spatial requirements for such Mediterranean cities' foodsheds.

From an empirical point of view, however, a foodshed assessment based on estimating distances in terms of a radius around the city has difficulty accounting for the precise consideration of the land use given and the diversity of existing crops. This information is needed in order to encourage farmers to change land use and inform decisions concerning food planning and urban food strategies, such as initiatives aimed at developing short supply chains for specific food products (in the case of Avignon, beef for the school canteens). The assessment is made by aggregating all of the different agricultural products used in the diet in one homogeneous foodshed area centered on the city. This yields a large foodshed containing too many diversified food production areas (in our case, mainly plant-based). In addition, an extended foodshed radius in high urban density areas may generate overlaps due to the city's competing procurement needs [29,38], with inefficient results from a public action perspective. In summary, from an empirical perspective, it might not be suitable to extend the size of the circle limitlessly. Two main arguments should be kept in mind.

Firstly, there is a negative relationship between the distance from the city and the likelihood of a production context to be favorable to the development of a local food supply chain [52]. Foodsheds for big cities may be so large that they incorporate highly varied production contexts, including some farms oriented toward local supply. The density of locally-oriented farms tends to decrease with their distance from the city [54], whereas monocultures and intensive production farms devoted to the global market are located mainly in areas that are not under urban influence [5].

Secondly, there is a negative relationship between the urban density at a regional level and the likelihood of finding a production area targeting only one market location [55,56]. Agricultural areas in a polycentric setting (i.e., a dense network of cities) tend to combine all of the demand from local markets into a food chain that supplies several cities [57]. Thus, when big cities' foodsheds are as large as a region, it is highly unlikely that a production area can be allocated to a single city [58].

4.2. *Foodshed and Self-Sufficiency Assessment: From the Isotropic Circle to the Archipelago Foodshed*

An alternative method of supporting strategic discussions and decision-making based on empirical evidence to allocating agricultural areas and land use in order to enhance regional self-sufficiency would be to create multiple foodsheds according to the main food production types. The foodshed pattern would not necessarily be centered on the city: geomorphological and pedoclimatic criteria do not necessarily select areas in physical contact with the city, and socioeconomic and cultural habits may determine market chains geographically. The breeders supplying meat to Avignon, for example, are located mainly in the surrounding mountains (predominantly in the Southern Alps), where pastoral resources are naturally available (Figure 7). This is a common Mediterranean city model, in which cities are often located on a dry piedmont of mountains with more humid climates, but are historically integrated within the same economic and social territory [59].

There is, therefore, a major scientific challenge involved in shifting from a size (an isotropic circle) to a spatial configuration of the foodshed, which would certainly imply a discontinuous assembly of interconnected parts, which we call the 'archipelago foodshed'; some of them can already be perceived in our foodshed assessment after considering the pedoclimatic and geomorphological constraints (Figure 8). Our research perspectives are founded on a well-known concept from ecological sciences and planning approaches: the 'Biogeography of Islands' theory [60,61]. This states that the specific richness of an island is correlated to its size and the distance from other islands or continental sources of new species. Reasoning by analogy, when physical contact between urban and agricultural areas is not possible, the most appropriate production areas for connection with the city are those closest and large enough to provide sufficient agricultural produce to supply a food chain. In the landscape, urban, and regional planning field, the archipelago is a visual metaphor for an anisotropic space defined by the dimension of the islands (i.e., the different parts of the foodshed) and the distances between them [62–64]. Moreover, in regional economics, the archipelago notion highlights the relational efficacy of production processes, depending on the location of the production units [65]. Additionally, this socio-geographic concept could be enriched by linking it with others that are already existent, such as 'Functional Urban Areas' (https://ec.europa.eu/eurostat/web/cities/spatial-units)—defined as the city and its commuting zone—and by including or prioritizing those farms that apply sustainable management practices, such as agroecology [53,66,67], in the context of assessing and improving the environmental sustainability of the food system [9].

From this perspective, theoretical food self-sufficiency assessments considering the site-specific conditions of metropolitan city regions—such as the one presented in this study—become a suitable starting point for the definition of the size of the foodshed realistically, improving the knowledge of the current state of the food system, or informing policy-makers [4,29,33,34]. However, avenues for further research include rethinking the foodshed concept as an archipelago of parts, the barycenter of which might be located in the production area that is socially, historically and/or agroecologically connected most closely with the city. The challenge is, therefore, to provide a robust and unambiguous indicator—or set of indicators—connecting the city with the farming areas within this archipelago foodshed. Further research could usefully select a set of production areas to meet the food self-sufficiency objective, by developing the 'reserve-site-selection approach' [68,69]. This would involve selecting, firstly, the most efficient area in terms of foodshed supply, and then the second best choice if necessary, continuing the procedure until the objective is fulfilled. In addition to mapping a more realistic foodshed pattern that is appropriate to guide public action and usable in decision-making, a multi-criterion indicator of the agricultural areas' connection with the city could inform policy, for instance, by showing how the foodshed pattern is impacted when the prices of environmentally-friendly food products are positively weighted.

Finally, such an assessment could be used as a decision support methodology concerning the land use and food planning—such as, for example, the *Plans Alimentaires Territoriaux* (https://agriculture.gouv.fr/comment-construire-son-projet-alimentaire-territorial)—being developed in the region, or the specific urban food strategies promoting local agriculture and short food supply chains (e.g., the initiative promoted by the municipality of Avignon to serve local and organic beef in the menus of the canteens in public schools).

4.3. Potential Application of the Methodology to Other Study Cases

The MFSS model has a relatively high versatility, because the lack of data can be compensated for by applying some extrapolations and using default values. However, the lack of regional-scale data represents a trade-off with the accuracy of the assessment, as the model and the methodology are designed to be applied regionally. However, we highlight two variables that are of high importance for obtaining reliable and accurate results. Firstly, the availability of statistical data on regional crop yields, which leads us to indirectly include the site-specific pedoclimatic conditions in the assessment. These data are usually available because crop production and productivity are key agronomic data for farmers. Secondly, the share of UAA is also critical. However, this depends strongly on the pedoclimatic conditions and land cover. The selection can be performed, for example, with remote-sensing, land cover maps, and/or crop maps. The combination of these maps with other data and maps—such as edaphic properties—can be very helpful to improve the accuracy of the UAA selection. The availability of the data to estimate the UAA might be low, especially in developing countries where land cover maps may not be as precise as they are in other regions (e.g., the Corine Land Cover map in Europe).

Moreover, the MFSS model can be applied in city regions to develop scenarios considering dietary shifts (plant- vs. animal-based), population growth, the reduction of food losses and waste, producing systems (organic vs. conventional) [29], or to evaluate future policy targets (i.e., backcasting methodology) [34] and propose specific pathways to increase food self-sufficiency levels.

4.4. Limitations of the Study

The study provides a novel foodshed approach in order to improve the foodshed concept. However, some limitations must be highlighted. Firstly, the data on food consumption is from national estimations from the FAO; therefore, the differences between regions are not considered. Some differences in the dietary patterns might be expected because the area belongs to the Mediterranean part of France. The values of the regional food consumption in the area are estimated by some surveys, but they are shown in terms of dishes and processed food, making the translation to basic ingredients and crops very complex. Another limitation regarding the food consumption data is that they do not consider food wastage. Secondly, the assessment of the current food self-sufficiency is based on the use of the LPIS database and RPG map, which are constructed from cadastral data related to CAP payments, and, therefore, some plots might not be included in the assessment, leading to an underestimation of the real cultivated area. Thirdly, socioeconomic and cultural variables have not been considered to propose the expanded foodsheds, which are based only on biophysical constraints. Finally, the study does not cover explicitly potential overlaps with neighboring foodsheds.

5. Conclusions

Our study demonstrates that a quantitative food assessment combined with maps showing specific biophysical information might be a suitable and realistic approach for assessing food self-sufficiency at a regional level. These studies can become a suitable starting point for the definition of the size of the foodsheds, and in order to improve the knowledge of the current state of the food system. In this line, our results proposing two different foodsheds – one for animal products and another one for plant-based – have been demonstrated to perform realistically.

However, we believe that rethinking the foodshed concept is needed. For instance, to recognize that food supply and demand might be a result of social, historical, cultural and/or agroecological issues and, therefore, other concepts like the *archipelago* foodshed should be considered. In this regard, future studies should address the combination of the biophysical with the socio-cultural dimensions.

Supplementary Materials: The following are available online at https://www.mdpi.com/2077-0472/11/2/143/s1. Table S1: Surface and population of the communes of Bouches-du-Rhône, Table S2: Surface and population of the communes of Gard, Table S3: Surface and population of the communes of Vaucluse. Table S4: Food consumption and yields, Table S5: Area demand, current crop area, and current self-sufficiency.

Author Contributions: Conceptualization, A.P., J.L.V.-V., E.S.-S., C.N. and M.M.; validation, E.S.-S., C.N., M.M. and J.L.V.-V.; formal analysis, J.L.V.-V. and E.S.-S.; resources, A.P.; writing—original draft preparation, J.L.V.-V., E.S.-S. and C.N.; writing—review and editing, E.S.-S., and A.P.; supervision, J.L.V.-V. and E.S.-S. All authors have read and agreed to the published version of the manuscript.

Funding: This study is framed within the 'FOODSHIFT2030' project, and recieved funding from the European Union's Horizon 2020 research and innovation program under grant agreement number 862716.

Institutional Review Board Statement: Not applicable.

Informed Consent Statement: Not applicable.

Data Availability Statement: Not applicable.

Acknowledgments: The authors want to thank the anonymous reviewers for their diligent work and their helpful advice.

Conflicts of Interest: The authors declare no conflict of interest.

References

1. Deverre, C.; Traversac, J.-B. Local Food: A Concrete Utopia. Métropolitiques. 2012. Available online: https://metropolitiques.eu/Local-food-a-concrete-Utopia.html (accessed on 19 January 2021).
2. Hinrichs, C.C. Regionalizing food security? Imperatives, intersections and contestations in a post-9/11 world. *J. Rural Stud.* **2013**, *29*, 7–18. [CrossRef]
3. Moragues-Faus, A. Distributive food systems to build just and liveable futures. *Agric. Hum. Values* **2020**, *37*, 583–584. [CrossRef] [PubMed]
4. Carey, J. *Who Feeds Bristol? Towards a Resilient Food Plan*; Bristol City Council: Bristol, UK, 2011.
5. Pradhan, P.; Kriewald, S.; Costa, L.; Rybski, D.; Benton, T.G.; Fischer, G.; Kropp, J.P. Urban food systems: How regionalization can contribute to climate change mitigation. *Env. Sci. Technol.* **2020**, *54*, 10551–10560. [CrossRef] [PubMed]
6. Vicente-Vicente, J.L.; Piorr, A. Can a shift to regional and organic diets reduce greenhouse gas emissions from the food system? A case study from Qatar. *Carbon Balance Manag.* **2021**, *16*, 2. [CrossRef]
7. Doernberg, A.; Horn, P.; Zasada, I.; Piorr, A. Urban food policies in German city regions: An overview of key players and policy instruments. *Food Policy* **2019**, *89*, 101782. [CrossRef]

8. Piorr, A.; Zasada, I.; Doernberg, A.; Zoll, F.; Ramme, W. *Research for AGRI Committee—Urban and Peri-Urban Agriculture in the EU*; European Parliament: Brussels, Belgium, 2018.
9. Béné, C.; Oosterveer, P.; Lamotte, L.; Brouwer, I.D.; de Haan, S.; Prager, S.D.; Talsma, E.F.; Khoury, C.K. When food systems meet sustainability—Current narratives and implications for actions. *World Dev.* **2019**, *113*, 116–130. [CrossRef]
10. European Union. *From Farm to Fork Strategy: For a Fair, Healthy and Environmentally-Friendly Food System*; European Commission: Brussels, Belgium, 2020.
11. Puma, M.J.; Bose, S.; Chon, S.Y.; Cook, B.I. Assessing the evolving fragility of the global food system. Environ. *Res. Lett.* **2015**, *10*, 024007. [CrossRef]
12. SAPEA. *A Sustainable Food System for the European Union*; SAPEA: Berlin, Germany, 2020; pp. 161–194.
13. Wezel, A.; Herren, B.G.; Kerr, R.B.; Barrios, E.; Gonçalves, A.L.R.; Sinclair, F. Agroecological principles and elements and their implications for transitioning to sustainable food systems. A review. *Agron. Sustain. Dev.* **2020**, *40*, 40. [CrossRef]
14. Willett, W.; Rockström, J.; Loken, B.; Springmann, M.; Lang, T.; Vermeulen, S.; Garnett, T.; Tilman, D.; DeClerck, F.; Wood, A.; et al. Food in the Anthropocene: The EAT–Lancet Commission on healthy diets from sustainable food systems. *Lancet* **2019**, *393*, 447–492. [CrossRef]
15. Pradhan, P.; Lüdeke, M.K.B.; Reusser, D.E.; Kropp, J.P. Food self-sufficiency across scales: How local can we go? *Environ. Sci. Technol.* **2014**, *48*, 9463–9470. [CrossRef]
16. Macleod, G.; Jones, M. Territorial, scalar, networked, connected: In what sense a 'regional world'? *Reg. Stud.* **2007**, *41*, 1177–1191. [CrossRef]
17. Clancy, K.; Ruhf, K. Is local enough? Some arguments for regional food Systems. *Choices* **2010**, *5*, 5.
18. Moreno Perez, O.; de Benito, C.; Sanz Sanz, E.; Simón Rojo, M.; Yacamán Ochoa, C. DIVERCROP Project—Deliverable 5.2: Possibilities to Reconnect Urban Consumption to Peri-Urban Agriculture. Avignon, France, 2019.
19. Bala, B.K.; Alias, E.F.; Arshad, F.M.; Noh, K.M.; Hadi, A.H.A. Modelling of food security in Malaysia. *Simul. Model. Pract. Theory* **2014**, *47*, 152–164. [CrossRef]
20. Godenau, D.; Caceres-Hernandez, J.J.; Martin-Rodriguez, G.; Gonzalez-Gomez, J.I. A consumption-oriented approach to measuring regional food self-sufficiency. *Food Sec.* **2020**, *12*, 1049–1063. [CrossRef]
21. Hara, Y.; Murakami, A.; Tsuchiya, K.; Palijon, A.M.; Yokohari, M. A quantitative assessment of vegetable farming on vacant lots in an urban fringe area in Metro Manila: Can it sustain long-term local vegetable demand? *Appl. Geogr.* **2013**, *41*, 195–206. [CrossRef]
22. Karg, H.; Drechsel, P.; Akoto-Danso, E.K.; Glaser, R.; Nyarko, G.; Buerkert, A. Foodsheds and city region food systems in two West African cities. *Sustainability* **2016**, *8*, 1175. [CrossRef]
23. Moschitz, H.; Frick, R. *City Food Flow Analysis. A New Method to Study Local Consumption*; Renewable Agriculture and Food Systems; Cambridge University Press: Cambridge, UK, 2020; pp. 1–13. [CrossRef]
24. Osei-Owusu, A.K.; Kastner, T.; de Ruiter, H.; Thomsen, M.; Caro, D. The global cropland footprint of Denmark's food supply 2000–2013. *Glob. Environ. Chang.* **2019**, *58*, 101978. [CrossRef]
25. Tedesco, C.; Petit, C.; Billen, G.; Garnier, J.; Personne, E. Potential for recoupling production and consumption in peri-urban territories: The case-study of the Saclay Plateau near Paris, France. *Food Policy* **2017**, *69*, 35–45. [CrossRef]
26. Rothwell, A.; Ridoutt, B.; Page, G.; Bellotti, W. Feeding and housing the urban population: Environmental impacts at the peri-urban interface under different land-use scenarios. *Land Use Policy* **2015**, *48*, 377–388. [CrossRef]
27. Saha, M.; Eckelman, M.J. Growing fresh fruits and vegetables in an urban landscape: A geospatial assessment of ground level and rooftop urban agriculture potential in Boston, USA. *Landsc. Urban Plan.* **2017**, *165*, 130–141. [CrossRef]
28. Schreiber, K.; Hickey, G.M.; Metson, G.S.; Robinson, B.E.; MacDonald, G.K. Quantifying the foodshed: A systematic review of urban food flow and local food self-sufficiency research. *Env. Res. Lett.* **2020**, *16*, 023003. [CrossRef]
29. Zasada, I.; Schmutz, U.; Wascher, D.; Kneafsey, M.; Corsi, S.; Mazzocchi, C.; Boyce, P.; Doernberg, A.; Sali, G.; Piorr, A. City, culture and society food beyond the city—Analysing foodsheds and self-sufficiency for different food system scenarios in European metropolitan regions. *City Cult. Soc.* **2019**, *16*, 25–35. [CrossRef]
30. Joseph, S.; Peters, I.; Friedrich, H. Can regional organic agriculture feed the regional community? A case study for Hamburg and North Germany. *Ecol. Econ.* **2019**, *164*, 106342. [CrossRef]
31. Kurtz, J.E.; Woodbury, P.B.; Ahmed, Z.U.; Peters, C.J. Mapping U.S. food system localization potential: The impact of diet on foodsheds. *Env. Sci. Technol.* **2020**, *54*, 12434–12446. [CrossRef]
32. Orlando, F.; Spigarolo, R.; Alali, S.; Bocchi, S. The role of public mass catering in local foodshed governance toward self-reliance of metropolitan regions. *Sustain. Cities Soc.* **2019**, *44*, 152–162. [CrossRef]
33. Świader, M.; Szewrański, S.; Kazak, J.K. Foodshed as an example of preliminary research for conducting environmental carrying capacity analysis. *Sustainability* **2018**, *10*, 882. [CrossRef]
34. Tavakoli-Hashjini, E.; Piorr, A.; Müller, K.; Vicente-Vicente, J.L. Potential bioenergy production from Miscanthus × Giganteus in Brandenburg: Producing bioenergy and fostering other ecosystem services while ensuring food self-sufficiency in the Berlin-Brandenburg region. *Sustainability* **2020**, *12*, 7731. [CrossRef]
35. Wegerif, M.C.A.; Wiskerke, J.S.C. Exploring the staple foodscape of Dar Es Salaam. *Sustainability* **2017**, *9*, 1081. [CrossRef]
36. Liao, F.H.; Gordon, B.; DePhelps, C.; Saul, D.; Fan, C.; Feng, W. A land-based and spatial assessment of local food capacity in Northern Idaho, USA. *Land* **2019**, *8*, 121. [CrossRef]

37. Porter, J.R.; Dyball, R.; Dumaresq, D.; Deutsch, L.; Matsuda, H. Feeding capitals: Urban food security and self-provisioning in Canberra, Copenhagen and Tokyo. *Glob. Food Sec.* **2014**, *3*, 1–7. [CrossRef]
38. Kriewald, S.; Pradhan, P.; Costa, L.; Ros, A.G.C.; Kropp, J.P. Hungry cities: How local food self-sufficiency relates to climate change, diets, and urbanisation. *Environ. Res. Lett.* **2019**, *14*, 094007. [CrossRef]
39. Kinnunen, P.; Guillaume, J.H.A.; Taka, M.; D'Odorico, P.; Siebert, S.; Puma, M.J.; Jalava, M.; Kummu, M. Local food crop production can fulfil demand for less than one-third of the population. *Nat. Food* **2020**, *1*, 229–237. [CrossRef]
40. Sanz Sanz, E.; Martinetti, D.; Napoléone, C. Operational modelling of peri-urban farmland for public action in Mediterranean context. *Land Use Policy* **2018**, *75*, 757–771. [CrossRef]
41. Sanz Sanz, E.; Napoléone, C.; Hubert, B. A systemic methodology to characterize peri-urban agriculture for a better integration of agricultural stakes in urban planning. *Espace Geogr.* **2017**, *46*, 174–190. [CrossRef]
42. FranceAgriMer. Chiffres-Clés 2016: Les Filières des Fruits et Légumes—Données 2016. 2017. Available online: https://www.franceagrimer.fr/fam/content/download/55358/document/chiffres%20cl%C3%A9s%202016%20FL.pdf?version=9 (accessed on 19 January 2021).
43. Agreste Provence-Alpes-Côte d'Azur. *Mémento de La Statistique Agricole de La Forêt et Des Industries Agroalimentaires*; La préfecture et les services de l'État en région Provenc: Marseille, France, 2019.
44. Agreste Occitanie. *Memento de La Statistique Agricole: Région Occitanie*; Préfet de la Région Occitaine: Toulouse, France, 2019.
45. European Soil Bureau Network European Commission. *Soil Atlas of Europe; Joint Research Centre*; European Soil Data Centre (ESDAC): Brussels, Belgium, 2005.
46. Copernicus Land Monitoring Service. Corine Land Cover Map 2018. Available online: https://land.copernicus.eu/pan-european/corine-land-cover/clc2018 (accessed on 19 January 2021).
47. Institut National de l'Information Géographique et Forestière. *Registre Parcellaire Graphique*; Institut National de l'Information Géographique et Forestière: Paris, France, 2014.
48. Institut national de La Statistique (INSEE) et Des Études Économiques Bilan Démographique. Available online: https://www.insee.fr/fr/accueil (accessed on 20 September 2019).
49. Panagos, P. The European Soil Database. *GEO Connexion* **2006**, *5*, 32–33.
50. Copernicus Land Monitoring Service EU-DEM. 2017.
51. QGIS. org QGIS Geographic Information System. **2020**.
52. Plakias, Z.T.; Klaiber, H.A.; Roe, B.E. Trade-offs in farm-to-school implementation: Larger foodsheds drive greater local food expenditures. *J. Agr. Resour. Econ.* **2020**, *45*, 232–243. [CrossRef]
53. Zazo-Moratalla, A.; Troncoso-González, I.; Moreira-Muñoz, A. Regenerative food systems to restore urban-rural relationships: Insights from the Concepción metropolitan area foodshed (Chile). *Sustainability* **2019**, *11*, 2892. [CrossRef]
54. Zasada, I.; Fertner, C.; Piorr, A.; Nielsen, T.S. Peri-urbanisation and multifunctional adaptation of agriculture around Copenhagen. *Geogr. Tidsskr.—Danish J. Geogr.* **2011**, *111*, 59–72. [CrossRef]
55. Sinclair, R. Von Thünen and urban aprawl. *Ann. Am. Assoc. Geogr.* **1967**, *57*, 72–87. [CrossRef]
56. Capt, D.; Schmitt, B. Economie spatiale et agriculture: Les dynamiques spatiales de l'agriculture contemporaine. *RERU* **2000**, *3*, 385–406.
57. Filippini, R.; Marraccini, E.; Houdart, M.; Bonari, E.; Lardon, S. Food production for the city: Hybridization of farmers' strategies between alternative and conventional food chains. *Agroecol. Sustain. Food Syst.* **2016**, *40*, 1058–1084. [CrossRef]
58. Aubry, C.; Kebir, L. Shortening food supply chains: A means for maintaining agriculture close to urban areas? The case of the French metropolitan area of Paris. *Food Policy* **2013**, *41*, 85–93. [CrossRef]
59. Braudel, F. *La Méditerranée et Le Monde Méditerranéen à l'époque de Philippe II*, 2nd ed.; Armand Colin: Paris, France, 1949; (revised 1966).
60. MacArthur, R.H.; Wilson, E.O. *The Theory of Island Biogeography*; Princeton University Press: Princeton, NJ, USA, 1967; ISBN 0-691-08049-6.
61. MacArthur, R.H.; Wilson, E.O. An equilibrium theory of insular zoogeography. *Evolution* **1963**, *17*, 373–387. [CrossRef]
62. Chapuis, J.-Y.; Viard, J. *Rennes, La Ville Archipel: Entretiens Avec Jean Viard*; Aube: La Tour-d'Aigue, France, 2013; ISBN 978-2-8159-0754-5.
63. Ungers, O.M.; Koolhaas, R. *The City in the City: Berlin: A Green Archipelago*; Hertweck, F., Marot, S., Eds.; Lars Müller Publishers: Zürich, Switzerland, 2013; ISBN 978-3-03778-326-9.
64. Viganò, P. *Les Territoires de L'urbanisme: Le Projet Comme Producteur de Connaissance*; Mētispresses: Genève, Switzerland, 2014; ISBN 978-2-940406-88-3.
65. Veltz, P. *Mondialisation, Villes et Territoires: L'économie D'archipel*, 2nd ed.; PUF: Paris, France, 2014; ISBN 978-2-13-062776-0.
66. Cardoso, A.S.; Domingos, T.; de Magalhães, M.R.; de Melo-Abreu, J.; Palma, J. Mapping the Lisbon potential foodshed in Ribatejo e Oeste: A suitability and yield model for assessing the potential for localized food production. *Sustainability* **2017**, *9*, 2003. [CrossRef]
67. Vaarst, M.; Escudero, A.G.; Chappell, M.J.; Brinkley, C.; Nijbroek, R.; Arraes, N.A.M.; Andreasen, L.; Gattinger, A.; De Almeida, G.F.; Bossio, D.; et al. Exploring the concept of agroecological food systems in a city-region context. *Agroecol. Sustain. Food Syst.* **2018**, *42*, 686–711. [CrossRef]

68. Ay, J.-S.; Napoléone, C. Efficiency and equity in land conservation: The effects of policy scale. *J. Env. Manag.* **2013**, *129*, 190–198. [CrossRef]
69. Kirkpatrick, J.B. An iterative method for establishing priorities for the selection of nature reserves: An example from Tasmania. *Biol. Conserv.* **1983**, *25*, 127–134. [CrossRef]

Article

The Impact of Crop Diversification on the Economic Efficiency of Small Farms in Poland

Agnieszka Kurdyś-Kujawska *, Agnieszka Strzelecka and Danuta Zawadzka

Faculty of Economics, Department of Finance, Koszalin University of Technology, Kwiatkowskiego 6e, 75-343 Koszalin, Poland; agnieszka.strzelecka@tu.koszalin.pl (A.S.); danuta.zawadzka@tu.koszalin.pl (D.Z.)
* Correspondence: agnieszka.kurdys-kujawska@tu.koszalin.pl; Tel.: +48-94-3436-160

Abstract: Crop diversification finds an important place in the strategy of dealing with risk and uncertainty related to climate change. It helps to increase the resilience of farmers, significantly improving their income stability, but at the same time, it can lower the economic efficiency of small farms. The aim of the article is to identify the determinants of crop diversification and the impact of crop diversification on the economic efficiency of small farms in Poland. This article first provides a critical review of the literature on crop diversification, its role in stabilizing agricultural income and its impact on economic efficiency in small farms. Secondly, the level of crop diversification was determined and empirical research was conducted considering the economic, social and agronomic characteristics of farms. Thirdly, the economic efficiency of farms diversifying crops was compared with farms focused on one type of production. The research material consisted of small farms participating in the Polish system of collecting and using farm accountancy data (FADN) in 2018. The level of diversification was determined using the Herfindahl-Hirschman Index. The factors influencing crop diversification were identified using the logit regression model. The Mann–Whitney *U* rank sum test was used to assess the significance of the differences in distributions. The research results indicate an average level of crop diversification in small farms in Poland and its regional differentiation. In addition, a statistically significant positive impact on the probability of crop diversification in small farms in Poland was found of variables such as the level of exposure of agricultural production to atmospheric and agricultural drought and the location of the farm in the frost hardiness zone and a statistically significant negative impact of the variable: value of fixed assets. The existence of significant differences in the level of economic efficiency of farms diversifying crops and farms focused on one profile of agricultural production was proved. The study is an important voice in the discussion on increasing measures to strengthen support for small farms that diversify crops so as to ensure their greater stability and economic efficiency.

Keywords: crop diversification; small farms; economic efficiency; HHI-Index; Poland

Citation: Kurdyś-Kujawska, A.; Strzelecka, A.; Zawadzka, D. The Impact of Crop Diversification on the Economic Efficiency of Small Farms in Poland. *Agriculture* **2021**, *11*, 250. https://doi.org/10.3390/agriculture11030250

Academic Editors: Claudia Di Bene, Rosa Francaviglia, Roberta Farina, Jorge Álvaro-Fuentes and Raúl Zornoza

Received: 1 February 2021
Accepted: 7 March 2021
Published: 16 March 2021

Publisher's Note: MDPI stays neutral with regard to jurisdictional claims in published maps and institutional affiliations.

1. Introduction

Diversification of agricultural holdings consists in transforming homogeneous agricultural production into diverse. It is one of the possible farm development strategies aimed at stabilizing income and securing against risks, mainly climatic and natural [1]. Diversification influences the differentiation and often increases income, which is made independent from one source. In a situation of fragmentation of agriculture, as is the case, inter alia, in Poland, the issue of diversification of small farms becomes more important. In 2018, there were 1.4 million farms in Poland, of which more than half (53.1%) had arable lands up to 5 ha and $\frac{3}{4}$ entities (75.2%) farmed less than 10 ha [2]. A significant barrier in conducting research is the lack of an unambiguous definition of a small farm, which is often emphasized in the literature on the subject [3–6]. Among the classification criteria, on the basis of which small farms are distinguished, the following are taken into account: agricultural land [7–10], economic size [10–15] and the links between a farm and the market [16]. Various approaches to defining small farms mean that the results of research conducted among farms located in different regions/countries often cannot be directly comparable. Considering the above limitations, some researchers adopt several criteria that must be met simultaneously for a given farm to be considered small. For example, Hornowski et al. [6] selected small farms on the basis of the utilized agricultural area (from 1 to 15 ha) and the economic size not exceeding 25 thousand EUR Standard Output (SO determined in accordance with the Farm Accountancy Data Network—FADN methodology). In turn, Ardakani, Bartolini and Brunori [17] used an innovative approach to defining small farms in their research and proposed a composite index of farm structure, which took into account the average values of the following categories: area of holdings (ha), livestock units of holdings (LSU), labor force of holdings (AWU) and standard output of holdings (EUR). In their study, they assumed that a small farm is one that is not productive enough (considering inputs and results).

The research results confirm the important role of small farms in the food system and their importance for food security [18–20]. For example, Rodrigues Fortes et al. [18], conducting research in this area, showed that small farms are important for greater availability of food in the region in which they operate. The results of the research by Rivera et al. [19] also proved the important role of small farms in food supply, especially in regions where such farms dominate the agrarian structure. The results of these studies also indicate that the importance of small farms in relation to local food availability is closely related to non-market distribution channels. Galli et al. [20] have found that small farms ensure food and nutrition security for a household at local, regional and global levels.

Guarínet al. [21] indicate that the importance of small farms in Europe depends, inter alia, on the type of farms located in the area. They proposed the division of small farms based on the analysis of features relating to farmer's histories and motivations, farm production, assets and labor, market linkages and access to support. In the course of the research, taking into account the above characteristics, they distinguished the following five types of small farms: (a) farms with a relatively weak commercial orientation: peasant farms, part-time farms and (b) farms with a relatively strong commercial orientation: diversified businesses, specialized businesses and new enterprises. The results of these studies indicate that among small farms, apart from units with low economic strength, with relatively weak commercial orientation, focused on self-supply, there are also entities that are characterized by entrepreneurship (farmer), strong connections with the market, innovation and production specialization. It should be added that small farms are characterized by relatively low profitability—both of assets and equity, relatively high cost-consumption and a strong dependence of agricultural income on the amount of financial support for operating activities, compared to larger units from the agricultural sector [22,23].

The diversity of crops in farming systems is essential to help farmers adapt to increasing climate variability in the future [24,25]. By diversifying crops, small farms are less exposed to losses in production and are more resistant to environmental changes [23]. By diversifying crops, it is possible to reduce the risks associated with low income from agricultural production, food insecurity and nutrition insecurity [26]. Diversification can be an effective system for securing the financial situation of farmers and integrating them more effectively into local outlets. Researchers emphasize that diversification can contribute to the sustainable development of rural areas by strengthening the links between agriculture and other sectors of the economy [27]. Sustainable agriculture is also based on the use of technology in the pursuit of maximizing productivity while striving to minimize the negative impact on the environment. Diversification therefore, enables farmers to be involved in the implementation of the SARD (Sustainable Agricultural and Rural Development) concept [28]. In the Resolution of the European Parliament of 27 October 2016 on how the Common Agricultural Policy can improve job creation in rural areas (2015/2226 (INI)), it was stated that the diversification of agricultural activities would encourage young generations to return to rural areas and will support entrepreneurship as well as focus on innovation and promotion of products typical for given areas.

Diversification of crops may improve the economic efficiency of small farms [29–37]. By protecting them, inter alia, against an economic downturn [38]. The greater economic efficiency of small farms that diversify crops provides them with relative income stabilization [39]. However, as the research results indicate, crop diversification may have negative effects on the economic efficiency of farms [32,40,41]. This means that decisions to diversify crops can represent a trade-off between productivity and resilience (income volatility) for small farmers [41].

Crop diversification may be determined by both internal factors—related to the characteristics of farmers and the farm structure and external factors—related to territorial features, including regional and spatial patterns [42]. By analyzing the results of research on internal factors, it was found that the following socio-economic characteristics of a farm as well as farmer's household may have an impact on crop diversification: farmer's gender, farmer's age, level of education, household size and income level, fixed assets, livestock or technological limitations [25,42–51]. On the other hand, the most important external factors, as evidenced by the results of empirical research conducted in this area, include location of the farm, cultivation intensity, technical infrastructure, climatic conditions and access to credit and advisory services [42–46,48,50–52].

The results of the research concerning the influence of socio-economic features of an agricultural holding on the degree of crop diversification make it impossible to adopt a uniform approach to determining the relationship between these features and the studied phenomenon. The farmer's age has a positive effect on the level of diversification [45,47]. This means that with age, the probability that the farmer will diversify his crops increases. Similarly, with an increase in the level of education, the probability of crop diversification increases [53–56], although in some developing countries the literacy rate is also important. As evidenced by the results of Geethu and Sharma [48], the degree of literacy can reduce crop diversification. The research also established that there is a positive relationship between the size (number) of an agricultural household and crop diversification [25,45,47,49]. This means that a larger number of people in the farmer's household may contribute to the diversification of crops on the farm. With regard to farm resources, a positive impact of agricultural land on decisions concerning crop diversification was established. Thus, larger resources of land owned by a farmer may favor the diversification of crops [43,45,47,49,51]. Fixed assets are the basic component of the technical equipment of farms. They constitute the material and technical basis of the production capacity. The amount of these resources determines the way of organizing production. Kumar [50] proves that the diversification of crops is determined, inter alia, by the mechanization of a farm (farm equipment with tractors, electric trailers). The research conducted by Kołoszko-Chomentowska [57] shows that farms with low value of assets are characterized by relatively good equipment in buildings, while they are less equipped with machines and devices. On the other hand, farms with high value of fixed assets usually have modern machines and accompanying equipment. As suggested by Mańko and Płonka [58], the specificity of farms related to the orientation of production depends to a greater extent on the necessary equipping of farms with fixed assets than on the efficiency of their use. Moreover, a high value of fixed assets may mean that they have a high share in the property structure of a farm. According to Strzelecka [59], this proves a significant immobilization of farm assets and their low flexibility. Hence, the change in the activity profile and adaptation to climate change in these farms is difficult.

The risk in agriculture from uncertain factors such as the weather can result in variable returns (income) on decisions made in a given year. Hence, crop diversification is seen as a self-insurance strategy used by farmers to protect against risk [60]. Sarwosri and Mußhoff [61] considered the farmer's risk attitudes and time preferences of the farmer and examined the effect of these factors on crop diversification. They found that risk-averse farmers were more likely to diversify their crops, indicating that they found this option safer. As indicated by Auffhammer and Carleton [62], crop diversification increases the resilience of the entire production and farm income in the event of unfavorable climatic conditions. Ashok et al. [63] indicate that climatic factors significantly explain the probability of a change in the crop model. Additionally, they suggest that awareness of climate change increases the likelihood of changes in the crop pattern. Huang et al. [64] prove that farmers' decisions to diversify their crops are influenced by past experiences of extreme weather events. This is also confirmed by the results of the research by Mulwa and Visser [51], who proved that past exposure to climatic shocks and availability of climate information are factors that influence farmers' decisions to diversify their crops. In turn, Kurdyś-Kujawska [65] indicates that the diversification of crops is characteristic of farms with a high exposure to weather hazards. Diversification is the logical answer to the risks associated with bad weather and price volatility. Some crops are more resistant to drought, for example, than others, but may offer worse economic benefits. A diversified product portfolio should ensure that agricultural production is not completely destroyed in bad weather. According to Di Falco et al. [66] crop diversification as a form of insurance is the basis of modern portfolio theory. The use of crop diversification has increased in recent decades due to protracted droughts and other extreme events that have been exacerbated by climate and weather variability [67]. The diversity of crops in farming systems is essential to help farmers adapt to increasing climate variability in the future. By diversifying crops, small farms are less exposed to losses in production and are more resistant to environmental changes [68].

Small farmers use various adaptation strategies to increase income stability, guided by their resources, information, intrinsic values and motivation. Consequently, crop diversification is one of the decisions made to spread risk and make economically sound choices. Understanding what influences these decisions can help identify the appropriate support programs for which it is important to ensure an adequate level of income and to stabilize it in small farms. This study will contribute to the emerging but still ambiguous research on the determinants of crop diversification and its impact on the economic efficiency of small farms. The aim of the research is to identify factors determining crop diversification and to determine the impact of crop diversification on the economic efficiency of small farms in Poland.

Hypotheses 1 (H1). *The factor determining crop diversification in small farms is the level of exposure to climate risk.*

Hypotheses 2 (H2). *Small farms diversifying crops are characterized by lower economic efficiency than small farms focused on one agricultural production profile.*

The rest of the paper is organized as follows. Materials and methods are described in the next section. The results of the study were then presented, which included two main stages. First, the crop diversification was assessed among the surveyed group of farms, considering its regional differentiation and it was determined whether there are differences between the economic results obtained by farms diversifying crops and the results characterizing the second group of farms included in the analysis. In the next stage, the factors influencing crop diversification (using a logistic regression model) were identified and assessed. The last section concludes.

2. Materials and Methods

2.1. Materials

The survey is based on a dataset from the national system for the collection and use of Farm Accountancy Data Network (FADN). In the study, small farms were defined on the basis of their economic size and agricultural area. Taking the above into account, the study covered 1612 farms from the FADN sample with an area of up to 10 ha of agricultural land, the economic size of which in 2018 does not exceed 8 thousand euro. To separate small farms diversifying crops from the sample, the FADN criterion for grouping farms according to agricultural types was used (TF14). The research assumed that farms diversifying crops are mixed farms, as well as mixed crops and livestock. Among the analyzed group of small farms, 34.30% diversified their crops (F_CD). The others specialized in COP (cereals, oilseeds and protein crops), other field crops, horticulture, orchards-fruits, olives, permanent cops combined, milk, sheep and goats, granivores and mixed livestock (F_N_CD).

2.2. Methods

The Herfindahl–Hirschman Index (HHI), one of the most commonly used measures of concentration, was used to assess the differentiation (diversification) of small farm crops. The HHI is the sum of the squares of the share of acreage of individual types of crops in relation to the total area of crops and is determined by the equation [69–71]:

$$\text{Herfindahl–Hirschman Index (HHI)} = \sum_{I=1}^{N} \mathrm{P_i}^2 \tag{1}$$

where: $\mathrm{P_i}$ represents acreage proportion of the i-th crop in total cropped area.

As the level of diversification increases, the sum of squares of the proportions of individual crops in the total area decreases, and thus, the indicators (HHI). The Herfindahl–Hirschman index is one when there is a specialization. Its value approaches zero when there is diversification.

The logistic regression model was used to identify the factors influencing crop diversification in small farms and to verify the research hypothesis (H1) adopted in the article. It allows to study the influence of many independent variables $X_1, \ldots, X_k$ on the dependent variable Y, which is a dichotomous variable and can take one of the two values: 1 or 0. The value of the variable $Y = 1$ means that the given event occurs. Otherwise, this variable takes the value of 0 [72]. The regression analysis process allows to determine which factors are most important for the occurrence of a given event, which can be ignored and how they affect each other [73]. The logistic regression model is based on the logistic function. Its values are in the range $\langle 0;1 \rangle$. The function has the shape of the letter S. The analytical form of the logistic function used in logistic regression is defined by the equation [74]:

$$f(z) = \frac{e^z}{1+e^z} = \frac{1}{1+e^{-z}},\ z \in R \tag{2}$$

The logistic regression model, therefore, applies to two-categorical dependent variables, taking only two values: 0 and 1. The expected value of the dependent variable has been replaced with the conditional probability that the dependent variable Y will assume the value 1 for the independent variables $X_1, X_2, \ldots, X_k$. The logistic regression model for the dichotomous variable Y determines the conditional probability of assuming the distinguished value by this variable and is expressed by the following relationship [75]:

$$P(Y = 1/\ X_1, \ldots, X_k) = \frac{e^{\alpha_0 + \alpha_1 X_1 + \ldots + \alpha_k X_k}}{1 + e^{\alpha_0 + \alpha_1 X_1 + \ldots + \alpha_k X_k}} \tag{3}$$

where $_{0},\ _{1}, \ldots, _{k}$ they are parameters of the model, $X_1, \ldots, X_k$ independent variables that may have both the qualitative and the quantitative character.

Due to the non-linearity of the model with respect to independent variables and parameters, in a logistic regression model the regression coefficients do not represent a measure of the relationship between the variables. Therefore, logarithmization transforms a logistic model into a linear model. For this purpose, the concept of the *Odds Ratio* is introduced. The concept of chance is understood as the ratio of the probability that a given phenomenon will occur to the probability that a given phenomenon will not occur [76], that is:

$$\frac{P(Y = 1/\ X_1, \ldots, X_k)}{1 - P(Y = 1/\ X_1, \ldots, X_k)} = \frac{e^{\alpha_0 + \alpha_1 X_1 + \ldots + \alpha_k X_k}}{1 + e^{\alpha_0 + \alpha_1 X_1 + \ldots + \alpha_k X_k}} : \frac{1}{1 + e^{\alpha_0 + \alpha_1 X_1 + \ldots + \alpha_k X_k}} = e^{\alpha_0 + \alpha_1 X_1 + \ldots + \alpha_k X_k} \tag{4}$$

The odds ratio is a measure of the relationship between exposure and outcome. It provides an estimate (with a confidence interval) of the relationship between two binary variables ("yes" or "no"). It also allows to study the influence of other variables on this relationship using logistic regression [77]. The natural logarithm of the odds ratio is linear in relation to independent variables and considering the model parameters, which facilitates estimation to a high degree [53,74,78]:

$$\text{logit}P = \ln \frac{P(Y = 1/\ X_1\ , \ \ldots,\ X_k)}{1 - \ P(Y = 1/\ X_1\ , \ \ldots,\ X_k)} = \ _0 + \sum_{i=1}^{k} \alpha_i X_i \tag{5}$$

The boundary value α is established as the share [fraction] of "ones" in the sample. Then, the evaluation of the correctness of the estimated model can be carried out, counting correctly and mistakenly the classified cases.

The quality of the constructed logistic regression model can be assessed using the R^2_{count} measure, which takes values from the range $\langle 0, 1 \rangle$ defined as follows [75]:

$$R^2_{count} = \frac{n_{11} + n_{22}}{n_{11} + n_{12} + n_{21} + n_{22}} \tag{6}$$

The closer to one value of this measure the better adjustment of the logistic model to the empirical data of the studied phenomenon. R^2_{count} indicates the percentage of correctly classified cases. The model works well in forecasting the studied phenomenon when $R^2_{count} > 50\%$. This means that the classification based on the model is better than the random one [73]. The quality of the constructed logistic regression model can also be assessed on the basis of the Hosmer–Lemeshow test [72], it compares the values of the estimated probability and the observed values of the occurrence of the phenomenon under study (the null hypothesis indicates a good fit of the model). Additionally, the classification quality of the model is illustrated by the ROC curve [54] and, more specifically, the area under this curve (AUC). The ROC curve is built based on the value of the dependent variable and its predicted probability. When the ROC curve coincides with the $y = x$ diagonal, then the decision to assign a case to a selected class (+) or (−), made on the basis of the model, is synonymous with a random division of the studied cases. Each point on this curve has coordinates (1—specificity, sensitivity). Sensitivity means the ability to detect units without a distinguished feature and specificity is the ability to detect units with a distinguished feature [79].

The area under the ROC curve (AUC) is a measure of the quality of the method in such a way that the field 0.5 is a classification quality comparable to a random coin toss and the area 1.0 is a perfect, error-free classification. The classification quality of the model is good when the curve is significantly above the diagonal $y = x$, i.e., when the area under the ROC curve is significantly greater than 0.5 [55]. If the chances of the occurrence of the of the studied phenomenon, the so-called optimal cut-off point, i.e., the value of k from the interval $(0; 1)$ that if $y < k$, then the object is assigned to the class coded by −, otherwise, when $y \geq k$, to the class coded by + [80].

The variables adopted for the model were quantitative and qualitative. The selection of the variables was based on the available database and the analysis of the research conducted so far in the field of diversification of crops of small farms and the analysis of correlation between the variables. The model uses a set of explanatory variables and cultivation diversification (Y) was assumed as the dependent variable. There were 17 explanatory variables used in the model and they related to the socio-economic and agronomic characteristics of a farm (Table 1).

The non-parametric Mann–Whitney U test was used to verify the research hypothesis (H2) adopted in the work. The essence of this test is to weaken the impact of atypical values on the result and to make this result independent of the type of distribution of the studied variables. The Mann–Whitney U test was used to verify the hypothesis about the compatibility of distributions in two compared populations, which have distributions with continuous distributions $F(x)$ and $G(y)$. In this test, the hypotheses were formulated as follows:

Hypotheses 3 (H3). $F(x) = G(y)$ *The distributions of the selected variables in the two populations have the same distribution.*

Hypotheses 4 (H4). *$F(x) \neq G(y)$ The distributions of the selected variables in the two populations do not have the same distribution.*

Table 1. Set of variables adopted to determine the factors determining diversification in small farms.

Variables	Category	Expected Impact of the Variable
X_1	age (years)	+/−
X_2	farmer's education level: 0—primary education, vocational education 1—secondary education, tertiary education	+/−
X_3	farmer's education: 0—nonagricultural 1—agricultural	+/−
X_4	size of the family (number)	+
X_5	number of family members working on the farm	+
X_6	utilized agricultural area (UAA) (ha)	+
X_7	share of leased land in total UAA (%)	+
X_8	soil valuation index	−
X_9	value of non-current assets (PLN '000)	−
X_{10}	access to credits (1—yes; 0—no)	+
X_{11}	gross value added (PLN '000/ha)	−
X_{12}	income from non-agricultural activities: 0—primary education 1—nonagricultural basic vocational education	−
X_{13}	cash flows from operating activities (PLN '000/ha)	+
X_{14}	labor profitability (PLN '000/AWU)	-
X_{15}	land productivity (PLN '000/ha)	-
X_{16}	the level of exposure to atmospheric and agricultural drought [based on data IMWM-NRI] (1—yes; 0—no)	+
X_{17}	location in the hardiness zone [according to USDA zone] (number: 1—low chance of frost; ...; 4—greatest chance of frost)	+

Note: Predicted impact of the variable on the basis of: [42,46,48]; IMWM-NRI—Institute of Meteorology and Water Management—National Research Institute, Poland.

This test is performed on the basis of the sum of the ranks of the variables, not their mean values [74]. The test for this test is the statistic defined by the formula:

$$U = n_1 n_2 + \frac{n_1(n_1+1)}{2} - R_1$$

or

$$U = n_1 n_2 + \frac{n_2(n_2+1)}{2} - R_2 \quad (7)$$

where n_1, n_2—sample sizes, R_1, R_2—ranks sums for samples.

When the sample size for each sample is greater than 20, use the statistic that approximates the normal distribution:

$$Z = \frac{R_1 - R_2 - (n_1 - n_2)(n+1)/2}{\sqrt{n_1 n_2 (n+1)/3}} \quad (8)$$

where $n = n_1 + n_2$—total number of observations. The Z statistic has an approximately normal distribution.

The selection of diagnostic variables adopted for the analysis was based on the available database and the analysis of the research conducted so far in the field of economic efficiency of farms. The set of variables and their characteristics are presented in Table 2.

Table 2. Diagnostic variables included in the study of economic results of small farms.

Variables	Category
	Productivity and Profitability of the Land
Land productivity [PLN '000/ha]	The variable defining the productivity will change the agricultural use. The index level was established as the relation of the total production produced by an agricultural holding to the area of agricultural land.
Land profitability [PLN '000/ha]	Variable specifying profitability of agricultural land. The indicator was calculated as the relation of the family farm income to the arable land area.
Gross farm income [PLN '000]	Includes total production less intermediate consumption and adjusted for the balance of subsidies and taxes related to operating activities.
	Work Efficiency and Profitability
Total labor profitability [PLN '000/AWU]	The variable defining the total profitability of work. The level of the indicator was established as the relation of the net value added to the number of full-time employees.
Own labor profitability [PLN/h]	The variable determining the profitability of own work. The level of the indicator was established as the relation of the family farm income to the working time as part of the operating activities of unpaid persons (mainly family members).
	Asset Financing Sources
Total liabilities [PLN '000]	The value of all outstanding debt obligations and short-term.
Farm net income [PLN '000]	The fee for the involvement of own factors of production in the operational activity of the farm and the fee for the risk taken by the farm operator in the accounting year.
Total subsidies—excluding on investments [PLN '000]	Value of operating subsidies less investment subsidies.
Total support for rural development [PLN '000]	Value of agri-environmental subsidies, subsidies to areas with unfavorable conditions for agricultural production and other subsidies for rural development.
	Financial Indicators
Return on assets [%]	The variable describing the profitability of total assets. The level of the index was established as the relation of the family farm income (reduced by own labor costs) to the average total assets.
Return on equity [%]	This ratio allows to assess the effectiveness of using equity in the enterprise. The level of the index was established as the relation of the family farm income (less own labor costs) to the average equity.
Total assets debt ratio [%]	The variable specifying the share of all liabilities in financing the property. It provides the most general picture of the financing structure of an agricultural holding's assets. The ratio was set as the ratio of total liabilities to the average total assets. A low level of the ratio indicates financial independence, while a high level indicates excessive credit risk.
Cash flow (1) [PLN '000]	They show the ability of an agricultural holding to self-finance its activities and create savings within operating activities. Cash flow is the sum of products sold, other income, sales of animals less o the cost of purchasing animals, the balance of subsidies and taxes relating to operating activities and the balance of subsidies and taxes relating to investments.
Gross investments [PLN '000]	Value of purchased and produced commodity assets, less the value of fixed assets sold and transferred free of charge in the accounting year + change in the value of the livestock.

3. Results and Discussion

3.1. Characteristics of Small Farms Diversifying Crops

The average area of the analyzed group was 7.41 ha. In most farms, the land was owned by the farmer. On average, 10.91% was the share of leased agricultural land. The structure of agricultural land was dominated by cereals, fodder crops and other field crops such as potatoes, sugar beet, herbs, oilseed and fiber, hops, tobacco and other industrial crops. The soil valuation index was at the level of 0.79, which means arable soils of average quality, which may periodically be too dry or too moist and which are very susceptible to fluctuations in groundwater levels. The average value of fixed assets in the analyzed farms was PLN 372,100. These farms were characterized by a large diversification in terms of the value of fixed assets (coefficient change: 56.14%) and a very large differentiation (coefficient change: 108.82%) in terms of the farm's ability to self-finance its activities and create savings as part of its operating activities. The amount of cash flows from operating activities averaged PLN 23,570. The increase in the value of goods produced in small farms (the so-called gross added value) was at the level of PLN 28,160. Income from a family farm amounted to PLN 12,830 on average. An important source of income for small farms were subsidies to operating activities, their value oscillated on average at the level of PLN 11,030. In the analyzed group, 21.33% of farmers obtained income from activities other than agriculture. 9.94% of small farmers had liabilities to financial institutions. Small farmers mainly used loans for day-to-day operations. The average level of current liabilities was PLN 89,90. On the other hand, loans taken for a period longer than one year amounted to PLN 42,670 on average. In the analyzed period, farmers did not have a valid crop insurance policy. The average number of people employed full-time in a farm was 1.42. The maximum number of full-time employees was 5.7. The average age of the farm manager was 48 years. One fourth of the surveyed farmers were over 56 years old. Experience in agricultural production is related to age. The farmer's experience influences the effectiveness of decisions regarding the achieved income and its stability. Most of the small farmers had secondary (43.40%) and vocational (39.24%) education. The smallest group were farmers with primary education (6.50%). A total of 10.85% of the surveyed farmers had higher education. It should be noted that almost half of the farmers (47.20%) had agricultural education. The family of the analyzed farmers was mostly not very large. Two- (25.85%) and three-person (29.65%) households prevailed. The smallest group were farms where the number of family members was higher than 7 (1.08%). On average, two people of working age, who were members of the farmer's family, worked on a farm. In 25.68% of farms, the farmer's family members were of retirement age.

3.2. *The Level of Crop Diversification*

The intensity of diversification in the group of researched farms was low. The diversification index (HHI) averaged 0.59. In terms of the intensity of diversification of crops, agricultural holdings were characterized by an average differentiation (change: 32.67%). For comparison, in the group of other small farms, the HHI index was on average 0.70. Units with the HHI index equal to 1 prevailed. The minimum value of the HHI index in the analyzed group of small farms was 0.22, while the maximum value was 0.94. The median value of the HHI index was 0.55. However, small farms with the HHI value above the median prevailed slightly (Figure 1).

In terms of the size of the crop diversification index, Poland can be divided into two parts. In the north-western part there are small farms with a lower level of crop diversification. The HHI value in these regions of Poland exceeded the value of 0.60. Only in 25% of small farms the value of the HHI index oscillated between 0.40 and 0.59. In turn, in the rest of the country, the value of the cultivation diversification index in small farms was below 0.59 (Figure 2). In 25% of small farms in this part of the country, the HHI index fluctuated at the level of 0.60. In most of them (75%), it reached the value of 0.40. It should be noted that the regional differentiation of the cultivation diversification index in small farms is influenced by the organizational and economic conditions of farms, topography or natural conditions. Regions with a high diversity of crops are characterized by low land productivity and low labor productivity. These regions have high employment in agriculture. They are also characterized by worse production parameters, i.e., the predominance of weaker soils with low agricultural culture [81]. Greater crop diversity is rational behavior for farmers in these regions, as they have to adapt to the existing agrotechnical conditions, which can create many production niches.

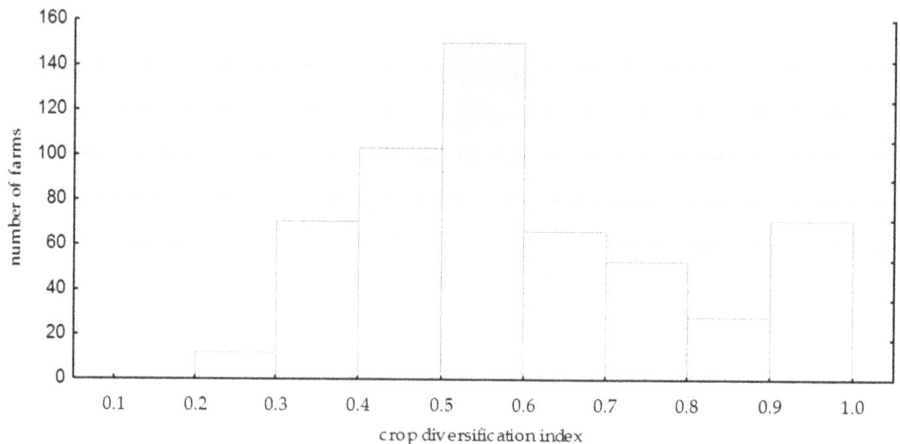

Figure 1. Distribution of the crop diversification index. Source: the authors' own analysis based on FADN data.

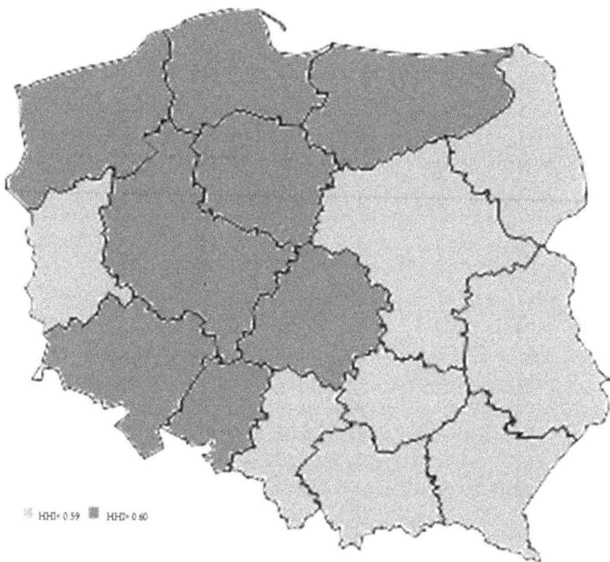

Figure 2. Regional differentiation of the cultivation diversification index (HHI) of small farms in Poland. Source: the authors' own analysis based on FADN data.

3.3. Economic Results of Small Farms Diversifying Crops versus Others

Small farms diversifying crops were characterized by significantly lower productivity and profitability of land than farms focused on one production profile. This is evidenced by the lower average values of the analyzed indicators in both groups. The land productivity index showed the greatest differentiation in the group of small farms focused on one production profile. Small farms diversifying crops were characterized by much lower differentiation of land productivity. On the other hand, a significant dispersion indicates a high volatility of land profitability in this group of farms. The presented land profitability indicators indicate more efficient land use by small farms conducting targeted production. Considering the coefficient of variation, small farms diversifying crops were characterized by lower diversification of the increase in the value of goods produced on a farm. Comparing the average increase in the value of goods produced in a given farm in small farms diversifying crops, it was two times lower than in small farms focused on one production profile (Table 3). For all indices describing productivity and profitability will change, statistically significant ($p < 0.05$) differences in distribution have been demonstrated. As Katchova [82] suggests, diversified farms show lower efficiency because these farms support less profitable activities by cross-subsidizing them with more profitable activities or accept lower returns in exchange for risk reduction.

Labor profitability on a farm is an indicator that assigns all farm income to the labor factor used on a farm, while own labor profitability is an indicator that assigns all farm income to the involvement of a farmer and his family, excluding remuneration from land and capital. The farm owner is primarily interested in the total effect of his involvement in agricultural activity and he is the one who influences the decisions regarding the continuation or discontinuation of the activity [56]. It should be noted that the level of profitability of work is nowadays considered to be one of the basic factors determining the living standard of the agricultural population and one of the most important factors determining the competitive advantage of farms [83]. In the compared groups of small farms, labor profitability indicators were twice as high in the case of entities focused on one production profile, compared to farms diversifying crops. This is indicated by the average values of the discussed indicators of work profitability, as well as the results of the Mann–Whitney U test. In the group of small farms diversifying crops, the profitability indicators were characterized by relatively lower volatility, compared to small farms with targeted production.

The capital structure affects the economic results of entities from the agricultural sector. It is a fact, however, that farms are characterized by a high degree of self-financing and low propensity to incur debt [84]. This is especially true for small farms. It should be noted that the strategy of financing assets in farms is a function of many factors and in particular depends on the availability of a given source of capital, the cost of capital and production risk. As the analysis shows, in the group of small farms diversifying their production, the level of general debt was relatively low compared to the debt level of small farms with targeted production. Considering the coefficient of variation, small farms diversifying crops were characterized by lower differentiation of the average amount of total liabilities. The distribution of income from agricultural activity was similar, which also indicated the existence of significant differences in the average income values of both analyzed groups of small farms.

The income from agricultural activity of small farms diversifying their production was almost half lower in comparison to small farms targeting one production profile. The low level of profitability of small farms diversifying crops, with a relatively low level of use of financial leverage by this group of entities, may significantly inhibit the processes of technical modernization and thus reduce their effectiveness. In addition, as indicated by Wieliczko et al. [85], with low income, accumulation of capital from existing income is very difficult. The accumulation of capital as well as the development of the individual curve of the demand for agricultural production factors depend on the amount of economic surplus generated by agricultural activity.

Table 3. Economic results of small farms due to crop diversification.

Specification	Median	Min	Max	Lower Quartile	Upper Quartile	Gap	SD	CV [%]
Productivity and profitability of the land								
Land productivity [PLN/ha]								
F_CD	5320	360	70,090	3620	8220	69,720	6610	92.61
F_N_CD	7640	−250	810,900	4050	16,290	811,150	6360	271.69
Land profitability [PLN/ha]								
F_CD	1000	−5730	111,680	−40	2590	117,410	7600	316.83
F_N_CD	2250	−179,700	1,078,070	300,630	6680	1,257,780	71,330	451.59
Gross farm income [PLN]								
F_CD	2740	50	119,800	1630	4640	119,750	8310	179.87
F_N_CD	4830	30	3,398,460	2370	11,620	3,398,450	146,120	477.12
Profitability of labor								
Total labor profitability [PLN/AWU]								
F_CD	6350	−39,690	124,520	130	17,360	164,210	20,750	183.66
F_N_CD	14,830	−132,990	1,585,450	3810	35,500	1,718,440	78,540	255.63
Own labor profitability [PLN/h]								
F_CD	2630	−18,430	58,480	−90	7280	76,920	9400	198.09
F_N_CD	5690	−74,040	344,730	760	13,590	418,770	26,090	231.61
Income and sources of financing the property								
Total liabilities [PLN]								
F_CD	0	0	338,000	0	0	338,000	20,580	534.04
F_N_CD	0	0	5,129,920	0	0	5,129,920	270,930	704.22
Farm net income [PLN]								
F_CD	7940	−40,070	179,170	−310	19,090	219,240	25,240	196.59
F_N_CD	15,500	−416,120	1,823,860	2220	41,160	2,239,980	102,880	269.82
Total subsidies - excluding on investments [PLN]								
F_CD	9940	0	111,160	7420	12,810	111,160	8900	80.64
F_N_CD	8430	0	97,460	5590	11,980	97,460	9370	94.76
Total support for rural development [PLN]								
F_CD	1120	0	100,000	0	1710	100,00	7460	365.63
F_N_CD	370	0	81,580	0	1580	81,580	7060	387.86
Financial indicators								
Return on assets [%]								
F_CD	−9.94	−62.21	48.04	−16.27	−5.94	110.25	10.48	91.95
F_N_CD	−7.09	−91.59	93.38	−14.00	−1.54	184.97	14.56	199.39
Return on equity [%]								
F_CD	−10.03	−62.21	57.56	−16.35	−5.94	119.77	10.66	92.90
F_N_CD	−7.26	−91.59	475.66	−14.19	−1.55	567.25	21.33	315.21
Total assets debt ratio [%]								
F_CD	0.00	0.00	68.67	0.00	0.00	68.78	4.55	632.76
F_N_CD	0.00	0.00	95.72	0.00	0.00	95.72	8.13	400.94
Cash flow (1) [PLN]								
F_CD	17,260	−39,530	174,410	8690	30,220	213,940	25,660	108.82
F_N_CD	27,320	−88,200	1,922,020	13,440	58,880	2,010,230	114,950	196.24
Gross investments [PLN]								
F_CD	−223,830	579,000	−330	6500	802,830	34,880	342,550	−223.83
F_N_CD	−320,000	2,206,510	0	9890	2,526,510	106,880	585,500	−320.00

Note: F_CD—farms diversifying crops; F_N_CD—agricultural holdings without crop diversification. Source: the authors' own analysis based on FADN data.

Operating subsidies have a significant share in the income of farms, both in Poland and in other European Union countries [86]. Moreover, small farms in Poland are characterized by a higher production cost and a lower ability to generate income in the course of operating activities, compared to an average small farm in the European Union [22]. Kurdyś-Kujawska and Sompolska-Rzechuła [87] prove that agricultural subsidies affect not only an increase in farm income, but also an increase in the value of fixed assets and gross investments in farms. A large variation in the amounts of subsidies for rural development received by small farms and small differences in the amounts of subsidies received for operating activities were observed. The number of subsidies for rural development in small farms diversifying crops was three times higher than the number of subsidies received in the second of the analyzed groups of farms. As Wieliczko et al. [88] the possibility of using these subsidies allows small farms to adapt to EU requirements, improve the quality of agricultural products, or, as in the case of LFA (less favored areas) subsidies, compensate for lower incomes. Small farms are willing to use subsidies to improve overall farming performance, including increasing the productivity of agricultural activity. In small farms diversifying crops, a much higher value of received payments for operating production was also noted.

In both groups of small farms, the profitability ratios of total assets and the return on equity were negative. The distributions of the return on total assets and return on equity were significantly different. This is indicated by the average values of the discussed index, as well as the results of the Mann–Whitney *U* test. Small farms diversifying crops were characterized by much lower operating efficiency in terms of generating profits from owned assets. These entities were also characterized by a lower profitability growth potential. The ability of small farms to self-finance and create savings is an important aspect of the functioning of farms, as it allows farmers to have a direct impact on the development and changes in the field of activity, allowing, inter alia, to finance investments in future periods and is an important element of financial security in the event of unforeseen events [85,89]. Cash flow shows the farm's ability to self-finance its activities and create savings as part of its operating activities. The distribution of cash flow values in both groups is not uniform. Small farms diversifying crops were characterized by a much lower ability to self-finance and create savings than farms focused on one production profile. In both groups of small farms there was a negative balance of cash flows from operating activities, while in farms diversifying production the negative balance of cash flows was much lower than in other farms.

The distributions of gross investment value in small farms diversifying crops and targeting one production profile differed significantly. In both groups of farms, the average value of gross investment was negative, which means that the value of sold and free of charge fixed assets was higher than the value of purchased and manufactured fixed assets in a given year. Small farms diversifying crops had a relatively lower average gross investment value.

Based on the results of the non-parametric Mann–Whitney *U* test, the null hypothesis (at the significance level $p < 0.05$) was rejected about the insignificance of differences between the economic results of both groups of small farms that were subjected to the study. Thus, there is a statistically significant difference between the economic results of small farms that diversify crops and small farms focused on one agricultural production profile.

3.4. Determinants of Crop Diversification

In the initial model of the probability of crop diversification in small farms in Poland, all variables listed in Table 1 were considered. In accordance with the adopted methodology, only those variables that have a significant impact on the variable *Y*—crop diversification, using backward stepwise regression analysis were left. This means that from the list of potential dependent variables, the variables from the full model were gradually eliminated in such a way as to obtain the model with the highest value of the determination coefficient, while maintaining the significance of the parameters. The analysis of the results of the estimation of the parameters of the probability model of crop diversification by small farms in Poland showed the statistical significance of three variables: X_9—value of fixed assets; X_{16}—the level of exposure to atmospheric and agricultural drought; X_{17}—location in the frost resistance zone. The empirical results obtained from the estimation of the logit model are presented in Table 4.

Table 4. Evaluation of logit model parameters.

Variable	Variable Name	Parameter Evaluation	p-Value	Odds Ratio
	Constant	**0.3213**	**0.0083**	-
X_9	value of fixed assets	−0.0008	0.0006	0.9992
X_{16}	the level of exposure to atmospheric and agricultural drought	0.5593	0.0001	0.5716
X_{17}	location in the hardiness zone	1.1891	0.0001	0.3045

Source: the authors' own analysis based on FADN data.

The estimated logistic model is as follows:

$$p_i = P(y = 1) = \frac{e^{0.3213-0.0008x_9+0.5593x_{16}+1.1891x_{17}}}{1 + e^{0.3213-0.0008x_9+0.5593x_{16}+1.1891x_{17}}}$$

The correctness of the estimated model was assessed by counting the accuracy of the classification of the logit model, which is presented in Table 5.

Table 5. Accuracy of classification of the logit model.

Qualification of Small Farms Based on the Logit Model	Actual Affiliation		Overall Validity of the Classification
	$y_i = 1$	$y_i = 0$	
$\hat{y}_i = 1$	72	85	64.89%
$\hat{y}_i = 0$	481	974	
Sensitivity, specificity	13.02%	91.97%	

Source: the authors' own analysis based on FADN data.

Model quality was assessed based on the value of the coefficient R^2 *count* and the ROC curve. The degree of fit of the logistic model to empirical data is presented in Table 6.

Table 6. The degree of fit of the logistic model to empirical data.

Classification Relevance R^2_{count}	Hosmer-Lemenshow Test		Area Under the ROC Curve
	x^2	p	
64.89%	11.99	0.152	67.01%

Source: the authors' own analysis based on FADN data.

Based on the results in Table 5, it can be concluded that the logistic regression model is characterized by a fairly good fit to the empirical data. The results of the Hosmer-Lemenshow test show no significant differences between the empirical and theoretical numbers, which result from the estimated logistic regression models.

The field under the ROC curve is significantly greater than 0.5 (at the significance level greater than 0.000001), therefore, it is possible to classify farms on the basis of the constructed model (Figure 3).

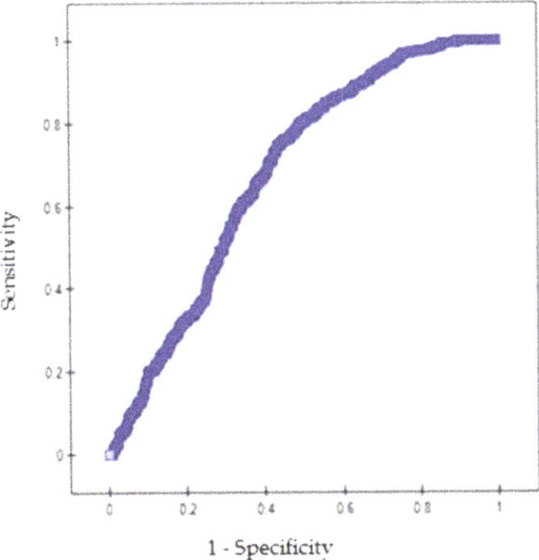

Figure 3. The ROC curve for the model.

In the model, the following factors have a positive, statistically significant influence on the dependent variable: X_{16}—the level of exposure to atmospheric and agricultural drought and X_{17}—the location of the farm in the hardiness zone. This means that the higher the values of these variables, the higher the probability of crop diversification. Interpreting the odds ratios for the *i*-th variable (assuming that the remaining variables included in the model remain unchanged), the following information is obtained: if the agricultural production is located in a region prone to atmospheric and agricultural drought, the chance for crop diversification will increase by 42.84%; the greater the likelihood of a harsher winter and, therefore, of frosts in the region where agricultural production is located, the greater the chance for crop diversification will increase by 69.55%. Similar results in terms of the influence of climatic factors on decisions regarding crop differentiation were obtained, among others, by Ashok et al. [63], Huang et al. [64], Kurdyś-Kujawska [65]. The adoption of a crop diversification strategy in the analyzed farms should be considered a rational behavior of farmers strongly exposed to weather uncertainty. Being highly dependent on rainfall or low temperatures, small farmers undertake ex ante actions minimizing losses resulting from the realization of production risk in the face of changing weather conditions. The results of the studies by Belay et al. [90] also prove that crop diversification is one of the strategies used by small farms to adapt to climate change. Climate change can cause large fluctuations in production and affect farmers' incomes and diversification can effectively stabilize them. Weather shocks such as drought, for example, can trap households in poverty [91]. By diversifying crops, the overall shortfall in income is reduced, by spreading the effects of climate risk across different crops, there is also a reduction in the average annual income volatility resulting from highly seasonal agricultural income flows and there is a reduction in the inter-year income volatility that results from the instability of production and the market [92]. Wan et al. [93] indicate that crop diversification can be seen as a strategy for managing the risk of drought, which is the greatest challenge for farmers worldwide. It is a deliberate ex ante strategy aimed at anticipating possible negative events and counteracting the expected failure in various income streams in the future. Small farmers do not generate enough savings, most of them do not participate in the crop insurance scheme [85], which means that they are not able to mitigate their consumption through ex post mechanisms. Therefore, they can allocate farm resources in such a way as to ensure a more stable income. In addition, to using crop diversification to deal with the risk of climate and market volatility [94–96], crop diversification optimizes crop production under heterogeneous agroecological conditions in marginal areas with heavy rainfall [97,98] or irregular frequency. These phenomena weaken farm yields, reducing food availability and lowering incomes [99].

The following variable had a negative, statistically significant influence on the dependent variable: X_5—value of non-current assets (PLN '000). If the value of fixed assets increases by PLN 1000, the chance for diversification of crops will decrease by 0.08%. This result shows that farmers with relatively larger fixed assets are less inclined to diversify their crops. Katchova [82] drew similar conclusions, indicating that farms with a high degree of diversification accumulate fewer assets than more specialized farms. This suggests that small farms with a higher value of fixed assets are better able to absorb or mitigate income shocks than farmers with fewer fixed assets. These farms may show greater ability and motivation to adopt new and improved production technologies necessary to increase and stabilize income. Since increasing the value of fixed assets is associated with investments, it can be concluded that small farms that do not diversify crops through investments in fixed assets, new technologies and innovative solutions may increase production and its efficiency. These smallholder farmers are most likely shifting from subsistence farming based on self-sufficiency to profit- and income-oriented decision making. Hence, the choice and degree of diversification may depend on the degree of commercialization of small farms [100].

4. Conclusions

This article presents the impact of crop diversification on the economic efficiency of small farms in Poland, as well as the factors conditioning crop diversification in these farms.

The results of our research indicate that, first of all, in the analyzed group of small farms in Poland 34.30% diversified their crops. The intensity of crop diversification was low. The majority of farms were farms with the HHI value above 0.55. Secondly, our research shows that the decisions of small farms in Poland in the field of crop diversification were determined by the value of fixed assets: the level of exposure to atmospheric and agricultural drought and location in the frost resistance zone. Farmers who have higher-value fixed assets are less likely to decide to diversify their crops. Furthermore, the location of a small farm in a region exposed to atmospheric and agricultural drought increases the chance of crop diversification by 42.84%, while severe winters and the risk of frost increase the probability of crop diversification by 69.55%. Diversification of crops was largely determined by the degree of exposure of small farms to climate risk, in particular to drought and frost. Thus, crop diversification helps increase farmers' resilience to changing weather conditions caused by climate change and stabilize their incomes. Therefore, the H1 hypothesis should be adopted, according to which the factor determining crop diversification in small farms is the level of exposure to climate risk. Thirdly, small farms diversifying crops were characterized by significantly lower productivity and profitability of land than farms focused on one production profile. The average increase in the value of goods produced in small farms diversifying crops was twice lower than in small farms focused on one production profile. The income from agricultural activity of small farms diversifying their production was almost 50% lower in comparison to small farms focused on one production profile. It was noticed that in the group of farms diversifying crops, the operating efficiency in terms of generating profits from the assets held was much lower than in other farms. Small farms diversifying crops were characterized by a much lower ability to self-finance and create savings than farms focused on one production profile. It should be emphasized that the positive effect of diversification of production by the examined small farms in relation to the number of subsidies for rural development. In the group of farms diversifying crops, this amount was three times higher than the amount of payments received in other farms. Despite relatively higher income support for farmers diversifying crops, these farms were characterized by lower economic surpluses. The obtained research results allow us to adopt the H2 hypothesis. Finally, the results of the study suggest that the choice of crop diversification involves a compromise between the efficiency and resilience (income volatility) of small farms. Maintaining crop diversification in small farms in Poland will largely depend on whether small farms will be able to maximize interactions and resolve trade-offs between crop diversification and economic efficiency and its increase.

Maintaining crop diversification in small farms in Poland will largely depend on whether small farms will be able to maximize interactions and resolve trade-offs between crop diversification and economic efficiency and its increase.

The future of agriculture and food production requires an integrated and coherent approach to risk prevention and management, complementary linking EU-level interventions with Member States' strategies and private sector instruments that address income stability and climate risk [101]. It becomes necessary to look for such solutions and such support programs that will provide small farmers who diversify crops with an increase in productivity and income. More targeted public support and policy responses are needed for small farms diversifying crops to minimize all expected and inevitable negative consequences of market volatility and income uncertainty [102]. Agricultural policy should focus on increasing the access of farmers from small farms to external sources of financing, which will enable farmers to invest in new plant varieties, more productive, resistant to changes in climatic and environmental conditions. This will ensure their income growth and at the same time increase their stability. It is also important to create an appropriate program of advisory services so that farmers from small farms increase their knowledge of the cultivation of new plant varieties. As Mzyece and Ng'ombe [41] points out crop diversification should be better promoted in conjunction with other strategies to increase farm productivity. These strategies can help offset or reduce the negative impact of crop diversification on small farm productivity.

This issue of crop diversification requires further research in terms of national and international (quantitative) as well as qualitative. This task is particularly important in the conditions of high fragmentation of agriculture and a relatively low level of profitability of this sector. The question arises to what extent decisions in the field of crop differentiation are aimed at protecting the potential/current income from agricultural production (striving to keep the income at an unchanged level) and to what extent are they determined by the desire to maximize income from the factors of production owned. In order to thoroughly analyze the problem of economic efficiency of small farms diversifying crops, further research is planned to extend the previous research and to compare the economic efficiency of small farms specializing in field crops with mixed farms using an alternative sample (for instance, recurring to matching techniques). In addition, future research will focus on identifying how the level of crop differentiation affects the economic efficiency of small farms.

Author Contributions: Conceptualization, A.K.-K., A.S. and D.Z.; methodology, A.K.-K., A.S. and D.Z.; software, A.K.-K.; validation, A.K.-K., A.S. and D.Z.; formal analysis, A.K.-K.; investigation, A.K.-K., A.S. and D.Z.; resources, A.K.-K.; data curation, A.K.-K.; writing—original draft preparation, A.K.-K., A.S. and D.Z.; writing—review and editing, A.K.-K., A.S. and D.Z.; visualization, A.K.-K., A.S. and D.Z.; supervision, A.K.-K., A.S. and D.Z.; project administration, A.K.-K., A.S. and D.Z.; funding acquisition, A.K.-K., A.S. and D.Z. All authors have read and agreed to the published version of the manuscript.

Funding: This research received no external funding.

Institutional Review Board Statement: Not applicable.

Informed Consent Statement: Not applicable.

Data Availability Statement: The data presented in this study are available on request from the corresponding author. Data was obtained from Polish Farm Accountancy Data Network.

Conflicts of Interest: The authors declare no conflict of interest.

References

1. Raj, T. Diversification of Small and Marginal Farms in Himachal Pradesh. *J. Agric. Econ.* **2010**, *7*, 7–16.
2. *Rural areas in Poland in 2018*; Statistics Poland, Statistical Office in Olsztyn: Warszawa, Olsztyn, 2020.
3. Lowder, S.K.; Skoet, J.; Raney, T. The Number, Size, and Distribution of Farms, Smallholder Farms, and Family Farms Worldwide. *World Dev.* **2016**, *87*, 16–29. [CrossRef]

4. Strzelecka, A. Zmiany dochodów drobnych gospodarstw rolnych w Polsce. (Changes in the incomes of small farms in Poland). *Probl. Small Agric. Hold.* **2018**, *1*, 73–91. [CrossRef]
5. Borychowski, M.; Poczta-Wajda, A.; Sapa, A. Small Farms in the World: Selected Issues. In *Small Farms in the Paradigm of Sustainable Development. Case Studies of Selected Central and Eastern European Countries*; Stępień, S., Maican, S., Eds.; Wydawnictwo Adam Marszałek: Toruń, Poland, 2020.
6. Hornowski, A.; Parzonko, A.; Kotyza, P.; Kondraszuk, T.; Bórawski, P.; Smutka, L. Factors Determining the Development of Small Farms in Central and Eastern Poland. *Sustainability* **2020**, *12*, 5095. [CrossRef]
7. Żmija, J.; Czekaj, M. Wspólna Polityka Rolna a rozwój drobnych gospodarstw rolnych. (Common Agricultural Policy and the Development of Small Farms). *Sci. J. Wars. Univ. Life Sci. SGGW. Eur. Policies Financ. Mark.* **2012**, *8*, 518–527.
8. Forgacs, C. Is Specialization a Way for Small Farms in Central and Eastern European Countries to Adjust? In Proceedings of the International Scientific Conference Economic Science for Rural Development, Jelgava, Latvia, 21–22 April 2016; p. 42.
9. Burkitbayeva, S.; Swinnen, J. Smallholder agriculture in transition economies. *J. Agrar. Chang.* **2018**, *18*, 882–892. [CrossRef]
10. Beluhova-Uzunova, R.; Hristov, K.; Shishkova, M. Small Farms in Bulgaria—Trends and Perspectives. *Agric. Sci. Agrar. Nauk.* **2019**, *11*, 59–65. [CrossRef]
11. Dzun, W. Drobne gospodarstwa w rolnictwie polskim—Próba definicji i charakterystyki. (Polish agricultures' small farms. An effort to define this category and its characteristics). *Village Agric.* **2013**, *2*, 9–27.
12. Hornowski, A.; Kryszak, Ł. Wyniki finansowe drobnych gospodarstw rolnych w świetle sprawozdań europejskiego FADN. (Financial results of small farms in the ligh to reports of European FADN). *Ann. Pol. Assoc. Agric. Agribus. Econ.* **2016**, *18*, 92–98.
13. Strzelecka, A.; Zawadzka, D. Does Production Specialization Have an Impact on the Financial Efficiency of Very Small Farms? In Proceedings of the 36th International Business Information Management Association, Granada, Spain, 4–5 November 2020.
14. Poczta-Wajda, A.; Sapa, A.; Stępień, S.; Borychowski, M. Food in security among Small-Scale Farmers in Poland. *Agriculture* **2020**, *10*, 295. [CrossRef]
15. Czyżewski, B.; Sapa, A.; Kułyk, P. Human Capital and Eco-Contractual Governance in Small Farms in Poland: Simultaneous Confirmatory Factor Analysis with Ordinal Variables. *Agriculture* **2021**, *11*, 46. [CrossRef]
16. Musiał, W.; Drygas, M. Dylematy procesu delimitacji drobnych gospodarstw rolnych. (Dilemmas in the Process of Marking Boundries of Small Farms). *Village Agric.* **2013**, *2*, 55–74.
17. Ardakani, Z.; Bartolini, F.; Brunori, G. New Evaluation of Small Farms: Implication for an Analysis of Food Security. *Agriculture* **2020**, *10*, 74. [CrossRef]
18. Rodrigues Fortes, A.; Ferreira, V.; Barbosa Simões, E.; Baptista, I.; Grando, S.; Sequeira, E. Food Systems and Food Security: The Role of Small Farms and Small Food Businesses in Santiago Island, Cabo Verde. *Agriculture* **2020**, *10*, 216. [CrossRef]
19. Rivera, M.; Guarín, A.; Pinto-Correia, T.; Almaas, H.; Arnalte-Mur, L.; Burns, V.; Czekaj, M.; Ellis, R.; Galli, F.; Grivins, M.; et al. Assessing the role of small farms in regional food systems in Europe: Evidence from a comparative study. *Glob. Food Secur.* **2020**, *26*, 100417. [CrossRef]
20. Galli, F.; Grando, S.; Adamsone-Fiskovica, A.; Bjørkhaug, H.; Czekaj, M.; Duckett, D.G.; Almaas, H.; Karanikolas, P.; Moreno-Pérez, O.M.; Ortiz-Miranda, D.; et al. How do small farms contribute to food and nutrition security? Linking European small farms, strategies and out comes in territorial food systems. *Glob. Food Secur.* **2020**, *26*, 100427. [CrossRef]
21. Guarín, A.; Rivera, M.; Pinto-Correia, T.; Guiomar, N.; Šūmane, S.; Moreno-Pérez, O.M. A new typology of small farms in Europe. *Glob. Food Secur.* **2020**, *26*, 100389. [CrossRef]
22. Strzelecka, A.; Zawadzka, D. Production Potential and Income of Very Small Farms in the European Union and Poland. In Proceedings of the 36th International Business Information Management Association, Granada, Spain, 4–5 November 2020.
23. Kurdyś-Kujawska, A.; Strzelecka, A.; Szczepańska-Przekota, A.; Zawadzka, D. *Dochody Rolnicze. Determinanty-Zróżnicowanie-Stabilizacja. (Agricultural income. Determinants-Differentiation-Stabilization)*; Wydawnictwo Politechniki Koszalińskiej: Koszalin, Poland, 2019.
24. Myeni, L.; Moeletsi, M.E. Factors Determining the Adoption of Strategies Used by Smallholder Farmers to Cope with Climate Variability in the Eastern Free State, South Africa. *Agriculture* **2020**, *10*, 410. [CrossRef]
25. Aribi, F.; Sghaier, M. Determinants and strategies of farmers' adaptation to climate change: The case of Medenine Governorate, Tunisia. *Agrofor Int. J.* **2020**, *5*, 122–129. [CrossRef]
26. Mango, N.; Makate, C.; Mapemba, L.; Sopo, M. The role of crop diversification in improving household food security in central Malawi. *Agric. Food Secur.* **2018**, *7*, 7. [CrossRef]
27. Iocola, I.; Angevin, F.; Bockstaller, C.; Catarino, R.; Curran, M.; Messéan, A.; Schader, C.; Stilmant, D.; VanStappen, F.; Vahove, P.; et al. An Actor-Oriented Multi-Criteria Assessment Framework to Support a Transition towards Sustainable Agricultural Systems Basedon Crop Diversification. *Sustainability* **2020**, *12*, 5434. [CrossRef]
28. Figurek, A.; Rokvić, G.; Vaśko, Ž. Diversification of the rural economy as the function of the sustainability of rural areas. *Agric. For.* **2012**, *58*, 51–61.
29. Ogundari, K. Crop diversification and technical efficiency in food crop production. *Int. J. Soc. Econ.* **2013**, *40*, 267–287. [CrossRef]
30. Banerjee, D.; Bhattacharya, U. Problems of Crop Diversification in West Bengal. In *Diversification of Agriculture in Eastern India. Indian Studies in Business and Economics*; Ghosh, M., Sarkar, D., Roy, B., Eds.; Springer: New Delhi, India, 2015. [CrossRef]
31. Salvioni, C.; Henke, R.; Vanni, F. The Impact of Non-Agricultural Diversification on Financial Performance: Evidence from Family Farms in Italy. *Sustainability* **2020**, *12*, 486. [CrossRef]

32. Lakner, S.; Kirchweger, S.; Hoop, D.; Brümmer, B.; Kantelhardt, J. The Effects of Diversification Activities on the Technical Efficiency of Organic Farms in Switzerland, Austria, and Southern Germany. *Sustainability* **2018**, *10*, 1304. [CrossRef]
33. McNamara, K.T.; Weiss, C. Farm Household Income and On-and Off-Farm Diversification. *J. Agric. Appl. Econ.* **2005**, *37*, 37–48. [CrossRef]
34. Chavas, J.P.; Kim, K. Economies of diversification: A generalization and decomposition of economies of scope. *Int. J. Prod. Econ.* **2010**, *126*, 229–235. [CrossRef]
35. Coelli, T.J.; Fleming, E. Diversification economies and specialization efficiencies in a mixed food and coffee small holder farming system in Papua New Guinea. *Agric. Econ.* **2004**, *31*, 229–239. [CrossRef]
36. Nguyen, H.Q. *Crop Diversification, Economic Performance and Household's Behaviours: Evidence from Vietnam*; MPRA Paper 59090; University Library of Munich: Munich, Germany, 2014.
37. Manjunatha, A.V.; Anik, A.R.; Speelman, S.; Nuppenau, E.A. Impact of land fragmentation, farm size, land owners hip and crop diversity on profit and efficiency of irrigated farms in India. *Land Use Policy* **2013**, *31*, 397–405. [CrossRef]
38. European Parliament Resolution of 27 October 2016 on How the CAP can Improve Job Creation in Rural Areas (2015/2226(INI). How the CAP can Improve Job Creation in Rural Areas. Available online: https://eur-lex.europa.eu/legal-content/EN/TXT/PDF/?uri=CELEX:52016IP0427&from=PL (accessed on 2 November 2020).
39. Vanden Berg, M.M.; Hengesdijk, H.; Wolf, J.; Ittersum, M.K.V.; Guangho, W.; Roetter, R.P. The impact of increasing farm size and mechanization on rural income and rice production in Zhejiang province, China. *Agric. Syst.* **2007**, *94*, 841–850. [CrossRef]
40. Haji, J. Production Efficiency of Smallholders' Vegetable-dominated Mixed Farming System in Eastern Ethiopia: A Non-Parametric Approach. *J. Afr. Econ.* **2007**, *1*, 1–27. [CrossRef]
41. Mzyece, A.; Ng'ombe, J.N. Does crop diversification involve a trade-off between technical efficiency and income stability for rural farmers? Evidence from Zombia. *Agronomy* **2020**, *10*, 1875. [CrossRef]
42. Mazzocchi, C.; Orsi, L.; Ferrazzi, G.; Corsi, S. The Dimensions of Agricultural Diversification: A Spatial Analysis of Itali Municipalities. *Rural Sociol.* **2020**, *85*, 316–345. [CrossRef]
43. Rehima, M.; Belay, K.; Dawit, A.; Rashid, S. Factors affecting farmers' crops diversification: Evidence from SNNPR, Ethiopia. *Int. J. Agric. Sci.* **2013**, *3*, 558–565.
44. Makate, C.; Wang, R.; Makate, M.; Mango, M. Crop diversification and livelihoods of smallholder farmers in Zimbabwe: Adaptive management for environmental change. *Springer Plus* **2016**, *5*, 1–18. [CrossRef]
45. Dembele, B.; Bett, H.K.; Kariuki, I.M.; LeBars, M.; Ouko, K.O. Factors influencing crop diversification strategies among smallholder farmers in cotton production zone in Mali. *Adv. Agric. Sci.* **2018**, *6*, 1–16.
46. Nayak, C.; Kumar, C.R. Crop diversification in Odisha: An analysis based on panel data. *Agric. Econ. Res. Rev.* **2019**, *32*, 67–80. [CrossRef]
47. Anuja, A.R.; Kumar, A.; Saroj, S.; Singh, K.N. The impact of crop diversification towards high-value crops on economic welfare of agricultural households in Eastern India. *Curr. Sci.* **2020**, *118*, 1575–1582. [CrossRef]
48. Geethu, P.G.; Sharma, H.O. Determinants of crop diversification in Kerala-a temporal analysis. *J. Trop. Agric.* **2020**, *58*, 99–106.
49. Bansal, H.; Sharma, S.; Kumar, R.; Singh, A. The Factors Influencing and Various Technological and Socio-Economic Constraints for Crop Diversification in Haryana. *Econ. Aff.* **2020**, *65*, 409–413. [CrossRef]
50. Kumar, C.R. Crop Diversification And Its Determinants: A Comparative Study Between Cuttack And Kandhamal Districts Of Odisha, India. *J. Glob. Resour.* **2020**, 6. [CrossRef]
51. Mulwa, C.K.; Visser, M. Farm diversification as an adaptation strategy to climatic shocks and implications for food security in northern Namibia. *World Dev.* **2020**, *129*, 104906. [CrossRef]
52. Dries, L.; Pascucci, S.; Gardebroek, C. Diversification in Italian Farm Systems: Are Farmers Using Interlinked Strategies? *New Medit* **2012**, *4*, 7–15.
53. Cramerm, J.S. *Logit Models from Economics and Other Fields*; Cambridge University Press: Cambridge, UK, 2003.
54. Hosmer, D.W.; Lemeshow, S. *Applied Logistic Regression*; John Wiley & Sons: New York, NY, USA, 2000.
55. Hoo, Z.H.; Candlish, J.; Teare, D. What is an ROC curve? *Emerg. Med. J.* **2017**, *34*, 357–359. [CrossRef]
56. Wojewodzic, T.; Jezowit-Jurek, M.; Rachwał, P. Remuneration for labour on agricultural commodity farms in the Małopolska and Pogórze macroregion. *Probl. Small Agric. Hold.* **2015**, *1*, 73–87. [CrossRef]
57. Kołoszko-Chomentowska, Z. Efektywność wykorzystania środków trwałych w gospodarstwach rolnych. (Efficiency in theuse of fixed assets on farm holdings). *Rocz. Nauk. Stowarzyszenia Ekon. Rol. I Agrobiz.* **2016**, *18/3*, 178–183.
58. Mańko, S.; Płonka, R. Struktura aktywów a wyniki działalności gospodarstw rolnych w świetle danych polskiego FADN. (The structure of assets vs performance of agricultural holdings in the light of Polish FADN Data). *Zagadnienia Ekon. Rolnej* **2010**, *4*, 134–145.
59. Strzelecka, A. Zmiany w strukturze majątkowo-kapitałowej przedsiębiorstw rolniczych w Polsce w latach 2004-2009. (Changes in the capital and asset structure of agricultural enterprises in Poland in the years 2004-2009). *Pr. Nauk. Uniw. Ekon. we Wrocławiu* **2011**, *166*, 702–715.
60. Mishra, H.S. El-Osta; Sandretto, C.L. Factors Affecting Farm Enterprise Diversification. *Agric. Financ. Rev.* **2004**, *64*, 151–166. [CrossRef]
61. Sarwosri, A.W.; Mußhoff, O. Are Risk Attitudes and Time Preferences Crucial Factors for Crop Diversification by Smallholder Farmers? *J. Int. Dev.* **2020**, *32*, 922–942. [CrossRef]

62. Auffhammer, M.; Carleton, T.A. Regional Crop Diversity and Weather Shocks in India. *Asian Dev. Rev.* **2018**, *35*, 113–130. [CrossRef]
63. Ashok, K.M.; Meti, S.K.; Vimenuo, S.K. Agricultural diversification and its impact on livelihood security of farmers. *J. Nutr. Health Sci.* **2017**, *1*, 1–10.
64. Huang, J.; Jiang, J.; Wang, J.; Hou, L. Crop diversification in coping with extreme weather events in China. *J. Integr. Agric.* **2014**, *13*, 677–686. [CrossRef]
65. Kurdyś-Kujawska, A. Dywersyfikacja upraw: Strategia zarządzania ryzykiem gospodarstw rolnych z Regionu Pomorza Środkowego. (Diversification of Crops: Risk Management Strategy of Farms Located in Central Pomerania). *Zarządzanie I Przedsiębiorczość* **2018**, *4*, 179–190.
66. DiFalco, S.; Mintewab, B.; Yesuf, M. Seeds for livelihood: Crop Biodiversity and Food Production in Ethiopia. *Ecol. Econ.* **2010**, *69*, 1695–1702. [CrossRef]
67. Dasmani, I.; Darfor, K.; Karakara, A. Farmers' choice of adoptation strategy against weather variation: Empirical evidence from the three agro-ecological zones in Ghana. *Cogent Soc. Sci.* **2020**, *6*, 1–17. [CrossRef]
68. Brenda, B.L. Resilience in Agriculture through Crop Diversification: Adaptive Management for Environmental Change. *Bio Sci.* **2011**, *61*, 183–193.
69. De Gioia, G. *A Decomposition of the Herfindahl Index of Concentration. Munich Personal RePEc Archive*; Paper No. 80360; University Library of Munich: Munich, Germany, 2017.
70. Nauenberga, E.; Basua, K.; Chandb, H. Hirschman-Herfindahl index determination under in complete information. *Appl. Econ. Lett.* **1997**, *4*, 639–642. [CrossRef]
71. Naldi, M.; Flamini, M. Dynamics of the Hirschman-Herfindahl Index uder new market entries, Economic Papers. *J. Appl. Policy* **2018**, 37. [CrossRef]
72. Hosmer, D.W.; Lemenshow, S.; May, S. *Applied Survival Analysis: Regression Modeling of Time to Event Data*; John Wiley & Sons, Inc.: New York, NY, 2008.
73. Strzelecka, A.; Kurdyś-Kujawska, A.; Zawadzka, D. Application of logistic regression models to assess household financial decisions regarding debt. *Procedia Comput. Sci.* **2020**, *176*, 3418–3427. [CrossRef]
74. Stanisz, A. *Modele Regresji Logistycznej. Zastosowanie w Medycynie, Naukach Przyrodniczych i Społecznych. (Logistic Regression Models. Application in Medicine, Natural and Social Sciences)*; Statsoft Polska: Kraków, Poland, 2016.
75. Maddala, G.S. *Introduction to Econometrics*, 3rd ed.; John Wiley & Sons, Inc.: New York, NY, 2001.
76. Stare, J.; Maucort-Boulch, D. Odds ratio, hazard ratio and relative risk. *Metodoloski Zv.* **2016**, *13*, 59–67.
77. Bland, J.M.; Altman, D.G. The Odds Ratio. *Br. Med J.* **2020**, *320*, 1468. [CrossRef]
78. Kleinbaum, D.G.; Klein, M. *Logistic Regression*; Springer: New York, NY, USA, 2002.
79. Sompolska-Rzechuła, A.; Świtłyk, M. Czynniki wpływające na prawdopodobieństwo poprawy przychodów gospodarstw rolnych specjalizujących się w produkcji mleka. (Factors affecting probability of income increase in agricultural holdings specialised in milk production). *Zagadnienia Ekon. Rolnej* **2016**, *4*, 107–121. [CrossRef]
80. Zweig, M.H.; Campbell, G. Receiver-operating characteristic (ROC) plots: A fundamental evaluation tool in clinical medicine. *Clin. Chem.* **1993**, *39*, 561–577. [CrossRef] [PubMed]
81. Kiryluk-Dryjska, E.; Więckowska, B. Territorial Clusters of Farmers' Interest in Diversification in Poland: Geospatial Location and Characteristics. *Sustainability* **2020**, *12*, 5276. [CrossRef]
82. Katchova, A.L. The farm diversification discount. *Am. J. Agric. Econ.* **2005**, *87*, 984–994. [CrossRef]
83. Gołaś, Z. Wydajność i dochodowość pracy w rolnictwie w świetle rachunków ekonomicznych dla rolnictwa. (Productivity and Profitability of Labour in Agriculture in the Light of the Economic Accounts for Agriculture). *Zagadnienia Ekon. Rolnej* **2010**, *3*, 19–42.
84. Strzelecka, A.; Kurdyś-Kujawska, A.; Zawadzka, D. Kapitał obcy a potencjał wytwórczy i wyniki produkcyjno-ekonomiczne towarowych gospodarstw rolnych. (Debt versus production potential as well as production and economic results of commodity farms). *Zesz. Nauk. Szkoły Głównej Gospod. Wiej. Warszawie. Probl. Rol. Światowego* **2019**, *19*, 110–119. [CrossRef]
85. Wieliczko, B.; Kurdyś-Kujawska, A.; Sompolska-Rzechuła, A. Savings of Small Farms: Their Magnitude, Determinants and Role in Sustainable Development. Example of Poland. *Agriculture* **2020**, *10*, 525. [CrossRef]
86. Runowski, H. Ekonomika rolnictwa—przemiany w gospodarstwach rolnych. (Economics of agriculture—Changes in farms). In *Rolnictwo, Gospodarka żywnościowa, obszarywiejskie—10 lat w Unii Europejskiej*; Drejerska, N., Ed.; Agriculture, food economy, rural areas—10 years in the European Union; SGGW: Warszawa, Poland, 2014.
87. Kurdyś-Kujawska, A.; Sompolska-Rzechuła, A. Public support for agriculture of EU countries under the CAP. Scale, dynamics and trends of changes. In *Subsidies versus Economics, Finance and Income of Farms (4)*; Monographs of Multi-annual Programme 2015–2019, 77.1; Soliwoda, M., Ed.; IAFE-NRI: Warsaw, Poland, 2018.
88. Wieliczko, B.; Sompolska-Rzechuła, A.; Kurdyś-Kujawska, A. Determinants of the Use of Subsidies for the Development of Rural Areas by Small Agricultural Holdings: Case of Poland. In Proceedings of the 2019 International Conference Economic Science For Rural Development, Jelgava, Latvia, 9–10 May 2018; p. 50.
89. Zawadzka, D.; Strzelecka, A.; Szafraniec-Siluta, E. Ukierunkowanie produkcji gospodarstwa rolnego a zdolność do samofinansowania nakładów inwestycyjnych—ujęcie porównawcze. (The type of farm's production and the ability to self-finance of investments—A comparative approach). *J. Manag. Financ.* **2014**, *12*, 289–305.

90. Belay, A.; Recha, J.W.; Woldeamanuel, T.; Morton, J.F. Small holder farmers' adaptation to climate change and determinants of their adaptation decisions in the Central Rift Valley of Ethiopia. *Agric. Food Secur.* **2017**, *6*, 1–13. [CrossRef]
91. Carter, M.R.; Cheng, L.; Sarris, A. Where and how index insurance can boost the adoption of improved agricultural technologies. *J. Dev. Econ.* **2016**, *118*, 59–71. [CrossRef]
92. Ellis, F. Household strategies and rural livelihood diversification. *J. Dev. Stud.* **1998**, *35*, 1–38. [CrossRef]
93. Wan, J.; Li, R.; Wang, W.; Liu, Z.; Chen, B. Income Diversification: A Strategy for Rural Region Risk Management. *Sustainability* **2016**, *8*, 1064. [CrossRef]
94. Asfaw, S.; Scognamillo, A.; DiCaprera, G.; Sitko, N.; Ignaciuk, A. Heterogenous impact of livelihood diversification on household welfare: Cross-country evidence from Sub-Saharan Africa. *World Dev.* **2019**, *117*, 278–295. [CrossRef]
95. McCord, P.F.; Cox, M.; Schmitt-Harsh, M.; Evans, T. Crop diversification as a livelihood strategy within semi-arid agricultural systems near Mount Kenya Land. *Use Policy* **2015**, *42*, 738–750. [CrossRef]
96. Bellon, M.R.; Kotu, B.H.; Azzarri, C.; Caracciolo, F. To diversify or not to diversify, that is the question. Pursuing agricultural development for smallholder farmers in marginal areas of Ghana. *World Dev.* **2020**, *125*, 104682. [CrossRef] [PubMed]
97. DiFalco, S.; Chavas, J.P. On crop biodiversity, risk exposure, and food security in the High lands of Ethiopia. *Am. J. Agric. Econ.* **2009**, *91*, 599–611. [CrossRef]
98. Kawa, N.C.; Clavijo-Michelangeli, J.A.; Clement, C.R. Household agrobiodiversity management on Amazonian Dark Earths, Oxisols, and Floodplain soils on the lower Madeira River, Brazil. *Braz. Hum. Ecol.* **2015**, *43*, 339–353. [CrossRef]
99. Guido, Z.; Zimmer, A.; Lopus, S.; Hannah, C.; Gower, D.; Waldman, K.; Krell, N.; Sheffield, J.; Caylor, K.; Evans, T. Farmer forecasts: Impact soft seasonal rainfall expectations on agricultural decision making in Sub-Saharan Africa. *Clim. Risk Manag.* **2020**, *30*, 100247. [CrossRef]
100. Ahmadzai, H. *Crop Diversification and Technical Efficiency in Afghanistan: Stochastic Frontier Analysis*; CREDIT Research Paper, No. 17/04; The University of Nottingham, Centre for Research in Economic Development and International Trade (CREDIT): Nottingham, UK, 2017.
101. European Union. Assessing How Policies Enable or Constrain the Resilience of Farming System in the European Union: Case Study Results. 2019. Available online: https://www.eca.europa.eu/ (accessed on 25 May 2020).
102. Giampietri, E.; Trestini, S.; Boatto, V. Toward the implementation of the Income Stabilization Tool: An analysis of factors affecting the probability of farm income losses in Italy. *New Medit* **2017**, *16*, 24–30.

 agriculture

MDPI

Article

Sunflower Husk Biochar as a Key Agrotechnical Factor Enhancing Sustainable Soybean Production

Agnieszka Klimek-Kopyra [1,*], Urszula Sadowska [2], Maciej Kuboń [3,4], Maciej Gliniak [5] and Jakub Sikora [5]

1 Institute of Plant Production, Faculty of Agriculture and Economy, Al. Mickiewicza 21, 31-120 Krakow, Poland
2 The Institute of Machinery Exploitation, Ergonomics and Production Processes, University of Agriculture in Krakow, Balicka 116B, 30-149 Krakow, Poland; urszula.sadowska@urk.edu.pl
3 Department of Production Engineering, Logistics and Applied Computer Science, University of Agriculture in Krakow, Balicka 116B, 30-149 Krakow, Poland; maciej.kubon@urk.edu.pl
4 Eastern European State College of Higher Education in Przemysl, Książąt Lubomirskich 6, 37-700 Przemysl, Poland
5 Department of Bioprocess Engineering, Power Engineering and Automation, University of Agriculture in Krakow, Balicka 116B, 30-149 Krakow, Poland; maciej.gliniak@urk.edu.pl (M.G.); jakub.sikora@urk.edu.pl (J.S.)
* Correspondence: agnieszka.klimek@urk.edu.pl

Abstract: Climate change has a decisive impact on the physical parameters of soil. To counteract this phenomenon, the ongoing search for more effective agri-technical solutions aims at the improvement of the physical properties of soil over a short time. The study aimed to assess the effect of biochar produced from sunflower husks on soil respiration (SR), soil water flux (SWF), and soil temperature (ST), depending on its dose and different soil cover (with and without vegetation). Moreover, the seed yield was assessed depending on the biochar fertilization. Field experiments were conducted on Calcaric/Dolomitic Leptosols (Ochric soil). SR, ST, and SWT were evaluated seven times in three-week intervals during two seasons, over 2018 and 2019. It was found that the time of biochar application had a significant effect on the evaluated parameters. In the second year, the authors observed significantly ($p < 0.005$) higher soil respiration (4.38 μmol s^{-1} m^{-2}), soil temperature (21.2 °C), and the level of water net transfer in the soil (0.38 m mol s^{-1} m^{-2}), compared to the first year. The most effective biochar dose regarding SR and soybean yield was 60 t ha^{-1}. These are promising results, but a more comprehensive cost-benefit analysis is needed to recommend large-scale biochar use at this dose.

Keywords: biochar; sunflower husk; soil respiration; soybean

Citation: Klimek-Kopyra, A.; Sadowska, U.; Kuboń, M.; Gliniak, M.; Sikora, J. Sunflower Husk Biochar as a Key Agrotechnical Factor Enhancing Sustainable Soybean Production. *Agriculture* **2021**, *11*, 305. https://doi.org/10.3390/agriculture11040305

Academic Editors: Claudia Di Bene, Rosa Francaviglia, Roberta Farina, Jorge Álvaro-Fuentes and Raúl Zornoza

Received: 12 February 2021
Accepted: 27 March 2021
Published: 1 April 2021

Publisher's Note: MDPI stays neutral with regard to jurisdictional claims in published maps and institutional affiliations.

1. Introduction

Soil respiration is an important indicator of soil fertility [1,2]. It includes diversified proportions of both autotrophic (root respiration) and heterotrophic components (microbial and soil fauna respiration), depending on soil type and growing season. The source of CO_2 emitted to the atmosphere from the soil surface is mainly root respiration, as well as decomposition of some root residues, soil organic matter, and plant litter [3,4]. The heterogeneity of the vegetation cover and physical properties of the soil contribute to the spatial variability of soil respiration [5,6]. Soil respiration also depends on the adopted farming system [7,8]. Many researchers argue that the farming system directly affects CO_2 emissions in soil and the content of C, and thus, the impact on global warming [9–11]. Switching from traditional to conservation tillage, including no-tillage (NT) cultivation, can reduce CO_2 emissions [12]. Soil management and changes in organic matter content are among the factors controlling CO_2 emissions [13]. Hence, it seems that determining the adaptability of the soil to the changing climatic conditions—reduced precipitation and temperature increase—would allow for safe and optimized soil management to ensure a higher yielding of plants while reducing CO_2 emissions. Kong et al. [14] proved the relationship between soil respiration, its temperature, and the amount of organic matter in the soil in the form of straw. The authors showed that straw retention in the soil is an effective method of conserving soil water and increasing carbon levels by reducing soil respiration. These studies are important in terms of the large-scale use of biochar as a source of cheap organic matter needed to improve soil retention properties. However, various scientific communities have thus far been unable to indicate the type of biomass that would indisputably and effectively, in a relatively short time, stabilize the physical parameters of the soil, as highlighted in previous studies by Liu et al. [15] and Ameloot et al. [16]. There are many sources of biomass, including wood and its waste, crops and their waste, municipal waste, food processing waste, as well as aquatic plants and algae [17–19]. Among the mentioned biomass sources, agricultural waste and energy crops are described as good precursors for the production of biogas, biofuel oil, and biodiesel [20,21]. A by-product of sunflower oil extraction from seeds demonstrates several benefits and possibilities in terms of biofuel production, especially bio-diesel [22]. In the past, the use of sunflower as a source of biomass was limited due to the unidirectional sales trend, mainly as animal feed. Recent attempts to diversify the use of sunflower in the energy industry have focused on the use of the husk as a raw material for the production of biofuels and other valuable chemical products. Sunflower husks are a promising alternative biomass source, offering numerous benefits and opportunities in biofuel research, in particular, in the production of biodiesel, biogas, and biochar [20,23]. Sunflower husks consist mainly of fibrous substances, nitrogen-free extractive proteins, oil, and ash. Its structural composition (cellulose, hemicellulose, and lignin) is diversified, impacted by environmental factors. On the other hand, according to Haykiri-Acam and Yaman [24], the sunflower husk contains 8.1% moisture, 76.4% volatile matter, 12.2% carbon, 3.3% ash, and its gross calorific value is 16.1 MJ/kg.

Biochar is produced by pyrolysis from various organic materials, including plants and organic waste. Its use on poorer or degraded soils has gained recognition as a strategic element in mitigating climate change due to its long-term and readily available carbon source [25–27]. The use of biochar on agricultural land is important for the improvement of degraded soils as it improves the physicochemical and soil properties [28–30]. According to some authors, biochar limits the absorption of heavy metals by plants, acting as a specific buffer [31]. Moreover, it is resistant to microbial degradation and remains in the soil for longer periods, thus providing a long-term benefit to soil fertility [32] and reducing the leaching of nutrients from the soil, to improve the nutrient life cycle.

Biochar made from various types of biomass sources can react in various ways depending on the type of soil to which it has been applied and broadly understood environmental conditions. This may be why, in some studies, biochar was reported to increase soil respiration and in other studies, to reduce it.

Thus far, no field studies have been conducted to assess the impact of the dosage of sunflower husk biochar on soil respiration and plant yield, although it was reported that the consequence of biochar addition on plant productivity depends on the amount added [23]. Although there is evidence on the relation between the biochar dose and its effect, the existing data gap prevents drawing general recommendations. Moreover, biochar materials can vary greatly in their characteristics; hence, the nature of the particular biochar material (e.g., pH and ash content) can also impact the application rate. Several studies have reported a positive effect of using biochar on crop yields at 5–50 tonnes per hectare with appropriate nutrient management [33]. The experiments conducted by Rondon et al. [34] resulted in a decrease in crop yield in a pot experiment with nutrient-deficient soil amended with biochar at 165 tonnes per hectare. Thus, controlling the biochar application rate is necessary to prevent its negative impact.

The study aimed to assess the effect of biochar produced from sunflower husks on physical soil properties (soil respiration, soil water flux, and soil temperature) and seed yield, depending on its dose and different soil cover (with and without vegetation).

2. Material and Methods

2.1. Field Experiment

The experiments were conducted on the experimental field of the University of Agriculture in Krakow (50°04′ N, 19°51′ E, 211 m MSL, slope 2°). The soil was characterized as Calcaric/Dolomitic Leptosols (Ochric), according to World Reference Base for Soil Resources [35]. The soil was mostly composed of sand (56.7%), silt (32%), and clay (10.4%) with a gravel fraction (0.9%).

2.2. Experiment Design

Two field experiments were conducted in the years 2018–2019. The experiments were established in a randomized block design with four replicates.

2.2.1. Experiment-1

The single-factor experiment tested the effects of four biochar doses, i.e., 0, 20, 40, and 80 t ha^{-1} applied on bare soil in March 2018. The biochar was incorporated and mixed into the topsoil layer (30 cm depth) to obtain a uniform mass.

In the first week of March, dragging was carried out to prevent evaporation. Then, after 3 weeks, cultivation was carried out with an active rototiller aggregate up to a depth of 20 cm to loosen the topsoil before applying the biochar to the experimental plots. This was done by hand and then the biochar was mixed with a manual rotary cultivator up to a depth of 20 cm. Each treatment had four replications. Each plot's size was 3 m^2.

2.2.2. Experiment-2

In 2019, a two–factor experiment was conducted to compare the effects of four doses of biochar application (i.e., 0, 20, 40, and 80 t ha^{-1}) on two different soil covers: with and without the plants (soybean).

Each treatment had four replications. The plot size was 3 m^2 each. Soybean was sown in the second week of April at a standard planting rate (80 seeds m^{-2}), followed by standard NPK mineral fertilization (30 kg N, 70 kg P_2O_5, 100 kg K_2O). Prior to sowing, the soybean seeds were inoculated with *Bradryzobium japonicum* bacteria. No pesticides were applied during plant vegetation; weeds were controlled mechanically. In the phase of full maturity, the soybean yield and the height of the first pod deposition were assessed based on the yield structures, as an important parameter of the plants' adaptation to the habitat conditions.

2.3. Biochar Characterization

The biochar was produced from sunflower husks by pyrolysis, at 450–550 °C [36,37]. It was prepared for scanning electron microscope (SEM) by thorough crumbling. Next, the sample was transferred under vacuum and imaged using SEM (Zeiss Ultra Plus, Microscopy GmbH, Potsdam, Germany) at 5 kV.

The obtained biochar's water content is 0.49%, ash 8.08%, volatile particles 11.56%, and fixed carbon 79.87%. Its elemental composition is as follows: C—85.32%; H—2.99%; N—1.06%; S—0.058%; O—2.01%; pHKCl—9.2.

The biochar is characterized by specific porosity (Figure 1): average pore radius is 0.24 µm, the total pore area is 19.01 m^2 g^{-1}, and the total porosity is 75.92%.

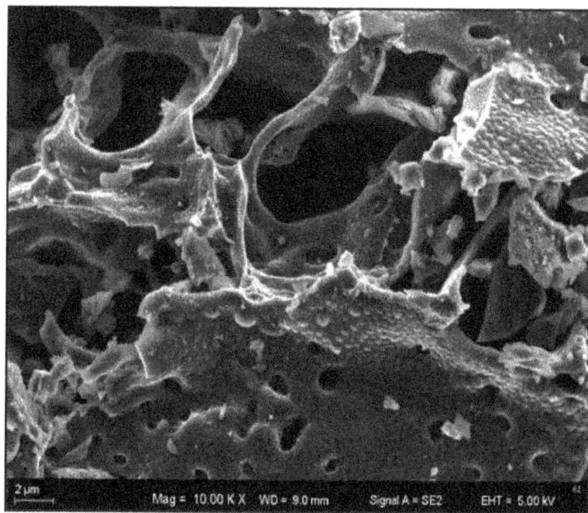

Figure 1. SEM image of the biochar porosity.

2.4. Soil Analysis

The chemical properties of soil were determined by standard methods and conducted in the second year of the study. The pH was measured potentiometrically in 1 M KCl after 24 h in the liquid/soil ratio of 10. Total organic carbon (TOC) was determined by TOC-VCSH (Shimadzu) with Solid Sample Module SSM-5000.

Measurements of soil respiration were conducted with the SRS-SD 1000 m (by ADC BioScientific Ltd., Hoddesdon, UK). Due to the specificity of the SRS-SD device (by ADC BioScientific Ltd., Hoddesdon, UK), CO_2 readouts in the soil were registered and recorded after 15 min from the moment the measurement was started. To reduce the measurement errors, readouts were made at the same time of day with similar atmospheric conditions. Measurements were not carried out during or shortly after precipitation. Prior to the measurements, the speed of gas flow was determined at 200 µmol s^{-1}, which guaranteed that the balance inside a measurement chamber was achieved after 15 min of active operation of the meter (SRS-SD 1000). The soil respiration, soil temperature, and water flux were measured 7 times during each season in three-week intervals during the two seasons.

Soil respiration (net molar flow of CO_2 in/out of the soil; µmol mol^{-1}) is:

$$Ce = u\,(-\Delta c), \tag{1}$$

where u is the molar air flow in mol s^{-1}; Δc is the difference in CO_2 concentration throughout the soil chamber, µmol mol^{-1}; $\Delta c = C_{ref}$—Can, where C_{ref} is the CO_2 flowing into the soil chamber, µmol mol^{-1}; and C_{an} is CO_2 flowing out from the soil chamber, µmol mol^{-1}.

The net H_2O Exchange Rate (Soil Flux) W_{flux} (m mol s^{-1} m^{-2}) is:

$$W_{flux} = \Delta eus/p, \tag{2}$$

where us is the molar flow of air per square meter of soil, m mol m^{-2} s^{-1}; Δe is the differential water vapor concentration, m Bar; and p is the atmospheric pressure, mBar.

2.5. Statistical Analyses

Results were statistically analyzed. The assumption of normality was checked and based on it, the statistical analysis was conducted. The one- and two-way analysis of variance (ANOVA) tests were performed at $\alpha = 0.05$, followed by an HSD Tukey's test. The Pearson coefficient of correlation between traits was calculated.

3. Results

3.1. Meteorological Conditions

The course of the weather was similar in the studied growing seasons; however, the distribution of rainfall changed in time (Figure 2). In 2018, heavy rainfall (over 59 mm) occurred in July, while in 2019, it occurred in April, May, and September (Figure 2A). In the analyzed period, there were periods without rainfall (June), but also numerous periods of drought (Figure 2B). More rainfall occurred in 2019; most days with rainfall occurred in May, and the least in June.

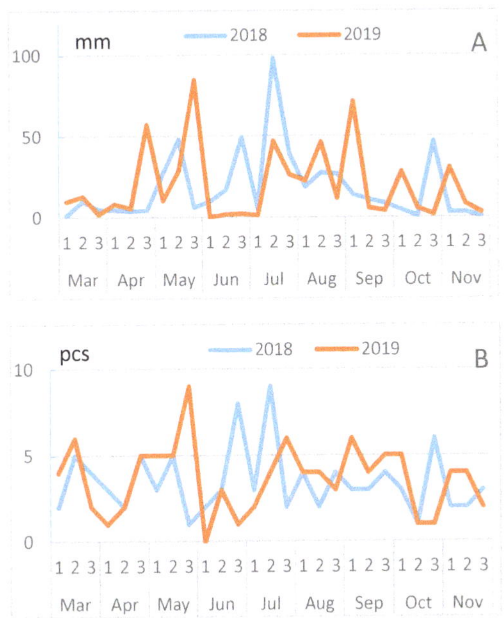

Figure 2. Rainfall distribution (**A**) and the number of days with rainfall (**B**) in 10-day intervals during the vegetation seasons.

3.2. Impact of Biochar Application on Selected Soil Parameters in the Year of Application and after One Year

The significantly positive conditional correlation obtained between soil water flux (SWF), soil respiration (SR), and soil temperature (ST) was related to the date of application of biochar (Table 1). Smaller correlations of other factors of parameters were visible in the first year of the study, which was impacted by the physical properties of the soil, e.g., looseness due to recent biochar application. In the first year of the study, the most significant relationship was found between SR and SWF. As the flow of water between the soil and the atmosphere increased, an increase in soil respiration was observed. In the second year after biochar application, the relationship between SR and SWF increased to ultimately prove the strongest mean correlation (r = 0.76) over the years. Along with the increase in respiration, the water flow in the soil increased significantly. On the other hand, a weaker correlation was found between soil respiration and temperature (r = 0.55) and between temperature and water flow in the soil (r = 0.44).

3.3. Impact of Biochar Application on Selected Soil Parameters in the First and in the Second Year on Bare Soil

The biochar-amended soil was characterized by higher pH and TOC compared to control soil. The pH increased proportionally to the biochar rate (Table 1). The TOC increase was proportional to the increase of biochar rate mainly in treatments of bare soil. No significant differences of TOC were revealed (Table 2).

Without the use of a protective plant, the analyzed soil parameters significantly varied between seasons (Table 3). The lower efficiency of the respiration process identified in the first year of the study was due to the physical properties of the soil, probably related to the lack of compactness resulting from the timing of biochar application. The water content in the soil was the result of the amount of rainfall and the number of days with rainfall. Higher precipitation was recorded in 2019, as confirmed by the significantly higher values of the obtained soil water flux index. The amount of biochar used significantly impacted the soil respiration process. The best effects were observed in the test objects with 60 t ha^{-1} biochar applied compared to control. Moreover, the use of biochar significantly improves the water flow in the soil compared to the control object.

Table 1. Pearson coefficient of correlation between soil water flux (SWF), soil respiration (SR), and soil temperature (ST).

	2018			2019			Mean		
	SWF	**SR**	**ST**	**SWF**	**SR**	**ST**	**SWF**	**SR**	**ST**
SWF	1	0.76 *	0.14	1	0.82 *	0.71 *	1	0.76 *	0.40
SR	0.76 *	1	0.42 *	0.82 *	1	0.73 *	0.76 *	1	0.55
ST	0.14	0.42 *	1	0.71 *	0.73 *	1	0.40 *	0.55 *	1

* Significant at the 0.05 probability level.

Table 2. Soil pH and total organic carbon (TOC) in soil after the second year from biochar incorporation.

Dose of Biochar (t ha^{-1})	pH $_{KCl}$		Total Organic Carbon (TOC) %	
	Bare Soil	**Soybean**	**Bare Soil**	**Soybean**
0	6.3	6.3	0.9	0.9
40	7.4	8.0	1.3	1.3
60	7.5	8.3	1.4	1.3
80	7.6	8.1	2.0	1.3
p-value	ns	ns	ns	ns

N = 4. Means labelled with different letters were significantly different for Tukey's as per test at $p < 0.05$, ns—not significant at the 0.05 probability level.

Table 3. Soil respiration, average soil temperature, and water vapor flow in the soil in the studied years (2018–2019), in plots without plants (bare soil).

Factor	Soil Respiration—SR (μmol s^{-1} m^{-2})	Soil Surface Temperature—ST (°C)	H_2O Exchange Rate (Soil Water Flux) = SWF (m mol s^{-1} m^{-2})
Year (Y)			
2018	2.94 b	22.2 a	0.36 b
2019	4.38 a	21.2 b	0.38 a
p-value	0.002	0.04	ns
Biochar dose t ha^{-1} (B)			
0	1.55 b	20.3 b	0.31 b
40	4.25 a	21.4 ab	0.38 a
60	4.99 a	22.3 a	0.39 a
80	3.87 a	22.7 a	0.40 a
p-value	<0.001	<0.002	<0.001
p-value Y × B	ns	ns	ns

N = 4. Means labelled with different letters were significantly different for Tukey's as per test at $p < 0.05$. ns—not significant at the 0.05 probability level.

Upon analyzing the soil respiration process throughout the growing season, significant object-related differentiation was found, depending on the dose of biochar used (Figure 3). The respiration process fluctuated depending on temperature and humidity. The significantly higher soil temperature in the summer months significantly increased soil respiration. The highest activity of soil respiration, irrespective of the dose of biochar used, was found in August. Biochar had a significant impact on the soil respiration process, which resulted in high readings in objects with a dose of 60 t ha^{-1} (18 µmol s^{-1} m^{-2}).

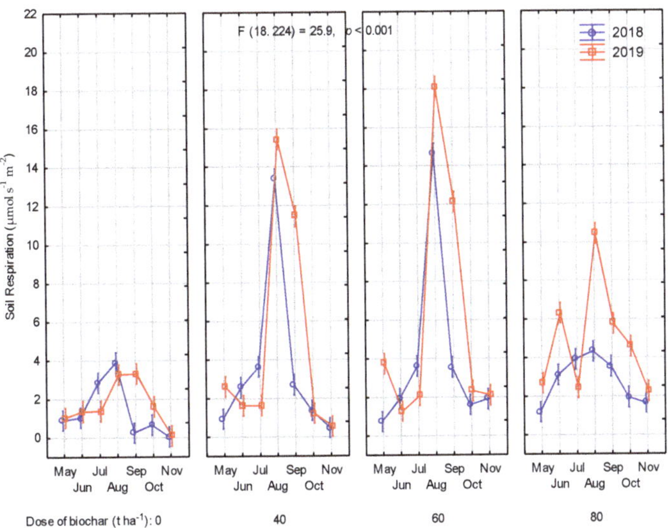

Figure 3. Distribution of soil respiration throughout the growing seasons in objects fertilized with biocarbon, without a protective plant (bare soil).

3.4. Impact of Biochar Application on Selected Soil Parameters in the Second Year Depending on the Soil Protection Variant

The use of a protective plant in the second year of the study had no significant effect on the soil respiration process and water flow in the soil (Table 4). However, a significant impact of the applied biochar dose on the soil respiration process and soil temperature was observed. Application of an average dose of biochar (60 t ha^{-1}) resulted in a significant increase in soil respiration compared to the control. This test object also obtained a slightly higher soil temperature and an increased water flow rate in the soil. The in-depth statistical analysis showed a significant convergence of the analyzed factors on the soil respiration process (Table 3, Figure 4). The use of biochar significantly decreased the respiratory activity of the soil, especially in the 40 t ha^{-1} dose. However, applying a higher dose did not increase soil respiration.

The course of soil respiration in the analyzed period (May–October) depended on the adopted soil cover variant (Figure 5). The lack of plant cover slightly increased the respiratory activity of the soil in May–July, but it significantly increased it in the summer months, i.e., August–September. Application of an average dose of biochar (60 t ha^{-1}) resulted in a significant increase in soil respiration compared to the control.

The minimal soil cover and characteristic of plants in the juvenile phase (June) resulted in a slight increase in soil respiration after the use of biochar (Figure 5b). A significant observation in soil respiration was found in August and September (during the period of intensive growth of plant biomass and roots) in objects with a high dose of biochar. The biochar used had a significant impact on the soybean yield (Figure 6). Soybean yields were significantly higher in the object where the average dose of biochar was applied (60 t ha^{-1}) compared to the control. However, no significant variation in the plant morphotype was found. The height of the first fruiting node on plants was similar regardless of the biochar dose used (Figure 7).

Table 4. Soil respiration activity, average soil temperature, and water vapor flow in the soil in the second year after biochar application, depending on the soil protection variant.

Factor	Soil Respiration—SR (μmol s^{-1} m^{-2})	Soil Surface Temperature—ST (°C)	H_2O Exchange Rate (Soil Water Flux) = SWF (m mol s^{-1} m^{-2})
Soil protection variant (SV)			
Bare soil	4.43 a	21.8 a	0.39 a
Soybean	4.32 a	21.6 b	0.59 a
p-value	ns	<0.05	ns
Biochar dose t ha^{-1} (B)			
0	2.21 c	20.1 c	0.32
40	4.74 b	21.1 b	0.37
60	5.56 a	22.7 a	0.84
80	4.98 b	23.1 a	0.43
p-value	<0.001	<0.001	ns
p-value SV × B	<0.001	<0.001	ns

N = 4. Means labelled with different letters were significantly different for Tukey's as per test at $p < 0.05$; ns—not significant at the 0.05 probability level.

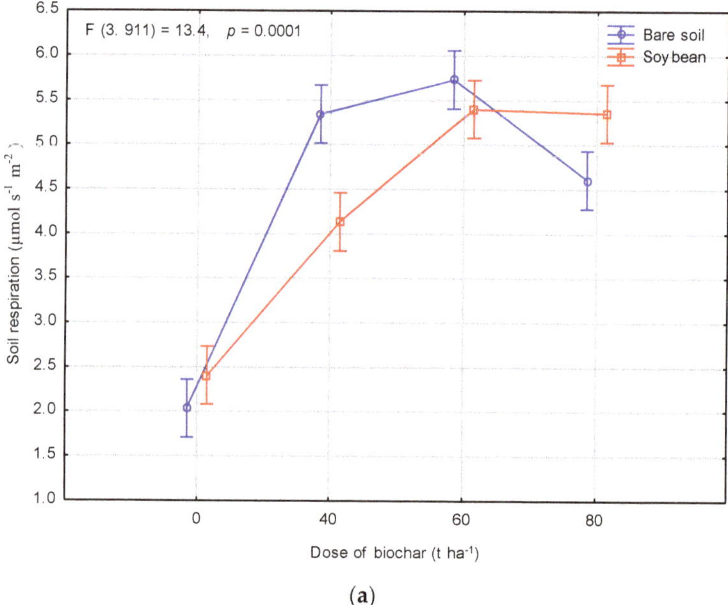

(a)

Figure 4. *Cont.*

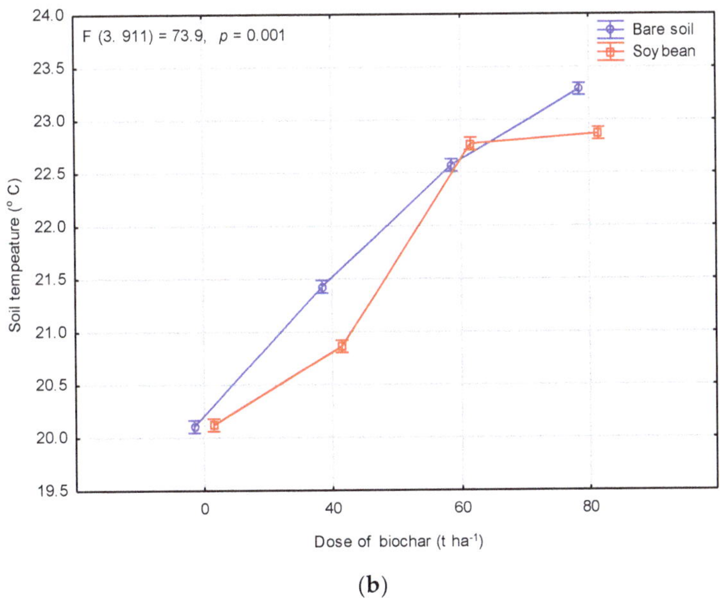

(**b**)

Figure 4. Effect of factor convergence on (**a**) soil respiration and (**b**) soil temperature.

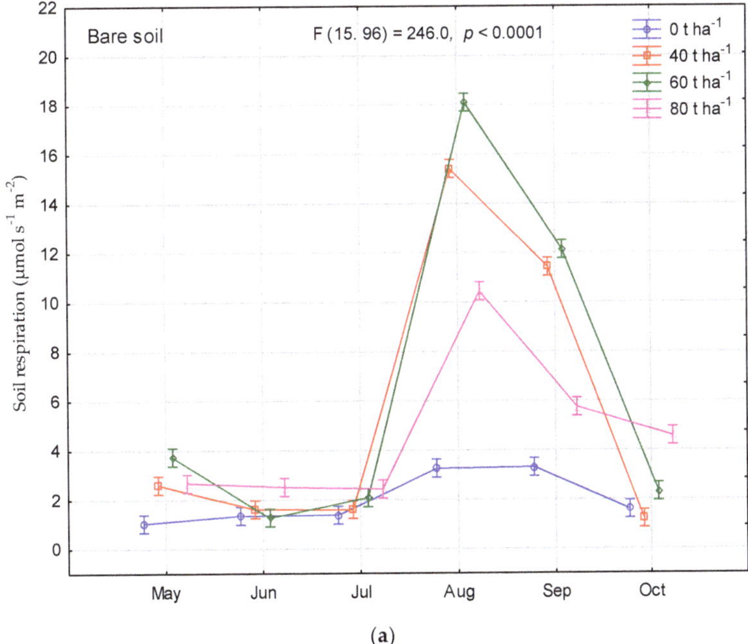

(**a**)

Figure 5. *Cont.*

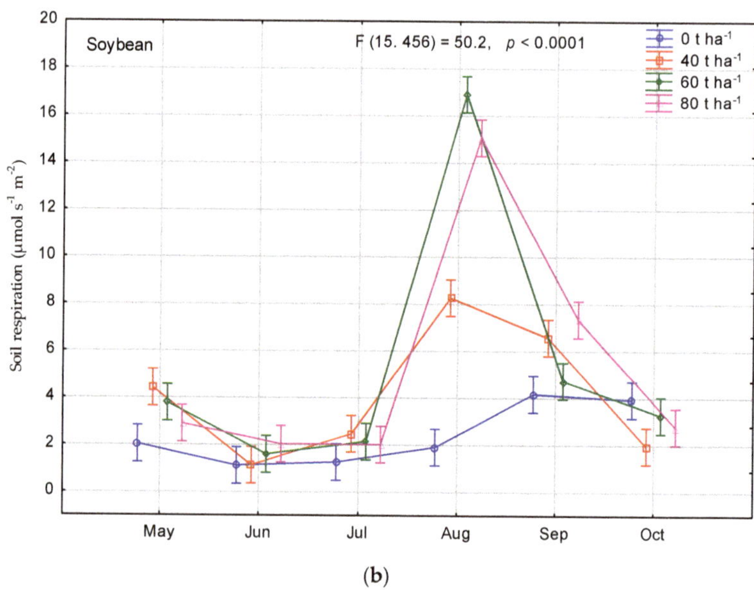

(**b**)

Figure 5. Soil respiration as a convergent effect of the dates of measurements and the dose of biochar in the objects measured: (**a**) without a protective plant, and (**b**) with a protective plant.

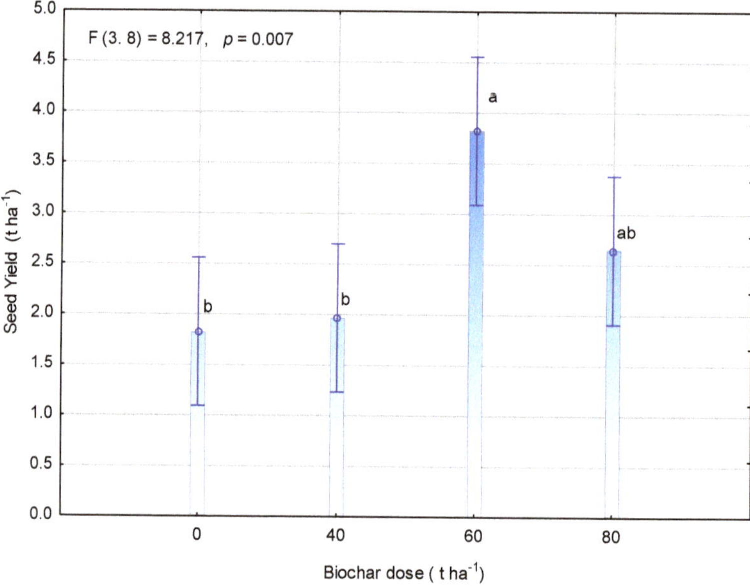

Figure 6. Soybean yield (t ha^{-1}) depending on the level of biochar fertilization. Means labelled with different letters were significantly different for Tukey's as per test at $p < 0.05$. Error bars indicate one standard error.

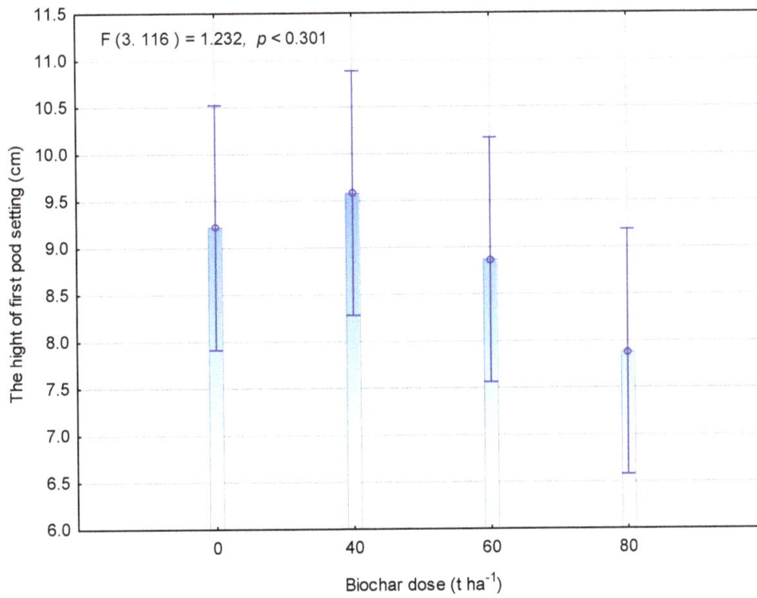

Figure 7. Height of the first pod setting depending on the level of biochar fertilization.

4. Discussion

Our results showed that biochar application increased soil respiration compared with the control treatment, which is in contradiction with several studies based on short-term incubation [1,37,38]. According to Lu et al. [25], the effects of biochar on soil respiration are varied because of differences in biochar type, soil type, soil moisture and temperature conditions, and crop planting. There was a significant negative correlation between soil respiration and soil moisture [25]. Their results indicated that rainfall during the maize-growing season suppressed soil respiration and limited the effects of biochar. The effect of soil temperature on soil respiration was greater than that of soil moisture, and soil respiration due to biochar incorporation was more sensitive to the soil temperature than that of control treatments. The research confirmed the above results since seasonal variations in soil respiratory activity, conditioned by the course of the weather, were shown. The lower efficiency of the respiration process was found in the first year of the study, which was impacted by the physical properties of the soil, e.g., lack of compactness due to recent biochar application. Moreover, the soil respiration activity was found to be highly dependent on the water flow rate and temperature. The significantly higher soil temperature in the summer months significantly increased soil respiration. The highest activity of soil respiration, irrespective of the dose of biochar used, was found in August. The presented results have been partially supported by the research of Rutigliano et al. [32], who observed that the speed of respiration was growing within the first 3 months and was statistically higher than the control, but after 14 months, there was no difference between the samples.

Lu et al. [25] analyzed soil respiratory activity in the following four years of consecutive application of straw biochar. The authors highlighted that application of straw biochar neither increased nor inhibited soil respiration throughout the entire maize-growing season compared to the control. In our own research, the authors showed that the use of biochar has a positive effect on the soil respiration process, but it depends on the soil protection variant. In the case of biochar application without soil protection, the positive effect of soil respiration was noticed regardless of biochar dose differentiation, compared to control. The differences in soil respiration between biochar treatments were significant in the experiment with soil protection. The use of biochar (up to 60 t ha^{-1}) in the experiment with soybean as a soil protector significantly increased the respiratory activity of soil compared to the control.

Zhang et al. [39] proved that the soil respiration of fields treated with returned wheat straw was 547 kg C ha^{-1} year^{-1} higher than in fields without residue in the same region. In the experiment, the authors proved relevant differences in respiration of soil conditioned by biochar compared to control conditions. However, the biochar application in different doses did not change soil respiration significantly. Shah et al. [38] tested the effect of different doses of biochar (5, 10, 20 t ha^{-1}) on soil respiration. The authors showed that with the increase in the dose of biochar, the soil respiratory activity increased. Similar conclusions were presented by Kubaczyński et al. [37], who stated that in short-term incubations, soil respiration was positively correlated with increasing biochar dose, while during long-term (several years) observation, the impact of biochar dose on the amount of emitted CO_2 was not so significant. It is worthwhile to conduct short- and long-term field studies in this area. In our own research, the authors showed that the soil respiratory activity increased proportionally to biocarbon fertilization. The best results were obtained in an object with 60 t ha^{-1} biochar, beyond which the soil respiratory activity slightly decreased.

Seremesic et al. [40] tested the effect of biochar at various doses (12.5, 25.75, 125 t ha^{-1}) and different soil types (Alluvium (A), Chernozem (C), and Humogley) on the biometric parameters of soybeans. The authors showed that soybean shoot biomass was significantly affected by soil type and biochar level. Soil types had less effect on morphological trait manifestation in soybeans. Sun et al. [41] suggested that biochar incorporation to brown soil can benefit soybean production by N retention in the soil and enhanced microbial turnover that resulted in P and K feedback. Results obtained by Seremesic et al. [40] correspond with a study of Yin et al. [42] on acid black soil, in which soybean yield increased by 35.97% compared to the control. Significant effects of biochar application on the soybean shoot were observed on Humogley soil compared to soybean height that was observed on Chernozem. Regarding shoot biomass, Humogley significantly influenced its formation compared to Alluvial soil. The obtained result could be explained with an improved water retention capacity of Humogley.

The obtained results of the soil tests for Calcaric/Dolomitic Leptosols prove that high soybean yields can be obtained with appropriate biocarbon fertilization. The authors showed that the soybean yield was significantly differentiated as impacted by the applied doses of biochar. Significantly higher soybean yields were obtained in the object with a dose of 60 t ha^{-1} biochar compared to control. However, the biochar application resulted in no significant difference in the formation of the first fruiting node on plants. Only slightly lower-placed pods were observed in test objects with a high dose of biochar. Upon analyzing the impact of biochar application on the soil respiration process throughout the growing season of soybean, the authors showed a significant difference between the objects. A significant observation in soil respiration was found in August and September (during the period of intensive growth of plant biomass and roots) in objects with a high dose of biochar.

Yooyen et al. [43] compared the effects of different doses of *Blachia siamensis Gagnep.* biochar (10, 20, 30 t ha^{-1}) on soybean yield. Growth and yields of soybean, including stem height, number of nodes, dry matter of stems, dry matter of leaves, dry matter of pods, and dry matter of seeds in the biochar treatments, show statistically significant differences at $p < 0.05$ compared to control (BC 0). The most significant result obtained in this study was the statistically significant increase of pods and seeds ($p < 0.05$). Moreover, according to the results, treatments with 20 t ha^{-1} and 30 t ha^{-1} of biochar yielded seeds 28.0 percent and 36.8 percent heavier, respectively, compared to the untreated control. In our own research, the authors showed that the biochar application increased the seed yield of the soybean, but the impact on the height of the first pod was not relevant. The highest yield (3.8 t ha^{-1}) was obtained in an object with 60 t ha^{-1} biochar, and with a higher dose, the yield slightly decreased.

5. Conclusions

The respiration process fluctuated depending on temperature and humidity. The significantly higher soil temperature in the summer months significantly increased soil respiration. The highest activity of soil respiration, irrespective of the dose of biochar used, was found in August. Biochar had a significant impact on the soil respiration process, which resulted in high readings in objects with a dose of 60 t ha^{-1} (18 µmol s^{-1} m^{-2}). The use of a protective plant in the second year of biochar application had no significant effect on the soil respiration process and water flow in the soil. However, a significant impact of the applied biochar dose was observed on the correlation between soybean cultivation on the soil respiration process and soil temperature. Among the compared treatments, a significantly higher soil respiration activity was found in the object after the application of 60 t ha^{-1} biochar, which increased soybean yield by an average of 2 t ha^{-1} compared to the control. The dose of 60 t ha^{-1} of biochar from the sunflower husk can be recommended for soybean cultivation since it increases the physical properties of sandy soil.

Author Contributions: Conceptualization, A.K.-K., U.S., M.K., M.G. and J.S. Methodology, A.K.-K. and U.S. All authors have read and agreed to the published version of the manuscript.

Funding: This work was financed by the National Centre for Research and Development (NCBiR) under Grant No. BIOSTRATEG3/345940/7/NCBR/2017.

Institutional Review Board Statement: Not applicable.

Informed Consent Statement: Not applicable.

Data Availability Statement: The data presented in this study are available on request from the corresponding author.

Conflicts of Interest: The authors declare no conflict of interest.

References

1. Bojarszczuk, J.; Księżak, J.; Gałązka, A. Soil respiration depending on different agricultural practices before maize sowing. *Plant Soil Environ.* **2017**, *63*, 435–441.
2. Niemiec, M.; Chowaniak, M.; Sikora, J.; Szeląg-Sikora, A.; Gródek-Szostak, Z.; Komorowska, M. Selected Properties of Soils for Long-Term Use in Organic Farming. *Sustainability* **2020**, *12*, 2509. [CrossRef]
3. Hanson, P.J.; Edwards, N.T.; Garten, C.T.; Andrew, J.A. Separating root and soil microbial contributions to soil respiration: A review of methods and observations. *Biogeochemistry* **2000**, *48*, 115–146. [CrossRef]
4. Högberg, P.; Nordgren, A.; Buchmann, N.; Taylor, A.F.S.; Ekblad, A.; Högberg, M.N.; Nyberg, G.; Lofvenius, M.O.; Read, D.J. Large-scale forest girdling shows that current photosynthesis drives soil respiration. *Nature* **2001**, *411*, 789–792. [CrossRef]
5. Epron, D.; Farque, L.; Lucot, E.; Badot, P.M. Soil CO_2 efflux in a beech forest: Dependence on soil temperature and soil water content. *Ann. Forest Sci.* **1999**, *56*, 221–226. [CrossRef]
6. Tang, J.; Baldocchi, D.D.; Qi, Y.; Xu, L. Assessing soil CO_2 efflux using continuous measurements of CO_2 profiles in soils with small solid-state sensors. *Agric. Forest Meteorol.* **2003**, *118*, 207–220. [CrossRef]
7. Sainju, U.M.; Jabro, J.D.; Stevens, W.B. Soil carbon dioxide emission and carbon content as affected by irrigation, tillage, cropping system and nitrogen fertilization. *J. Environ. Qual.* **2008**, *37*, 98–106. [CrossRef] [PubMed]
8. Lamptey, S.; Li, L.L.; Xie, J.H.; Zhang, R.Z.; Luo, Z.Z.; Cai, L.Q.; Liu, J. Soil respiration and net ecosystem production under different tillage practices in semi-arid Northwest China. *Plan Soil Environ.* **2017**, *63*, 14–21.
9. Liu, Y.; Hu, C.; Mohamed, I.; Wang, J.; Zhang, G.; Li, Z. Soil CO_2 Emissions and Drivers in Rice Wheat Rotation Fields Subjected to Different Long-Term Fertilization Practices. *Clean Soil Air Water* **2016**, *44*, 867–876. [CrossRef]
10. Yang, F.; Ali, M.; Zheng, X.; He, Q.; Yang, X.; Huo, W. Diurnal dynamics of soil respiration and the influencing factors for three land-cover types in the hinterland of the Taklimakan Desert, China. *J. Arid Land* **2017**, *9*, 568–579. [CrossRef]
11. Sikora, J.; Niemiec, M.; Szeląg-Sikora, A.; Gródek-Szostak, Z.; Kuboń, M.; Komorowska, M. The Impact of a Controlled-Release Fertilizer on Greenhouse Gas Emissions and the Efficiency of the Production of Chinese Cabbage. *Energies* **2020**, *13*, 2063. [CrossRef]
12. Álvaro-Fuentes, J.; López, M.V.; Cantero-Martínez, C.; Arrúe, J.L. Tillage effects on soil organic carbon fractions in Mediterranean dryland agroecosystems. *Soil Sci. Soc. Am. J.* **2008**, *72*, 541–547. [CrossRef]
13. Ginting, D.; Kessavalou, A.; Eghball, B.; Doran, J.W. Greenhouse gas emissions and soil indicators four years after manure and compost applications. *J. Environ. Qual.* **2003**, *32*, 23–32. [CrossRef]
14. Kong, D.; Liu, N.; Wang, W.; Akhtar, K.; Li, N.; Ren, G. Soil respiration from fields under three crop rotation treatments and three straw retention treatments. *PLoS ONE* **2019**, *14*, e0219253. [CrossRef]

15. Liu, Y.X.; Yang, M.; Wu, Y.M.; Wang, H.L.; Chen, Y.X.; Wu, W.X. Reducing CH_4 and CO_2 emissions from waterlogged paddy soil with biochar. *J. Soil Sedim.* **2011**, *11*, 930–939. [CrossRef]
16. Ameloot, N.; De Neve, S.; Jegajeevagan, K.; Yildiz, G.; Buchan, D.; Funkuin, Y.N.; Prins, W.; Bouckaert, L.; Sleutel, S. Short-term CO_2 and N_2O emissions and microbial properties of biochar amended sandy loam soils. *Soil Biol. Biochem.* **2013**, *57*, 401–410. [CrossRef]
17. Kraszkiewicz, A.; Niedziółka, I.; Parafiniuk, S.; Sprawka, M.; Dula, M. Assessment of selected physical characteristics of the English ryegrass (*Lolium perenne* L.) waste biomass briquettes. *Agric. Eng.* **2019**, *23*, 21–30. [CrossRef]
18. Denisiuk, W. Elements of precision agriculture in malting barley cultivation and the use of barley straw for energy purposes. *Agric. Eng.* **2020**, *24*, 21–27. [CrossRef]
19. Lucia, L.A.; Argyropoulos, D.S.; Adamopoulos, L.; Gaspar, A.R. Chemicals and energy from biomass. *Can. J. Chem.* **2006**, *84*, 960–970. [CrossRef]
20. Sikora, J.; Niemiec, M.; Szeląg-Sikora, A.; Gródek-Szostak, Z.; Kuboń, M.; Komorowska, M. The Effect of the Addition of a Fat Emulsifier on the Amount and Quality of the Obtained Biogas. *Energies* **2020**, *13*, 1825. [CrossRef]
21. Ozcimen, D.; Karaosmonglu, F. Production and characterization of bio-oil and biochar from rapeseed cake. *Renew. Energy* **2004**, *29*, 779–787. [CrossRef]
22. Perea-Moreno, M.A.; Manzano-Agugliaro, F.; Perea-Moreno, A.J. Sustainable energy based on sunflower seed husk boiler for residential buildings. *Sustainability* **2018**, *10*, 3407. [CrossRef]
23. Saleh, M.; El Refaey, A.; Mahmoud, A.H. Effectiveness of sunflower seed husk biochar for removing copper ions from wastewater: A comparative study. *Soil Water Res.* **2016**, *11*, 53–63. [CrossRef]
24. Haykiri-Acma, H.; Yaman, S. Interpretation of biomass gasification yield regarding temperature intervals under nitrogen steam atmosphere. *Fuel Proc. Technol.* **2007**, *88*, 417–425. [CrossRef]
25. Lu, N.; Liu, X.-R.; Du, Z.-L.; Wang, Y.-D.; Zhang, Q.-Z. Effect of biochar on soil respiration in the maize growing season after 5 years of consecutive application. *Soil Res.* **2014**, *52*, 505–512. [CrossRef]
26. Lehmann, J. A handful of carbon. *Nature* **2007**, *447*, 143–144. [CrossRef] [PubMed]
27. Gliniak, M.; Sikora, J.; Sadowska, U.; Klimek-Kopyra, A.; Latawiec, A.; Kubon, M. Impact of Biochar on Water Retention in Soil. *Earth Environ. Sci.* **2019**, *362*, 012046. [CrossRef]
28. Glaser, B.; Lehman, J.; Zech, W. Ameliorating physical and chemical properties of highly weathered soils in the tropics with charcoal—A review. *Biol. Fertil. Soil* **2002**, *35*, 219–230. [CrossRef]
29. Yamato, M.; Okimori, Y.; Wibowo, I.F.; Anshori, S.; Ogawa, M. Effects of the application of charred bark of *Acacia mangium* on the yield of maize, cowpea and peanut, and soil chemical properties in South Sumatra, Indonesia. *Soil Sci. Plant Nutr.* **2006**, *52*, 489–495. [CrossRef]
30. Gliniak, M.; Sikora, J.; Sadowska, U.; Klimek-Kopyra, A.; Latawiec, A.; Kubon, M. Impact of Biochar on Soil Water Content and Electrical Conductivity. *Earth Environ. Sci.* **2019**, *362*, 012044. [CrossRef]
31. Lehmann, J.; Rillig, M.C.; Thies, J.; Masiello, C.A.; Hockaday, W.C.; Crowley, D. Biochar Effects on Soil Biota—A Review. *Soil Biol. Biochem.* **2011**, *43*, 1812–1836. [CrossRef]
32. Rutigliano, F.A.; Romano, M.; Marzaioli, R.; Baglivo, I.; Baronti, S.; Miglietta, F.; Castaldi, S. Effect of biochar addition on soil microbial community in a wheat crop. *Eur. J. Soil Biol.* **2014**, *60*, 9–15. [CrossRef]
33. Liu, X.; Zhang, A.; Ji, C.; Joseph, S.; Bian, R.; Li, L.; Pan, G.; Paz-Ferreiro, J. Biochar's effect on crop productivity and the dependence on experimental condition—A meta-analysis of literature data. *Plant Soil* **2013**, *373*, 583–594. [CrossRef]
34. Rondon, M.A.; Lehmann, J.; Ramırez, J.; Hurtado, M. Biological nitrogen fixation by common beans (*Phaseolus vulgaris* L.) increases with bio-char additions. *Biol. Fertil. Soil* **2007**, *43*, 699–708. [CrossRef]
35. Hulisz, P.; Charzynski, P.; Giani, L. Application of the WRB classification to salt-affected soils in Poland and Germany. *Pol. J. Soil Sci.* **2010**, *43*, 81–92.
36. Gondek, K.; Mierzwa-Hersztek, M.; Kopeć, M.; Sikora, J.; Głąb, T.; Szczurowska, K. Influence of biochar application on reduced acidification of sandy soil, increased cation exchange capacity, and the content of available forms of K, Mg, and P. *Pol. J. Environ. Stud.* **2019**, *28*, 103–111. [CrossRef]
37. Kubaczyński, A.; Walkiewicz, A.; Brzezińska, M.; Usowicz, B. How does biochar affect soil respiration? *EGU Gen. Assem.* **2020**. [CrossRef]
38. Shah, T.; Shah, S.; Shah, Z. Soil respiration, Ph and EC as influenced by biochar. *Soil Environ.* **2017**, *36*, 77–83. [CrossRef]
39. Zhang, W.; Parker, K.M.; Luo, Y.; Wan, S.; Wallace, L.L.; Hu, S. Soil microbial responses to experimental warming and clipping in a tallgrass prairie. *Global Chang. Biol.* **2005**, *11*, 266–277. [CrossRef]
40. Seremesic, S.; Zivanov, M.; Milosev, D.; Vasin, J.; Ciric, V.; Vasiljevic, M.; Vujic, N. Effect of biochar application on morphological traits in maize and soybean. *J. Nat. Sci.* **2015**, *129*, 17–25. [CrossRef]
41. Sun, D.Q.; Jun, M.; Zhang, W.M.; Guan, X.C.; Huang, Y.W.; Lan, Y.; Chen, W.F. Implication of temporal dynamics of microbial abundance and nutrients to soil fertility under biochar application–field experiments conducted in a brown soil cultivated with soybean, north China. *Adv. Mater. Res.* **2012**, *518*, 384–394. [CrossRef]
42. Yin, D.W.; Meng, J.; Zheng, G.P.; Zhong, X.M.; Yu, L.; Gao, J.P.; Chen, W.F. Effects of biochar on acid black soil nutrient, soybean root and yield. *Adv. Mater. Res.* **2012**, *524*, 2278–2289. [CrossRef]
43. Yooyen, J.; Wijitkosum, S.; Sriburi, T. Increasing yield of soybean by adding biochar. *J. Environ. Res. Dev.* **2015**, *9*, 1066–1074.

MDPI
St. Alban-Anlage 66
4052 Basel
Switzerland
Tel. +41 61 683 77 34
Fax +41 61 302 89 18
www.mdpi.com

Agriculture Editorial Office
E-mail: agriculture@mdpi.com
www.mdpi.com/journal/agriculture